D1126508

Data Reduction

Data Reduction

Analysing and Interpreting Statistical Data

A. S. C. EHRENBERG
London Graduate School of Business Studies

A Wiley–Interscience Publication

JOHN WILEY & SONS
London · New York · Sydney · Toronto

Library of Congress Cataloging in Publication Data:

Ehrenberg, A. S. C.
Data reduction; analysing and interpreting statistical data.

"A Wiley–Interscience publication."
Bibliography: p.
1. Mathematical statistics. 2. Statistics. I. Title.

QA276.E34 519.5 74–3724

ISBN 0 471 23399 4

ISBN 0 471 23398 6 (pbk).

Printed in Great Britain by
J. W. Arrowsmith Ltd., Bristol BS3 2NT

2[ΓM]$^{-1}$

Preface

This is a basic textbook on how to handle and interpret statistical data. It is for anyone who comes up against statistical data in his ordinary work.

Many people feel a need to know more about statistics. They want to know how to deal with numerical data, how to judge other people's analyses, and how to choose between all the techniques that are on offer. This book is for such people: administrators and business executives, economists and technologists, social, natural, and biological scientists, and teachers and students in these areas.

There are already many books on statistics. This one differs in a number of respects. Firstly, it recognises that one is often faced with undigested tables of numbers where the author has in effect said "I do not understand these data, perhaps *you* can". The initial emphasis is therefore on how to reduce data to meaningful summaries, or how someone else should have done so in the first place. This is not adequately dealt with in most books.

Secondly, the book brings out the reasons behind statistical procedures. It emphasises *why* things are done, rather than merely giving techniques to be followed in a cookbook manner. Thirdly, although the book is primarily a basic text, it also evaluates more advanced techniques, sometimes with rather critical conclusions. These techniques are mostly employed by statistical specialists, but the general reader needs to learn something about the broad principles, so as to be able to use both the results and the specialists.

Next, many conventionally taught statistical concepts such as frequency-distributions, probability, sampling, and tests of significance are described and evaluated in the *latter* parts of the book, because they are relevant only in certain limited situations. Finally, the book recognises that results are only useful to the extent to which they *generalise*. Empirical generalisation is therefore the key concept.

The book concentrates on teaching the reader to be numerate, to see patterns and relationships that exist in numerical data and to reduce these to summaries that can readily be interpreted, used, and communicated. *The need is to let the data speak.*

Being numerate does not mean being mathematical. The use of mathematics in this book is minimal. The reader only has to be reasonably comfortable with an equation like $y = 2x + 5$, and occasionally with some simple algebra and a logarithm or a square root. The real role of mathematics is in the *later* stages of analysis, to help in integrating many different results. There is a natural progression from seeing the simpler pattern in one's data to developing more general models and theories, and this is touched on in various parts of the book.

To reduce the data to summary results one must see the form that the data take. The book has therefore been built around numerical examples. Some have been taken from work with which I have been personally involved and deal with ordinary, humdrum problems, not unlike those which the reader could also face.

Data Reduction takes a radical look at statistical methods and problems. It aims to make explicit what is best in everyday practice amongst statisticians and scientists. The basic principle throughout is that data analysis must produce results which are usable in subsequent work. Nothing could be easier in analysing new data than to see whether the new results agree with the old ones (e.g. *does* $y = 2x + 5$ hold again?). Such use of prior results also leads to empirical generalisations: findings which are known to hold under a wide range of different conditions of observation. That is how scientific knowledge and understanding are usually generated.

The Structure of the Book

The book is in five parts. Parts I and II deal with ways of seeing and communicating systematic patterns in different sets of data, primarily in terms of a single variable in Part I and relationships *between* variables in Part II.

Part III discusses the conventional statistical procedures used for describing irregular variability within a single set of readings, and gives a critical evaluation of correlation and regression methods. Stochastic models are also introduced in this part.

Part IV deals with sampling and statistical inference from sample data. (Statistical teachers wishing to use the book along conventional lines may be more comfortable starting with some of the chapters in Parts III and IV.)

Part V compares the observational and experimental approaches to data collection and discusses the link between the description and explanation of observable phenomena.

Each chapter ends with exercise questions and their discussion. These fill in or elaborate various points which are either too elementary for some readers or too advanced for others and would clutter up the main text (especially at a first reading).

Tables, Graphs and Teaching Aids

Almost all the tables in this book have been produced on a typewriter (by Mrs. Myra Davies of Typlan Ltd.) before being reduced by one-third. They therefore take a form which can readily be produced in any report, without using a printer's different type-faces. (Two luxuries are proportional spacing, which makes the headings look more attractive but is in fact complicated to use for tables, and a simple attachment for producing bold lettering by repeated overtyping.) Mr. Robin Hipps drew the figures.

Teachers wishing copies of the original tables or figures to make transparencies or slides should write to Aske Publications at 13 Dover Street, London W.1., or 1 Chase Manhattan Plaza, New York, N.Y. 10005. The originals are about twice as large as the tables and figures in the text (50% larger linearly).

Acknowledgments

The preparation of this book has been greatly aided by discussions with friends and statisticians over the years, especially my colleague, Mr. G. J. Goodhardt.

For comments on numerous drafts I am indebted to many including students, and especially to Prof. M. E. Beesley, Dr. T. K. Chakrapani, Mr. P. Charlton, Dr. C. Chatfield, Mr. M. A. Collins, Mr. S. M. Schaeffer, and Dr. H. Thomas. But I have been particularly aided by Mrs. Helen Bloom Lewis, without whom the book would have been more than twice as long and less than half as clear.

London,
January 1974 A.S.C.E.

Contents

Detailed Contents

PART I: DATA HANDLING

Data are often presented in a form that is not immediately clear. The reader can then either ignore the data, analyse them himself, or return them to their author for *him* to analyse. In the last two cases it helps to know what can be done to make the data easier to comprehend.

Basic procedures for clarifying a table of numbers are described in Chapter 1. Chapter 2 deals with the situation where one already has previous experience of similar data.

After the patterns of the data have been established, the results need to be *communicated.* Here summary figures are more effective than large bodies of numbers. This limits the role of tables and graphs, as is discussed in Chapter 3.

Numerical results cannot be interpreted in isolation but have to be compared with norms and theoretical concepts. This is illustrated in Chapter 4.

CHAPTER 1

Averages and Layout

We start with a table of undigested data. In this chapter we discuss ways of understanding and communicating such data better.

TABLE 1.1 Data in Four Areas and Eight Three-Month Periods in 1969-1970

	13-15	16-18	19-21	22-24	25-27	28-30	31-33	34-36
A	97.63	92.24	100.90	90.39	95.69	94.44	91.13	97.81
B	48.29	42.31	49.98	39.09	46.38	49.74	41.74	37.39
C	75.23	75.16	100.11	74.23	74.23	76.97	71.66	76.47
D	49.69	57.21	80.19	51.09	52.88	49.41	59.32	52.56

The aim is for the reader to be able to see the pattern in the data, which in the present form of the table is not clear. There are a number of useful procedures and guidelines for achieving this, mainly by

(a) improving the display of the data,
(b) developing an explicit summary or "model",
(c) checking on the deviations of the readings from this model.

The readings in the table are not identified so that we can at this stage concentrate on the numerical patterns—about the only case in this book to be treated like that—but the rows represent four geographical areas and the columns eight three-month periods in 1969 and 1970.

1.1 Initial Visual Scanning

The first step is to gain a quick visual impression of the data so as to see the wood for the trees and to be better prepared for a more detailed analysis.

However, Table 1.1 is not easy to take in. One difficulty is the lack of differentiation between the column headings (months 13–15, 16–18, etc.) and the figures in the body of the table. Drawing a separating line, as in Table 1.1a, makes the numbers easier on the eye. This is a common example of the improvements which are possible in the layout of data.

A second problem with the table is the lack of any visual or conceptual focus. What are we supposed to be comparing with what: one area with

TABLE 1.1a Separating off the Headings

	13-15	16-18	19-21	22-24	25-27	28-30	31-33	34-36
A	97.63	92.24	100.90	90.39	95.69	94.44	91.13	97.81
B	48.29	42.31	49.98	39.09	46.38	49.74	41.74	37.39
C	75.23	75.16	100.11	74.23	74.23	76.97	71.66	76.47
D	49.69	57.21	80.19	51.09	52.88	49.41	59.32	52.56

another, one period with the next, one *year* with the next, or what? All we can see is a jumble of varied numbers.

But suppose we calculate the average of all the readings, which is about 70. Bearing this one figure in mind, we now see that in the first column Region A is much higher than the average, Region B lower, Region C quite near, and Region D lower again. Looking at the second column, the same pattern seems to recur. We therefore begin to see that the figures in the rows are broadly similar. In Region A the figures are mainly in the 90's, those in Region B are much lower, mostly in the 40's, those in Region C are with one exception in the 70's, and in Region D the figures are mostly in the 50's. This rough summary has been reached by using only the first digit or so in each number, a normal process in the kind of mental arithmetic that is used to scan and interpret numerical data.

In analysing extensive data we can often start with some sub-group. We will use this selective approach here and look only at the data for 1969 in the remainder of this chapter, analysing that for 1970 in Chapter 2. With many kinds of data, one year tends to be like another. If this turns out to be untrue in the present case, our analysis of the 1970 data in Chapter 2 will show it up. At this point the reader may wish to pause and analyse the data in more detail before reading on.

1.2 Basic Lay-out

Table 1.2 is an intermediate working table of the 1969 readings that illustrates improvements in the physical layout of the data. It introduces better labelling and table grids and reduces the number of digits in the readings.

TABLE 1.2 Better Labels, Table Grids and Two Significant Figures

	Quarters (1969)			
Area	I	II	III	IV
No	98	92	101	90
So	48	42	50	39
Ea	75	75	100	74
We	50	57	80	51

Better Labels. The abstract area-codes and clumsy period-labels have been replaced with self-explanatory and memorable ones. Labels in tables should aim to be self-explanatory. Usually some mnemonic device is possible, such as t for time, h for height, and No for North (single letters are only essential for use in *equations*) instead of x, y and z.

Table Grids. To guide the eye, ruled lines and gaps of white space have been introduced between certain columns and rows. Usually table-grids are drawn to separate labels and summary figures from the body of the data, while white space on its own is reserved for lesser demarcations or groups of figures. It is worthwhile to draw lines even with rough working-tables and computer print-out because they make the data so much easier on the eye. (Vertical lines are expensive to set with printed matter but not with typed tables. Lines become especially helpful with typed tables because there is less scope for other variations in lay-out, e.g. the use of different type-faces.)

Significant Figures. Perhaps the most effective change made in Table 1.2 has been rounding the readings to two significant figures. In our first rough scrutiny of the data, we only looked at the first digit or so in each number, but a widely applicable rule is to work with about *two* significant digits: defined as digits that vary from one number to another.

Recording only two digits has the advantage of making the data clearer to the eye. Mental arithmetic becomes easier to do. Extra digits often hinder one in seeing the data because the least important part of a number is the part looked at or spoken last (e.g. one reads 97.63 as ninety-seven point six-*three*).

There is a loss of accuracy in rounding to two significant digits, but this is usually negligible. For example, in the original 1969 data, the total range of variation is from 39.09 to 100.90, or about 160%. The two decimal places account for only 1 or 2% of that variation. It is unlikely that anyone will be able to interpret such small differences, especially in the early stages of an analysis. The primary need is to see the main patterns in the data. Once the initial analysis is completed, one can see whether the results call for a greater degree of precision. If so, a re-run with an extra digit can be made. But this rarely occurs, the possible exceptions being when a completely new kind of data is analysed.

The effect of such changes of lay-out can be fairly startling. Not only is it easier to see which figures are higher and which lower in Table 1.2, but with hindsight even Table 1.1 begins to look clearer.

1.3 Summary Figures

Having improved the presentation of the data, we now turn to analysing the actual information contained in them. Table 1.3 introduces summary figures through row and column totals. These help us to see that QI and QII, for example, are very similar, both in their totals, 271 and 266, and in the individual figures, 98 and 92, 48 and 44, etc.

TABLE 1.3 Row and Column Totals

	Quarter (1969)				
Area	I	II	III	IV	Total
North	98	92	101	90	381
South	48	42	50	39	179
East	75	75	100	74	324
West	50	57	80	51	238
Total	271	266	331	254	1122

However, totals are clumsy because they are expressed on different scales of measurement than the original readings. For example, the grand total of 1,122 in the bottom right-hand corner is not directly comparable with any other figure. When analysing data it is best to stick to one type of figure, like the quarterly ones, and to avoid introducing what are effectively new variables. (If a gross annual total is required for some particular purpose, it is easy to multiply a quarterly rate by 4.)

The totals can be reduced to the original scale of measurement by forming *averages*—dividing the totals by the number of readings added together (here 16 for the overall total and 4 for the others). Because the averages have the same visual shape as the individual readings, it may be better to separate them off with full table grids, as in Table 1.3a, rather than merely with extra white space.

TABLE 1.3a Row and Column Averages

	Quarters				
Area	I	II	III	IV	Av.
No	98	92	101	90	95
So	48	42	50	39	45
Ea	75	75	100	74	81
We	50	57	80	51	60
Average	68	67	83	64	70

These averages give us a better feel of the data. Firstly, we see that none of the quarterly averages differs greatly from the overall average of 70, although Quarter III is relatively high at 83. Secondly, we note that the four area averages differ markedly, from a low of 45 to a high of 95. Thirdly, we can see that the area averages reflect quite well what occurs in most of the individual quarters. The Northern quarterly results are *all* about 95; the Southern ones are all about 45; and while the figures seem more variable in the East and West, *most* of them do not vary much from quarter to quarter.

1.4 More Lay-out: Rows, Columns, and Ordering by Size

Now that we are beginning to understand the data we can rearrange the lay-out accordingly. This is usually worth doing both as an intermediate step, when analysing the data further, and when presenting the data in final form to others.

It is easier to see the relative lack of quarter-by-quarter variation, e.g. 98, 92, 101, 90, and the average 95 in the North, if the figures are written in columns instead of in rows and Table 1.4 accordingly reverses the columns and rows.

TABLE 1.4 Approximately Constant Columns, with Exceptions

(Rows and Columns from Table 1.3a interchanged)

	Area				
1969	No	So	Ea	We	Av.
Q I	98	48	75	50	68
Q II	92	42	75	57	67
Q III	101	50	100	80	83
Q IV	90	39	74	51	64
Average	95	45	81	60	70

In running down the North column in Table 1.4, the eye can read off the consecutive digits in the "tens" column as 9, 9, 10, 9, 9, largely bypassing the figures in the "units" column. In contrast, reading the data for QI *horizontally* in Table 1.3a, the eye had to scan 9, 8, blank, 9, 2, blank, 10, 1, blank, and so on.

The new lay-out also makes any exceptions stand out more clearly. We have already noticed that the East and West figures are more variable, but now the table shows that there are in fact two exceptional figures in the data, the 100 in QIII for the East and the 80 in QIII for the West. Running down these columns in Table 1.4 we read 7, 7, **10**, 7, 8, and 5, 5, **8**, 5, 6.

This visual gain is lost again if the rows of the table are more widely spaced, as in Table 1.4a. The more luxurious lay-out with extra white space fails to guide the eye.

The numerical nature of the data can also help to determine the best *order* of the columns or rows of the table. Rather than keep to their predetermined

TABLE 1.4a The Rows in Double Spacing

	Area				
1969	No	So	Ea	We	Av.
Q I	98	48	75	50	68
Q II	92	42	75	57	67
Q III	101	50	100	80	83
Q IV	90	39	74	51	64
Average	95	45	81	60	70

and possibly accidental order, the columns have been rearranged in Table 1.4b by the size of the averages, i.e. as North, East, West and South.

TABLE 1.4b Columns Rearranged in Order of their Average Size

	Area				
1969	No	Ea	We	So	Av.
Q I	98	75	50	48	68
Q II	92	75	57	42	67
Q III	101	100	80	50	83
Q IV	90	74	51	39	64
Average	95	81	60	45	70

This gives the table a clearer structure. Knowing that the column *averages* decrease from left to right, we can readily check whether the individual rows also decrease. In our example we can see that they do and that a marginal case like the first two figures in QIII also stands out clearly, as would any real exception. Doing the same thing in Table 1.4 we would have had to check each row against the less memorable high-low-high-low pattern of the column averages 95–45–81–60. The quantitative detail would have been almost impossible to sort out.

The columns can be arranged either in decreasing or increasing order of size; i.e. either starting with the big values as here, or in line with common practice in plotting graphs. Apparently there is no strong perceptual argument either way for columns. But with the *rows* of a table it helps to put rows with larger numbers above rows with smaller ones, since it is easier to subtract mentally that way.

However, in our example there is no point in rearranging the order of the rows because there is little variation between the quarterly averages. The "natural" order of the four quarters here might in any case seem sacrosanct. But using the dimensions of a table to represent the previously unknown pattern of the data can be more useful to the reader than repeating the well-known order of some row or column labels (e.g. everyone already knows that QII follows QI).

There can, however, be other considerations in all this. For example, if there are many tables with the same format, usually the overriding concern is to keep the same lay-out to facilitate visual comparison between the different tables. This can also apply to the interchanging of rows and columns (although it may still be useful for the analyst to do this as an intermediate step).

1.5 Averages and Exceptions

Averages are the main tool for summarising extensive numerical data. However, the four area averages in Table 1.4b are not good summaries because the data are dissimilar. In the North and South all the readings are close to the averages of 95 and 45, but in the East and West they are not, mainly because of the two exceptional QIII readings. Therefore it is misleading to compare the four areas merely in terms of their averages.

When there is a small number of exceptional readings, the data are easier to describe if one excludes these exceptions from the main summary figures. This is done in Table 1.5, which gives "adjusted" averages for the East and West. We now have more effective summaries of the data—the readings are generally about 95 in the North, 75 in the East, 53 in the West, and 45 in the South, with two large exceptions for QIII in the East and West.

TABLE 1.5 **The Adjusted Area Averages**

(QIII in East and West excluded)

	Area				
1969	No	Ea	We	So	Av.
Q I	98	75	50	49	68
Q II	92	75	57	42	67
Q III	101	(100)	(80)	50	(83)
Q IV	90	74	51	39	64
Average	95	75*	53*	45	67*

* Excluding QIII

Having noted the exceptional readings, we have to start checking on the reasons for them. The most likely explanation is a computing or clerical error (Twyman's Law that "any figure that looks interesting or different is usually wrong"). In practice we would of course know what the readings in the table referred to, e.g. the sales of Product X, the number of working-days lost due to absenteeism, or the incidence of measles, and could therefore try to find out whether there had been anything special that year to cause such exceptions, such as a price-cut, a large strike, or an epidemic, and whether equally high readings also occurred the year before. (If we are completely new to the data, we can often ask old Joe next door who has been around for 20 years and knows everything.)

Whether or not we find an explanation, the exceptional figures are best excluded from the basic summary figures. This might seem misleading (as if the analyst were trying to mislead himself), but the QIII readings will be exceptional no matter how we report them. We can either give four comparable averages with two exceptions, or four averages that are *not* comparable and two exceptions. No one is trying to hide anything; the exceptional readings stand out even more from the new averages than from the old ones simply because they were excluded. (Chapter 2 will show whether these adjusted averages are successful when checking them against additional data. This is the ultimate test of the usefulness of any description of data.)

If the readings being summarized are not all more or less constant (as in each column here), allowing for exceptional figures will be more difficult. Thus the data in the rows are more complex than in the columns, but at least the three row averages for QI, QII and QIV also provide good summaries. The averages are similar and we can also see that the scatter of readings about each average takes the same form. These row averages therefore perform the useful role of showing where the distributions of readings quarter-by-quarter are the same and of making the exceptional QIII stand out clearly.

With the four area averages we now have a possible summary or "model" of the main features of the observed data. This model says that the quarterly figures are generally about 95 in the North, 75 in the East, 53 in the West and 45 in the South. (The model or "theory" here is obviously on a very low level, but this is because of the narrow range of the data and not because of the modelling process itself.)

1.6 Deviations from the Model—The Final Step

The final step is to examine the "fit" of this theoretical model, i.e. the differences between the observed readings and the area averages (e.g. 98 − 95 = 3 in QI in the North), as shown in Table 1.6.

Here once again we face an array of numbers that are new and largely undigested. But the data are already somewhat better organised than in Table 1.1 and we know something about their background. For example, we

TABLE 1.6 Deviations Between the Observed and Theoretical Figures

1969	Area			
	No	Ea	We	So
Q I	3	0	3	3
Q II	-3	0	4	-3
Q III	6	25	27	5
Q IV	-5	-1	-2	-6

know about the two exceptions in QIII, and that, compared with a range from an average of 95 in the North to one of 45 in the South, most of the deviations are small. Judging the deviations on their own, the variation of the numbers in Table 1.6 is however quite marked, e.g. from $+3$ to -3 to $+6$ to -5 in the first column, and so on. (Expressing these readings to more digits, e.g. as 2.6, -2.8, 5.9, -4.6, etc., let alone to a *second* place of decimals, would therefore have been pointless.)

In analysing the deviations we first work out row and column averages as in Table 1.6a, again excluding the exceptional values in QIII.

TABLE 1.6a Averages

1969	Area				
	No	Ea	We	So	Av.*
Q I	3	0	-3	3	1
Q II	-3	0	4	-3	0
Q III	6	(25)	(27)	5	6*
Q IV	-5	-1	-2	-6	-4
Average	0	0*	0*	0	0

*Excluding QIII in East and West

The *column* averages are all zero because we used the area averages as our model. Therefore positive and negative deviations must balance out. The area averages in Table 1.6a tell us only that the arithmetic is right.

The *row* averages are more variable. The QIII average of 6 reflects individual readings that are positive, and the QIV average of -4 reflects individual readings that are all negative. But there are no such patterns in the *first* two quarters: in QI, the readings in the North and South are positive and that in the West is negative, while the opposite occurs in QII, negative readings in the North and South and a positive one in the West. Not too much should therefore be made of the apparent pattern in QIII and QIV, since it has to be interpreted in conjunction with the *absence* of a pattern in QI and QII.

Concluding that there is no general pattern of *any* kind in the table would, however, be premature. The signs in the first column alternate systematically + − + −. The same occurs in the South, and there is something of this pattern even in the West (and virtually *no* variation in the East anyway). Successive readings over time—usually called a "time-series"—can contain a tendency to be specially related to each other. This is called serial correlation. In our example the serial correlation is negative. A relatively high reading is followed by a low one, and a low one is followed by a high. (For sales data, for example, this might reflect a tendency to over-sell in one quarter, leading to an excess of stock, and so followed by under-ordering in the next quarter.) With only 16 readings for 1969 it is impossible to tell whether the apparent pattern here is real. But when we analyse the 1970 data in the next chapter we can readily check whether the pattern generalises.

1.7 The Size of the Deviations

In analysing the deviations of the original readings from their averages, some are positive and some negative and they necessarily have an overall average of zero. This does not help to describe the data. Instead, we need a measure of the size of the deviations irrespective of their signs.

For example, in the North the straightforward average of the four deviations is 0 to the nearest whole number (3 − 3 + 6 − 5 = 1, divided by 4). But the average *ignoring the sign* is 4, i.e. (3 + 3 + 6 + 5 = 17 divided by 4). This tells us how *big* the deviations generally are. Table 1.7 gives the average size of the deviations for each area, ignoring the negative signs.

TABLE 1.7 **The Average Size of the Deviations**

	Area				
1969	No	Ea	We	So	Av.
Av. Size	4	0*	3*	4	3

*Excluding QIII in East and West

The overall average size of the deviations is 3 (again ignoring the two QIII exceptions). The areas do not vary much in this respect, except that the values in the East are low. Similarly, visual inspection of Table 1.6a shows that the average size of the deviations in each quarter is also about 3, except that in QIII even the two "normal" readings are on the large side.

Thus, apart perhaps from the high values in QIII and the low ones in the East, the average size of the deviations does not vary dramatically by area or by quarter. The overall average of 3 is a fairly good summary. (Such a measure of the average size of the deviations ignoring sign is usually known as the *mean deviation.* This and other measures of statistical scatter are discussed in Chapter 11.)

1.8 A Full Description

In Section 1.2 at the beginning of this chapter we started with the 16 quarterly readings for 1969, as reproduced in Table 1.8.

TABLE 1.8 **The Original Data for 1969**

	13-15	16-18	19-21	22-24
A (North)	97.63	92.24	100.90	90.39
B (South)	48.29	42.31	49.98	39.09
C (East)	75.23	75.16	100.11	74.23
D (West)	49.69	57.21	80.19	51.09

Now that we have achieved some understanding of the data, they seem fairly clear even in their original form. In addition, we have reduced the data to seven summary figures.

Four area averages: North 95, East 75, West 53, South 45.

An apparently irregular quarterly variation of about 3 units about these averages.

Two exceptionally high values in QIII (months 19–21) in the East and West, differing from the area averages by about 25 units.

Seven summary figures, the four area averages, one measure of average-scatter, and two large exceptions, may not seem like much of a reduction of 16 initial readings. One reason for this limited success is that the data analysed so far were not, in any case, very extensive. Another reason is that we still have not accounted for the main regularity that we have found, the large systematic differences between the four areas, nor for the residual scatter, i.e. the deviations of the individual readings. What the analysis *has* achieved is to bring into focus the fact that it is these features which need to be explained. Before attempting any deeper explanation of these features it is worth checking whether they generalise. The 1969 data started with month 13 so there should be at least one year's previous data; but since Table 1.1 already gave the *following* year's data, we shall analyse this in the next chapter.

1.9 Summary

Most data reach us in undigested form. In this chapter we have illustrated various analytic steps that can be used to see the pattern in a table of numbers and to communicate the results. The main guide-lines are as follows.

(i) See if the number of digits shown can be reduced. (Mental arithmetic is difficult with more than two significant digits, i.e. ones which vary.)
(ii) Use self-explanatory, memorable symbols and labels (e.g. t for time, h for heights, and No for North, not x, y and z).
(iii) Separate different types of items or sub-groupings in a table by grid-lines or white space.
(iv) Use averages to help focus the eye when examining any array of numbers.
(v) Ensure that figures that must be compared are close together.
(vi) Try to use the *columns* of a table to show figures that vary little since this makes both the regularities and the exceptions easier to see.
(vii) See if the columns or rows of a table can be ordered "graphically" to reflect their average size, thus making it easier to see patterns in the body of the table.
(viii) Avoid introducing new variables or scales whenever possible (e.g. use averages rather than totals as summary figures).
(ix) Note any dramatic exceptional values separately and exclude them from the main summary figures.
(x) Summarise irregular aspects of the data statistically, e.g. by a measure of the average size of the deviations.

In general, one should aim to present simple, well-digested patterns and summaries. The alternative amounts to leaving the analysis to the reader.

The analytic process which has been described here is organic rather than mechanical. One step is usually taken because of the patterns revealed by the previous steps. This makes the analysis laborious and slow, but helps to ensure that the final result models the actual structure of the data. Any subsequent analysis of similar data should then be relatively quick and straightforward. This is illustrated in Chapter 2.

CHAPTER 1 EXERCISES

Exercise 1A. Alternative Analyses of the Data

Discuss alternative ways of handling the original data analysed in the present chapter.

Discussion.

Popular approaches to such time-series data include the use of graphs, the calculations of percentage changes or shares, and the use of indices or moving averages. (A widely used alternative is to leave the raw data for the reader to sort out.)

Graphs. Figure 1.1 shows the 32 readings for 1969 and 1970 graphed. The QIII blips in the East and West show up clearly. They typically dominate the picture, but the generally steady levels and the tendency for slightly lower values in QIV can also be discerned.

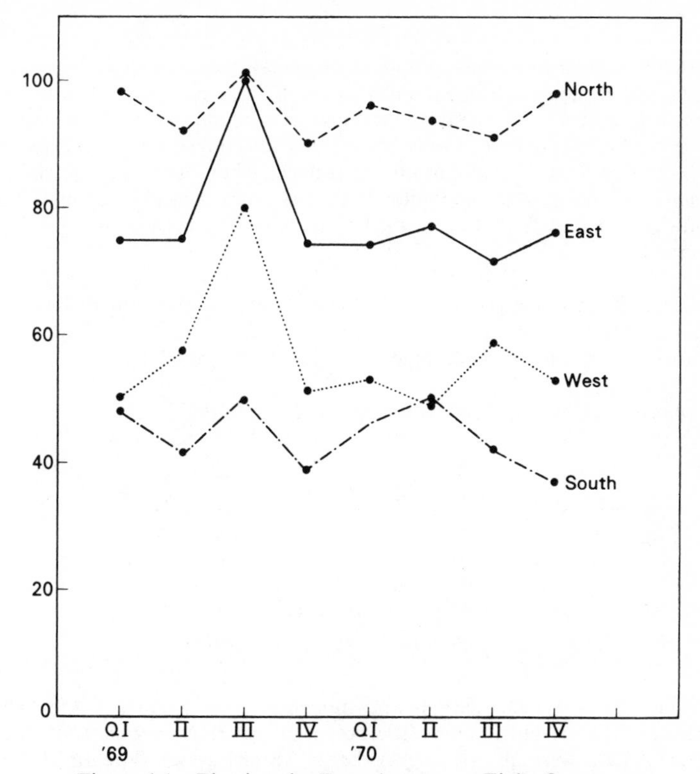

Figure 1.1 Plotting the Four Areas over Eight Quarters

Graphs are, however, relatively time-consuming to construct. Reading-off any detailed *quantitative* information is difficult, if not impossible. Nor does a graph provide a succinct and usable *summary* of the data.

Percentage Changes. Table 1.9 shows the data expressed as percentages of the preceding quarter (a popular alternative is to show percentage changes from the same quarter a year before).

TABLE 1.9 Quarter-by-Quarter Percentage Changes

Region	1969				1970		
	QI	QII	QIII	QIV	QI	QII	etc.
A (North)	-	94	109	89	107	99	.
B (South)	-	88	118	78	119	107	.
C (East)	-	100	133	74	100	104	.
D (West)	-	115	140	64	104	93	.

Any feeling for the actual data is lost in this form of presentation, e.g. that the North is bigger than the South. The exceptionally high values in QIII show up, but they also recur as exceptionally *low* values in QIV. With such percentages, every variation in the original figures shows up *twice*, once as up and once as down (or vice versa). The figures are therefore difficult to interpret. For example, the original figures in QIV 1969 were low, so that the percentage changes to QI 1970 are high, even though the actual figures (95, 76, 54, 46) in that quarter were almost exactly average.

Percentage changes can only be properly interpreted by going back to the original data. They are usually introduced when the original readings have been shown to so many digits that any quick mental scrutiny of the data is inhibited. But without further analysis they provide no effective summary.

Indices. Percentaging on a constant base avoids the excess up-and-down movements of the change-from-last-quarter type of percentages. Table 1.10 shows the readings as percentages using QI of 1969 as a base.

TABLE 1.10 The Readings Indexed on Quarter I = 100

Region	1969				1970		
	QI	QII	QIII	QIV	QI	QII	etc.
A (North)	(100)	94	103	93	98	97	.
B (South)	(100)	88	104	81	96	103	.
C (East)	(100)	100	133	99	99	102	.
D (West)	(100)	115	161	103	106	99	.
Total	(100)	99	122	94	99	100	.

This brings out fluctuations and trends over time, and also the two QIII exceptions, but it does little better in this respect than when the original data were shown to two digits. All feeling for the size of the original readings is once more lost (e.g. that the North is twice as big as the South). The choice of base figures is also often arbitrary, and becomes more so as time progresses.

Regional Shares. Table 1.11 gives each area as a percentage of the national total that quarter. (Such share figures can be meaningful with "additive" quantities like numbers of people, sales of a product or rainfall, but not with ages, prices or temperatures.)

TABLE 1.11 Regional Shares

Region	1969				
	QI	QII	QIII	QIV	Av.
	%	%	%	%	
A (North)	36	35	31	36	34
B (South)	18	16	15	15	16
C (East)	28	28	30	29	29
D (West)	18	21	24	20	21

The general stability of the figures is made clear. It is also easy to see that the North is just over twice as large as the South. Most of this visual clarity, however, is lost in the common practice of showing percentage figures to one decimal place (i.e. as "per mille"), as in Table 1.11a.

TABLE 1.11a The Percentages to One Decimal Place

Region	1969				
	QI	QII	QIII	QIV	Av.
A (North)	36.0	34.6	30.5	35.5	33.9
B (South)	17.8	15.9	15.1	15.3	16.0
C (East)	27.8	28.1	30.1	29.1	28.9
D (West)	18.3	21.4	24.2	20.1	21.2

Exceptional results do not show up well in either table because they influence the total on which the percentages are based. This can also influence other figures. For example, the North had its lowest share in QIII, 31%, but its highest *absolute* figure, 101 (see Table 1.8). Percentage shares are best used when there are large but proportional variations in the absolute figures from column to column or row to row. The shares are then more or less steady.

Moving Averages. These are a device for smoothing away some of the fluctuations that occur in time-series data. In our example one would first calculate each area average over the four quarters QI to QIV of 1969 (e.g. 95.4 in the North). Then one would calculate the area averages over the next four quarters QII 1969 to QI 1970 (e.g. 94.7 in the North), the area averages over QIII 1969 to QII 1970 (95.2 in the North), and so on.

The figures for each area are then very steady, but they are difficult to interpret because an increase from one reading to the next could be due to a low figure being dropped or a high figure being added in forming the next average. (A better alternative is to summarise the regularities in the data by an overall average or trend line, with deviations. The fitting of such lines is discussed in Part II.)

Exercise 1B. Another Example

What is the main pattern of the following four pairs of readings?

	1956	1957	1958	1959
X:	108	60	89	51
Y:	206	158	187	149

Discussion.

Applying the various guide-lines of Chapter 1 (e.g. arranging in order of size, averaging, and rounding) helps to bring out the pattern which is otherwise not immediately obvious.

For example, rearranging the X values in order of their size and using single spacing gives

	1959	1957	1958	1956
X:	51	60	89	108
Y:	149	158	187	206

This makes it easy to see that the X and Y values follow the same trend. Departing from the natural order of the years may appear peculiar but would be done without second thoughts in plotting the data as in Figure 1.2. (Plotting such a graph is more laborious than simply re-arranging the numbers in a new table, but it also shows clearly that X and Y vary together.)

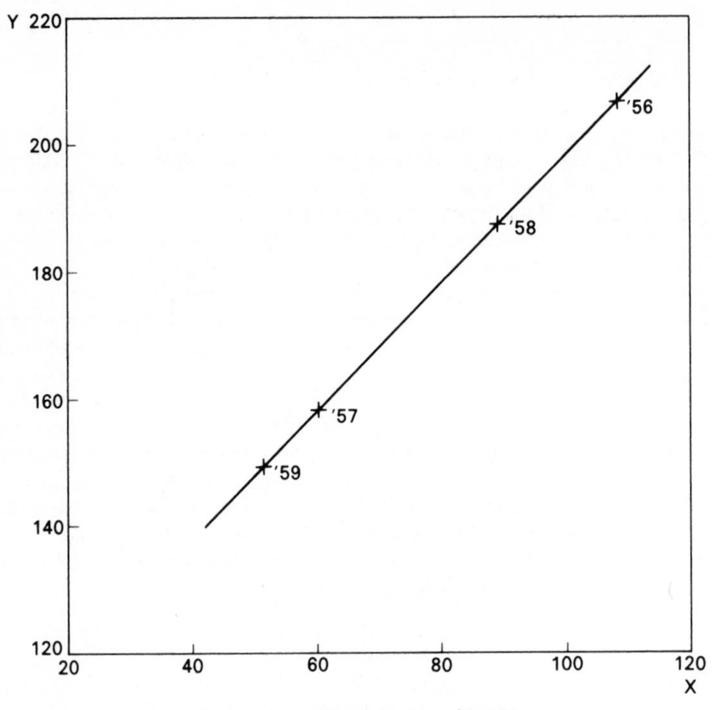

Figure 1.2 Plotting *Y* against *X*

Introducing averages shows up the *quantitative* form of the relationship:

	1959	1957	1958	1956	Av.
X:	51	60	89	108	77
Y:	149	158	187	206	175
Av.	100	109	138	157	120

The average values of Y and X differ by 98. With this figure in mind we can scan the individual years to see if the differences are higher or lower. We see that the differences are in fact all 98 (e.g. 149–151 in 1959), so that

$$Y = X + 98.$$

That is the pattern.

Rounding off to the nearest 10 can be done without serious loss of information and also vastly clarifies the pattern:

	1959	1957	1958	1956
X:	50	60	90	110
Y:	150	160	190	210

Approximately, therefore, X and Y differ by 100.

However, even after getting to know it, this result remains fairly difficult to see in the original data:

	1956	1957	1958	1959
X:	108	60	89	51
Y:	206	158	187	149

This is because we mostly find it hard to subtract large numbers from smaller ones written above them. Rearranging the rows with the larger numbers on top (and in single spacing once more) helps:

	1956	1957	1958	1959
Y:	206	158	187	149
X:	108	60	89	51

In conclusion, since the simple pattern in these four pairs of readings was relatively difficult to see in the original form, one wonders what regularities are being missed in *other* data due to clumsy lay-out and presentation.

Exercise 1C. The Current Economic Climate

Each week the American journal *Business Week* publishes certain "Figures of the Week". Discuss the production figures shown in Table 1.12 from the issue published April 24, 1971 (reproduced by permission).

TABLE 1.12 Business Week's "Figures of the Week" (April 24, 1971)

Production	1957-59 average	Year ago	Month ago	Week ago	Latest week
Raw steel *[Amer. Iron & Steel Inst., thous. of net tons]*	1,860	2,686	2,844	2,932	2,905
Automobiles *[Ward's Automotive Reports]*	102,264	139,838	190,160	146,400	150,941
Electric power *[Edison Elec. Inst., millions of kilowatt-hours]*	12,385	27,280	29,735	28,633	28,111
Crude oil, refinery runs *[Amer. Pet. Inst., daily av., thous. of bbl.]*	7,852	10,556	11,014	11,132	11,315
Bituminous coal *[Bureau of Mines, daily av., thous. of net tons]*	1,437	2,011	2,114	2,140	2,154
Paperboard *[Amer. Paper Inst., thous. of tons]*	284.3	521.0	513.5	523.8	503.0

Discussion.

The apparent purpose of the magazine's "Figures of the Week" is to help non-technical businessmen and administrators to judge the current economic climate with a quick indication of recent trends. However, the data are expressed to too many digits to be taken in easily by eye. (Such detail could not even influence the most painstaking economic analysis.)

Table 1.12a shows the data reduced to two digits. Decimal points are avoided by using suitable units, like 100,000 tons of steel instead of million tons. The use of odd-looking units (like 10,000 cars) could be obnoxious in other cases, but it does not seem to matter here since the units quoted from *Business Week* are not self-explanatory to the lay-user anyway: e.g. tons of paper board but *net* tons of steel, bulk barrels of oil (bbl), and "daily averages" without saying whether these are based on a 5- or 7-day week.

TABLE 1.12a The Figures Rounded and Ordered by Size

Weekly Production Figures	1957-59 average	Year ago	Month ago	Week ago	Latest week
Paperboard (in 10,000 tons)	28	52	51	52	50
Raw Steel (in 100,000 net tons)	19	27	28	29	29
Bitum. Coal (daily av., 100,000 tons)	14	20	21	21	21
Electricity (in billion kwh)	12	27	30	29	28
Automobiles (in 10,000)	10	14	19	15	15
Crude Oil (daily av. in million bbl)	8	11	11	11	11

In Table 1.12a the rows have been arranged in decreasing order of size, using the 1957–59 averages as the norm because these appear every week. This ordering depends on the arbitrary choice of measurement units, but that is unimportant. What matters is that it is now visually easier to concentrate on the figures.

For example, we can now see quite easily that since 1957–59 production of the different items has increased by between 50% and 100% and that the four figures for the last year are rather similar. Since it is better to place figures with slight variation in columns, we can interchange the columns and rows, as in Table 1.12b, and move the information about measurement units below the table.

TABLE 1.12b Rows and Columns Interchanged

	Paper	Steel	Coal	Electr.	Autos	Oil
1957-59 average	28	19	14	12	10	8
Year ago	52	27	20	27	14	11
Month ago	51	28	21	30	19	11
Week ago	52	29	21	29	15	11
Latest week	50	29	21	28	15	11

Paperboard in 10,000 tons; Raw Steel in 100,000 net tons;
Daily average of Bituminous Coal in 100,000 net tons;
Electricity in billion kwh; Automobiles in 10,000;
Daily average crude oil refinery runs in mill. bbl.

Now it is clear that the oil production figures have been 11 for the past year (but have they perhaps been rounded too much?), that auto production was exceptionally high a month ago (a printing error, the end of a strike, or what?), that the electricity figures fluctuated up and down, etc.

In this lay-out it is a matter of choice whether to give the 1957–59 base-line at the top or bottom of the table; the real question now is whether this base-line tells us anything useful anyway.

Exercise 1D. The National Income

Discuss the first five rows of National Income figures from the 1971 edition of the Government Statistical Service's "Britain's Economy in Figures" (a 3 × 5 inch pocket-card):

Government Statistical Service
BRITAIN'S ECONOMY IN FIGURES

		1964	1969	1970
Population and Manpower	000s			
Home population (June):				
United Kingdom	(94,216 sq. miles)	54,008	55,534	55,711
England	(50,333 sq. miles)	44,658	46,102	46,254
Wales	(8,017 sq. miles)	2,676	2,725	2,734
Scotland	(30,414 sq. miles)	5,206	5,195	5,199
N. Ireland	(5,452 sq. miles)	1,458	1,513	1,524
Working population (June)		25,849	25,802	25,637
Wholly unemployed (average)		405	582	619
Full-time students in higher education in Gt. Britain (Autumn of previous year)		239	410	427
National Income	£ million			
Gross national product at current factor cost		29,319	38,913	42,667
Consumers' expenditure		21,459	28,683	31,124
Gross fixed investment		5,857	8,024	8,742
Public		2,580	3,714	4,004
Private		3,277	4,310	4,700
Income from employment		19,702	27,141	30,246
Gross trading profits of companies		4,591	4,948	5,036
Economic growth (gdp at constant factor cost 1963 = 100)		105.6	118.4	120.4
Public Expenditure (years ending 31 March)	£ million			
Defence		1,811	2,245	2,206
Education	excl. N. Ireland	1,393	2,301	2,513
Health and welfare	excl. N. Ireland	1,107	1,770	1,931
Social security benefits	excl. N. Ireland	1,976	3,294	3,538
Roads and public lighting	excl. N. Ireland	365	590	648
Overseas aid		156	177	192

		1964	1969	1970
Production				
Index of industrial production	1963 = 100	108.3	122.9	124.0
Output per head	1963 = 100	106.5	124.9	128.1
Coal	m. tons	193.6	150.5	142.3
Gas—total available	m. therms	3,417	5,517	6,350
North Sea	m. therms	—	1,001	3,586
Electricity—total generated	m. kWh	165,445	219,087	228,894
From nuclear power	m. kWh	5,341	25,271	21,871
Crude oil processed	m. tons	58.5	90.3	100.3
Merchant ships under construction (end-year)	000 gross tons	1,616	1,617	1,612
Man-made fibres	million lb.	825	1,221	1,321*
Passenger cars	000s	1,868	1,717*	1,641
Transport (Great Britain)				
Index of vehicle miles travelled on roads	1963 = 100	112	144	152
Index of ton miles of inland goods (road and rail)	1963 = 100	111	127	132
Licences current:				
Total road vehicles	000s	12,370	14,753	14,951
Motor cars	000s	8,247	11,228	11,516
Mileage of motorways in use (at 1 April)		292	622	662
Prices, Incomes and Money	1963 = 100			
Retail prices		103.3	127.2	135.3
Average earnings		107.1	147.4	165.0
Real personal disposable income per head		103.2	109.4	112.8
Money supply (M3) (end-year)		105.4	145.3	159.0
Balance of payments	£ million			
Current account balance		−395	+437	+631
Visible		−519	−141	+3
Invisible		+124	+578	+628
Total currency flow		−695	+743	+1,287
Level of official reserves	end Dec.	827	1,053	1,178
Total external assets	end Dec.	13,145	19,375	n.a.
Total external liabilities	end Dec.	11,350	16,685	n.a.
Overseas Trade	£ million			
Total value of exports fob†		4,565	7,339	8,063
Sterling area		1,554	2,040	2,218
EEC		963	1,530	1,754
EFTA		641	1,080	1,277
North America		623	1,223	1,231

	1964	**1969**	**1970**
Total value of imports cif	5,696	8,315	9,051
Sterling area	1,874	2,404	2,460
EEC	941	1,609	1,822
EFTA	748	1,248	1,406
North America	1,109	1,639	1,857
Food, drink and tobacco	1,771	1,930	2,052
Raw materials and fuel	1,702	2,159	2,310
Semi-manufactures	1,324	2,302	2,509
Manufactured goods	837	1,835	2,070

* 53 weeks. † Excluding net adjustment for under-recording.

Further copies can be obtained from the Central Statistical Office, Great George Street, London S.W.1, or Regional Offices of the Department of Trade and Industry.

Prepared by the Central Statistical Office and the Central Office of Information, May 1971.

Printed in England for Her Majesty's Stationery Office by Stephen Austin and Sons Limited, Hertford.

(Reproduced by permission of The Central Statistical Office)

Discussion.

This particular card is a little out-of-date and has since been improved, but it illustrates how the business of presenting almost unintelligibly detailed figures has not always been restricted to business journals or to one side of the Atlantic.

For an "at-a-glance" card, figures to the nearest thousand million are enough and help one to see by simple mental arithmetic that GNP has increased by 50% from 1964 to 1970 (43 − 29 = 14, which is half of 28), and by 10% since 1969 (43 − 39 = 4, which is about 10% of 43).

National Income (£'000 million)	1964	1969	1970
Gross National Product	29	39	43
Consumers' Expenditure	72%	74%	72%
Gross fixed investment	6	8	9
Public sector	44%	46%	46%

The percentages have been introduced not because they are inherently a good thing to use with economic data, but because they remain almost constant despite the increases in GNP and gross investment, and thus reflect simple aspects of the data in a simple manner. We now have figures which are memorable, like consumers' expenditure being roughly a constant 73% of GNP, and public investment being roughly 45% of total capital investment. If it is felt that minor variations are important, such as public investment going up from 44 to 46%, then this can now at least be *seen*. (Because the pocket-card requires reasonably explicit descriptions of each item, rearranging the rows and columns seems in this case impossible, even if the pattern of the data suggests it.)

If we now move on to the more up-to-date information of this kind, the 1973 edition of the GSS card entitled "United Kingdom in Figures" (a 10 × 5 inch folding pocket-card), we find several changes.

Two are that the figures are shown to three digits only and that breakdowns are shown as percentages. Thus the comparable National Income figures, as given on the 1973 card, are

	1951	1966	1972
Gross Domestic Product (£'000 mn.)	12.6	32.8	52.6
Used for—Consumption %	87.5	80.6	79.8
—Investment %	12.5	19.4	20.2

Much has been gained since 1971 but the third digit still inhibits any simple visual interpretation, compared with the following version:

	1951	1966	1972
Gross Domestic Product (£'000 m)	13	33	53
Used for—Consumption %	88	81	80
—Investment %	12	19	20

Another change from the 1971 card is one of definitions, e.g. something called "Consumption" at over 80% instead of "Consumer expenditure" at 72% or so. Such problems over definitions (presumably *public* consumption has been added in), make it especially doubtful whether *three* digits on such a card can serve any useful purpose.

Exercise 1E. Elegant Variations

Improve the following report on the incidence of fatal accidents in air travel (from *The Guardian*, July 31, 1971):

> "In 1950 . . . , the world's airlines carried 30 million people. There were 27 fatal crashes that year, and just over three passengers were killed for every 100 million miles flown. . . . Ten years later the number of passengers had more than tripled, there were 33 crashes, and the rate in terms of miles had nearly halved. . . . The world's air travellers now number 300 million annually and each can reckon to fly 125 million miles before the statistics catch up with him."

Discussion.

The incidence of fatal accidents is compared at roughly ten-year intervals, but the results are expressed in a different form each time:

1950: Three passengers killed per 100 million miles flown.
1960: The rate (presumably of deaths) in terms of miles had nearly halved.
1971: Each air traveller "can reckon to fly 125 million miles before the statistics catch up with him."

The comparisons can be clarified by expressing each result in the same terms, giving approximately:

1950: 1 fatality per 30 million miles flown,
1960: 1 fatality per 60 million miles flown,
1971: 1 fatality per 120 million miles flown.

The results are simple: The fatality rate per mile has approximately halved every 10 years or so. (Similarly, the number of passengers has roughly *tripled* every 10 years.) *Now* one can add stylistic gloss to make for easier reading.

Exercise 1F. Taking One's Own Medicine

Few statistical writers discuss the virtues of working with rounded figures, but Robert Golde (1966a) has stressed the small loss in accuracy of simply "dropping-off" digits. He used the following example for the number 21,742:

Digits Dropped	New Number	Approximate Loss in Accuracy
2	21,740	0.01%
42	21,700	0.19%
742	21,000	3.41%

Discuss the example.

Discussion.

The accuracy lost in "dropping-off" the last three digits of the number 21,742 could have been reported as 3% instead of 3.41%! Neither the second nor the first decimal can matter. No one would *not* drop the digits if the loss were 3.5%, but do so if the loss were only 3.4%.

If Golde had used the normal procedure of rounding to the nearest number, 22,000 (instead of merely dropping off the last digits to 21,000), the error would have been about 250 out of 22,000, or only 1%:

Digits Dropped	New Number	Approximate Loss in Accuracy
2	21,740	.01%
42	21,700	.2%
742	22,000	1%

Although the lay-out in the last column above is unorthodox, it can be effective. Certainly the visual obfuscation of writing numbers less than 1 with an initial zero (e.g. 0.01) can often be avoided.

Exercise 1G. What's Wrong?

Golde has quoted elsewhere (1966b) the company president who said:

> "The first few times my controller sent me a report with figures rounded to the nearest $1,000 I felt very uneasy. The report seemed lacking and incomplete. Now that I have grown used to the shorter figures, I find I mentally round all figures I look at down to two or three digits."

Why did the company president feel uncomfortable?

Discussion.

Rounded figures like 95 in the North and 45 in the South seem naked because it is easy to see what they say (the figures in the North are about twice those in the South) but it is not obvious what we ought to do about it. In contrast, figures like 97.63 and 48.29 are comforting because although we may not know what they *mean*, they are at least very precise.

But why should anyone like the company president still be faced with figures which need to be rounded mentally? Figures ought to be presented in suitable form in the first place, consisting only of what matters. The reader's role is to return undigested figures to their originator with rude comments, rather than do his arithmetic for him.

Exercise 1H. Company Reports

Below are given the basic figures in a company's Annual Report, printed in "big, easily-read type and organised to highlight the guts of the business" (Foy, 1973).

But the data are still not easy to take in. How much higher are the 1972 sales than those the year before, and how does this increase compare with those in the other figures?

1972 AT A GLANCE

	1972	1971
Sales	$172,045,539	$153,220,890
Operating profit	$ 11,612,434	$ 8,790,576
As percent of sales	6.7%	5.7%
Net income	$ 6,009,155	$ 4,248,645
As percent of sales	3.5%	2.8%
As percent of investment	4.8%	3.5%
Earnings per share of common stock	$1.87	$1.32
Dividends paid per share of common stock	$1.03	$1.00
Taxes paid per share of common stock	$6.23	$5.24
Market price range of stock during year	$\$29\frac{7}{8}$–$17\frac{1}{8}$	$\$23\frac{1}{2}$–$14\frac{1}{4}$

Discussion.

Rounding helps as usual (including dropping the stock-market's archaic habit of quoting price-ranges in eighths of a dollar). Table 1.13 begins to make some mental arithmetic on the figures possible (e.g. one can see that sales have gone up by about 20 million, which is over 10%, but that operating profits and the range of stock-prices have gone up by over 20%).

The new table is smaller, but its lay-out generally helps to guide the reader's eye. It is not spread across the page, the rounding allows figures which are to be compared to be close to each other, single spacing is used to indicate which item belongs to which, "ditto" marks indicate the similarity of the "per share" ratios, and the % symbol triggers visual recognition.

Has some of the rounding perhaps been overdone? For an "At a glance" table, showing operating profit as growing from 6% to 7% tells a good story well. (Anyone who wants the percentages to a further place, 5.9% and 6.8%, can work them out from the dollar figures given.)

TABLE 1.13 An Improved Format

		1971	1972
Sales	($'000)	150,000	170,000
Operating Profit	($'000)	8,800	11,000
As % of sales		6	7
Net Income	($'000)	4,200	6,000
As % of sales		3	4
As % of investment		4	5
Earnings per share		$1.30	$1.90
Dividends " "		$1.00	$1.03
Taxes " "		$5.20	$6.30
Range of stock price		$14-23	$17-29

Nobody is saying that $2 or 3 million, e.g. the difference between $172 million and $170 million, isn't money. It ought therefore to be noted down somewhere, for the record (so that nobody makes off with it). But given a $20 million or so increase in sales, $2 million more or less does not affect any *conclusions* to be drawn from these figures.

The real problem is that two years' figures are not enough anyway. Was 1973 an exceptionally high year or 1972 exceptionally low? To see the full picture, a longer series of data is needed. Adequate digestion and clear presentation of much more extensive data will then be even more necessary.

Exercise 1I. Further Examples

From the viewpoint of this chapter examine the last memorandum involving numerical data which you prepared and/or received. Also examine the last scientific paper, technical report, or financial or technical article involving numbers which you read. Is there room for improvement? Would it be easy to achieve? Would it facilitate understanding and communication of the information?

CHAPTER 2

Using Prior Knowledge

Now that we know the results of the data analysed in Chapter 1, we can analyse any additional data for that variable with some prior knowledge.

The notion of using prior knowledge in analysing data is a fundamental one and runs through the remainder of this book. Since our aim is to provide information for future use, the use of prior information should be the common situation. The reason it may not always seem so is that previous data are often left in an undigested, and hence unusable, state.

2.1 New Data and Prior Knowledge

The new data to be analysed here are the readings for 1970 which were given earlier in Table 1.1. They are now set out in Table 2.1 with the lay-out and manner developed in Chapter 1.

TABLE 2.1 The 1970 Data in the Same Layout as the 1969 Data (Table 1.5)

1970	Area				
	No	Ea	We	So	Av.
Q I	96	74	53	46	67
Q II	94	77	49	50	68
Q III	91	72	59	42	66
Q IV	98	76	53	37	66
Average	95	75	54	44	67

If we remember the 1969 results, the new results follow quickly.

(i) The 1970 area averages are virtually the same as for 1969: 95, 75, 53, and 45 (as is also demonstrated in Table 2.1a below).

(ii) There are no systematic quarterly differences, just as in 1969.

(iii) The deviations of the individual quarterly figures from the area averages are about 3 units on average, again as in 1969.
(iv) There are no exceptionally high figures in Quarter III in the East and West, or at any point in 1970.

TABLE 2.1a The 1969 and 1970 Quarterly Averages for each Area

	No	Ea	We	So	Av.
1969	95	75*	53*	45	67*
1970	95	75	54	44	67

* Excluding QIII in 1969

The new results are accompanied by more conviction just because they are largely the same as in 1969. This analysis has also been much easier to do than the original one because we already know something about this kind of data. This is an advantage of being something of an expert.

It might seem that the analysis was easier only because there is so much agreement between the two sets of data. But it was equally easy to establish that the two exceptionally high values in the East and West in QIII of 1969 did *not* recur in 1970. (It was therefore a good decision to treat them separately in the last chapter. But it would have been a good decision even if the exceptions *had* recurred, because then they clearly would have to be treated separately!) Furthermore, it is also easy to establish that hardly any of the other smaller deviations in 1969 occur again in 1970, as we shall now see.

2.2 Irregular Deviations

Table 2.2 sets out the deviations of the 1970 quarterly figures from the area averages (i.e. 96 − 95 = 1 for QI in the North, etc.).

TABLE 2.2 Deviations in 1970 from the Area Averages

1970	Area				
	No	Ea	We	So	Av.
Q I	1	-1	-1	2	0
Q II	-1	2	-4	6	0
Q III	-3	-3	5	-2	-1
Q IV	3	1	1	-7	0
Average	0	0	0	0	0

There is no systematic quarter-by-quarter variation and of necessity the area averages are all zero. Therefore only the individual deviations need to be scrutinised further. At first sight these appear to be mainly irregular. However, some of the individual deviations may recur from year to year and be regular in that sense.

For the 1969 deviations (repeated here in Table 2.2a), we recall that they were rather high in QIII and consistently low in QIV, possible seasonal trends, and that there also seemed to be a + − + − pattern, or negative "serial correlation". A glance at the 1970 deviations in the previous table shows that none of this recurs. In the 1970 data the only apparent regularities are that the QI deviations are all small and that the *East* column has a − + − + pattern, and neither of these things occurred in 1969.

TABLE 2.2a The 1969 Deviations (Table 1.6a)

1969	Area				
	No	Ea	We	So	Av.
Q I	3	0	-3	3	1
Q II	-3	0	4	-3	0
Q III	6	(25)	(27)	5	6*
Q IV	-5	-1	-2	-6	-4
Average	0	0*	0*	0	0*

* Excluding QIII

To check whether other "local sub-patterns" are consistent from one year to another, we can average the deviations over the two years. Two deviations which are inconsistent (one negative and one positive) tend then to cancel and produce a relatively small average. In contrast, consistent results reinforce each other and show up even more strongly. (A common alternative form of this kind of analysis is to average the original readings over the two years

TABLE 2.2b The 1969 and 1970 Deviations Averaged
(From Tables 2.2 and 2.2a)

Average of '69 and '70	Area				
	No	Ea	We	So	Av.
Q I	2	0	-2	2	1
Q II	-2	1	0	2	0
Q III	2	-2*	3*	2	1
Q IV	-1	0	0	-6	-2
Average	0	0	0	0	0

* Excluding QIII '69

and then to examine *their* variability.) Table 2.2b shows the average deviations over the two years.

Only two of the readings are 3 or more and the overall average has dropped to 2 from the yearly averages of 3. This shows that most of the deviations were inconsistent. However, the reading of −6 for QIV in the South stands out strikingly as a consistent result. With this as the sole exception, the analysis has shown that the residual deviations from the basic model are irregular—not only from quarter to quarter and area to area, but also from year to year. It is this *irregularity* of the readings that appears to be generalisable.

2.3 Empirical Generalisation

The analyses of the data for 1969 and 1970 have led to the empirical generalisation that the quarterly readings are generally about 95 in the North, 75 in the East, 53 in the West, and 44 in the South, that the deviations from this model are apparently irregular, and that these deviations average about ± 3 ("plus or minus 3").

This model is now known to hold under a fairly wide range of circumstances.

(i) In each of two years (1969 and 1970), despite all the other things that varied from one year to the other.
(ii) In the different quarters of the year, irrespective of the season, etc.
(iii) For different numerical values of the variable ranging from 39 in the South to 101 in the North.
(iv) Despite the large and unrepeated 1969 deviation of about 25 units in Quarter III in the East and West.

The resulting generalisation is limited only by the range of conditions covered. In order to extend it, all we have to establish is whether the same model of 95 ± 3 in the North, 75 ± 3 in the East, etc. also holds for data in other years and other countries, and for other related variables. If it does not, we must determine what sort of generalisable differences there are. But we no longer need to tackle each new set of data from the beginning. Use of prior knowledge avoids having to re-invent the wheel every time.

We can also use these results for prediction, e.g. that the 1969/70 patterns will hold for other years. If the prediction is successful, we have a further generalisation covering the new conditions as well.

2.4 The Beginnings of Explanation

The results so far have shown which factors do not matter, such as the seasons of the year, the year itself, and so on. Such conclusions about things that are *not* related are very simple and powerful. But we also need to explain the main variation in the data, i.e. the systematic differences among the four area results 95, 75, 53, and 44. Why is it 95 in the North and only 44 in the South?

To determine this we need to relate the area results (which refer to millions of units) to some other characteristic of each area. One possibility that springs to mind is the *size* of each area, e.g. the number of potential users of Product X, or the number of employed. Perhaps the results in the North are higher than in the South because there are more people in the North. Table 2.3 shows that this is so. Given the size of the populations in the four areas, about 30, 25, 20, and 15 millions, we can see that the average incidence is about 3 units.

TABLE 2.3 **The Area Averages and Population Size**

In millions	Area				
	No	Ea	We	So	Av.
Average Quarter	95	75*	53*	44	67
Population	30	25	20	15	22
Incidence per capita	3	3	3	3	3

* Excluding QIII '69

We now have a very simple summary statement: 3 per capita. This relationship with population size accounts for the area difference. It also accounts for the relative stability of the quarter-by-quarter figures, given that populations are generally stable over time.

2.5 Summary

In this chapter, the data for 1970 have been analysed against the background of the 1969 results obtained in Chapter 1.

The analysis illustrates two kinds of results which can stem from this use of prior knowledge:

(i) Further generalisation of previous results: here the recurring lack of systematic quarterly differences, the same area averages, and same size quarterly deviations from the area averages, at about ± 3.

(ii) Contradiction of previous results: the two exceptional 1969 QIII results do not recur in 1970, and the deviations of the individual quarters do not generally recur in 1970. (Instead it is the *irregularity* of the deviations that generalises.)

Combining the results for 1970 with those of 1969 has led to increased understanding and conviction. This is particularly so for the factors which we have learnt do *not* affect the results, such as the season of the year, the year itself, and anything else that changed from one year to the other.

The analysis of the 1970 data has also been much simpler to perform than the "first-time" analysis of the 1969 data in Chapter 1. This is because we already knew the 1969 results. It is one advantage of being something of an expert.

CHAPTER 2 EXERCISES

Exercise 2A. Predicting from Two Readings

The results for QIII in the East were 100 in 1969 and 72 in 1970. Discuss what prediction can be made about QIII in the East for 1971 solely from these two readings.

Discussion.

Because the readings are so discrepant and there are only two of them there is no prediction that one can confidently make. For example, the 1969 and 1970 readings are both "100 to the nearest 100", so that one might predict that the 1971 result should also be "about 100". But equally one could predict

about 81 (the average of 100 and 72),
about 40 (a drop of 30 units per year),
about 150 (based on the notion of some increasing cyclical up-and-down movement over the years).

One's confidence or "strength of belief" in a prediction can be assessed by imagining its outcome. Suppose a reading of 150 were actually observed in 1971. This could not be interpreted as a major discrepancy because the two initial readings were not enough to establish a pattern in the first place. Similarly, a reading of 81 could not be judged to be "as expected" even though it is the average of the two previous results; neither of the given readings lies near this average, so there is no firm reason why the third reading should.

Now suppose we had a third reading to start with, e.g. for 1968. This would begin to provide more predictive information. Three examples are:

	Given Data			
	1968	1969	1970	Prediction for 1971
Case (a)	150	100	72	About 50
Case (b)	79	100	72	About 85 $\pm$ 15
Case (c)	20	100	72	Between 0 and 150

Any 1971 outcome markedly different from that predicted, for example 150 in Case (a) or 500 in Case (c), would seem discrepant with the apparent pattern of the results. But three readings are still a thin base for prediction.

We can see this because *with hindsight*, a sequence of 150, 100, 72 and 150 in Case (a) and even 70, 100, 72, 500 in Case (c), would not appear at all impossible.

Yet another reading, say for 1971, could strengthen the picture:

	Given Data				
	1968	1969	1970	1971	Prediction for 1972
Case (a)	150	100	72	50	About 30
Case (b)	79	100	72	78	About 82 ± 15
Case (c)	20	100	72	120	Between 0 and 150, say

If we *now* had an outcome of 150 in Case (a), it would look peculiar even with hindsight; something different has occurred to change the pattern.

Exercise 2B. More Prior Knowledge

What prediction can be made about QIII in the East if the 1969 and 1970 readings in all areas and all quarters are taken into account?

Discussion.

There is no trend in the general pattern of 1969–70 results. Instead there is an irregular quarterly scatter of ±3 about each area average, which in the East was 75. The 1969 QIII reading of 100 in the East was an exception.

Therefore for any other year we predict QIII in the East will be 75 ± 3. (Any result well outside these limits would be discrepant with the directly relevant data for the East and with the pattern in all the other areas as well.)

Although the only "normal" reading for QIII in the East was 72 this is not the best prediction for QIII. If the QIII Eastern values in other years were generally about 72, then QIII in the East would be consistently lower than the three other quarters. This would be discrepant with all the other available findings for the other areas. Hence the yearly average of 75 for the East is the appropriate prediction even in QIII.

Exercise 2C. *Business Week*'s "Latest Week"

In Exercise 1.C we discussed *Business Week*'s report of the latest week's production data, accompanied by four earlier figures as interpretative background. The magazine also gave the following commentary:

> "General Motors scheduled only light overtime while Ford shut down operations at four assembly plants. Small declines occurred in steel and electricity. Crude oil refinery runs rose slightly."

The figures are shown here in the layout used in the earlier exercise. How can the results given in the week of April 24 be interpreted?

TABLE 2.4 **Business Week's Figures, April 24 1970**

	Paper	Steel	Coal	Electr.	Autos	Oil
1957-59 average	28	19	14	12	10	8
Year ago	52	27	20	27	14	11
Month ago	51	28	21	30	19	11
Week ago	52	29	21	29	16	11
Latest week	**50**	**29**	**22**	**28**	**15**	**11**

Discussion.

These figures cannot be interpreted without additional information such as:

the "normal" week-by-week changes that occur,
whether the single week's figure "a year ago" was normal,
whether special events account for any of the *earlier* figures, e.g. the very high auto figure of 19 "a month ago",
whether the economy is sensitive to weekly changes such as a 2% increase in oil production or a 1% drop in steel, and if so, in what way.

Exercise 2D. More *Business Week* Data

From back issues of *Business Week*, a continuous sequence of weekly figures can be reconstructed, as illustrated in Table 2.5 for the first sixteen weeks of 1971. How does this additional information help in interpreting the latest week's figures?

TABLE 2.5 **Week-by-Week Figures in 1971**

1971 Week ending		Production (per week) Paper	Electr.	Steel	Coal	Autos	Oil
January	2	24	29	24	20	6	11
	9	43	31	24	20	17	11
	16	50	31	25	19	19	11
	23	51	32	26	21	19	11
	30	50	31	27	20	18	11
February	6	50	32	26	20	19	11
	13	50	32	27	18	18	11
	20	51	30	27	18	20	11
	27	51	30	28	19	19	11
March	6	52	30	28	20	18	11
	13	51	30	28	20	19	11
	20	51	30	28	21	19	11
	27	51	30	29	21	19	11
April	3	52	29	29	22	18	11
	10	52	29	29	21	15	11
"Latest"	17	50	28	29	22	15	11
Average		49	30	27	20	17	11

Discussion

The new table firstly helps us to judge what is normal. We can now see that the "Month Ago" Automobile figure of 19 in Table 2.4 was not exceptionally high (as was thought in Exercise 1.C). Instead, Table 2.5 shows that this figure (March 27) is normal and that it is the other three figures in Table 2.4 which were unusually low. (Presumably there were short working-weeks for Easter in the weeks of April 10 and 17, 1971, and Easter "a year ago".) Similarly, the Paper and Automobile figures for the week of January 2 are seen to be low and seem to be tied with Christmas working arrangements in these two industries (this also occurred in the previous year's figures).

Next we can try to see whether this fuller background can help the businessman judge the current economic climate from the latest week's figures ("Shall I invest in the new plant this week, next week, sometime, never?"). We consider three industries with different properties in the early part of 1971: Steel (which shows a steady upward trend), Automobiles (which fluctuate), and Oil (which does not vary).

Steel: the steady upward trend in 1971 is new—it did not occur the year before. The trend however does not predict anything similar for the other five industries, because none of *them* show an upward trend. The figures therefore show that more steel is being produced, but we are given no back-log of other such cases to indicate what else this implies. (It might also be that the latest figure itself signals a levelling-off, but if so, we do not know what *that* would imply.)

Automobiles: the figures fluctuate and show no trend. Suitably adjusted to account for fluctuations due to short working-weeks, strikes, inventory adjustments, heavy overtime coming on, heavy overtime being eased off, and other factors annotated in *Business Week*, the 1971 figures might well be steady. Therefore, it is not clear what the latest weekly figures tell us.

Oil: the figures to two digits do not vary in 1971 (nor did they do so in 1970). Thus, the latest week's figures tell us nothing new. (The figures might appear to have been unduly rounded, since there is *no* variation at all, but to one decimal place they still show no trend: e.g. the five 4-week averages for Table 2.5 are 10.8, 11.0, 11.1, 10.8 and 10.9.)

It follows that, even in the context of additional data, the latest week's figures say nothing very clear about the general economic climate. Changes of 2% up or 3% down in one week as commented on by *Business Week*, "small declines occurred in steel and electricity and crude oil refinery runs rose slightly", therefore seem uninterpretable. Showing the figures to 3 or more digits is pointless and even the oil figures have probably not been rounded too much.

The data are too complicated for a layman to interpret immediately. But it should not be too difficult for a professional analyst to derive some understanding to pass on to the reader. For example, the steady rise in steel production in the first four months of 1971 may have some long-term "lagged" implications. Looking at previous years and other countries, a skilled analyst might for example establish that such an increase, unaccompanied by similar increases in other industries, generally represents something like a build-up in stocks preparing for an expected stoppage during the wage-negotiations later in the year.

If, however, skilled economists have been unable to determine any such generalisable patterns, then it is useless to expect the lay reader of *Business Week* to discover them himself.

Exercise 2E. Prior Information in Practice

Examine how prior information was used in the last two memoranda involving numerical data that you either received or reported, and the last two scientific papers or technical reports that you read. Did they make any detailed numerical use of previous results? If so, did it lead to increased confidence and understanding? If not, how would it have helped?

Exercise 2F. Rounding to Two Digits

Most people would be fairly happy to round readings to *three* significant digits, but baulk at *two*. People fear that too much information might be lost. Is the difference important?

Discussion.

Unfortunately, the difference between two and three digits is a big one. Most of us can do mental arithmetic with two-digit numbers, and we can memorise them. But we cannot do this with three-digit numbers (except by first mentally rounding them down to two!).

A simple example is percentages expressed to one decimal place. Few of us can divide 35.2% by 17.8% in our heads. But almost all can see at a glance that 35% is almost twice 18%. Being able to see patterns, structures and relationships is what really matters in understanding and interpreting data.

What then should we do in practice—should we round to two digits despite our fears?

With data that are being looked at for the very first time (as in the main example of Chapter 1), we will not know beforehand what degree of precision is really required. So we can *record* the data to three or more digits, just in case they might be needed. But the first analysis, to get the *feel* of the data, can best be done with two digits.

After this initial analysis we can then go back and rework the data with three or more digits, now that we know what we are looking for. But, in practice, one finds that this is only rarely called for!

In all other cases, i.e. when there is already prior knowledge of that type of data, we should be able to judge *beforehand* what precision is needed. If, in the past, useful interpretations or conclusions have been made from the variations in the third digit (e.g. 35.2% as against 35.3%), then three or more digits are called for. But this is not the case if the third digit is used only because there might be something interesting.

Data can always be recorded or stored somewhere in great precision, in case they are needed for further, more sensitive, analysis and research. But in trying to *communicate* data, a balance has to be struck between the possibility that occasionally something of possible interest is missed through undue rounding and the likelihood of not seeing *anything* in the data if they are presented in too much detail.

Ultimately the decision is one for each user of the data and each analyst to make for himself.

CHAPTER 3

Tables and Graphs

We have seen in the two preceding chapters ways in which extensive data can be reduced to a few summary figures. Tables of data were used for working purposes, but no tables were needed to communicate the final results. The same applies in other situations. Indeed, the more complex the data, the less can the task of understanding the undigested data be left to the reader.

The rule "No report should contain tables—with some exceptions" means that if the analyst *does* present a table of extensive data, he should have a specific (i.e. exceptional) purpose in mind. Two such purposes are to illustrate a summary statement or to provide a record of the detailed data for possible *later* analysis.

Sometimes graphs are considered easier to understand than tables of numbers. But graphs also require a clear purpose. They are discussed in the later part of this chapter.

3.1 Tables to Illustrate

Communication of key numerical results can be aided by *displaying* the figures, e.g. as

North 95, East 75, West 53, South 44,

instead of hiding them in the text. Displays of this kind, or even a mini-table like

Average deviation: 3 units

can make the results more memorable and easier to find again later when skimming the pages.

Tables for Proof

Sometimes tables of detailed data are reported for *proof*, since people do not always trust the analyst. Did he add the figures correctly? Has he given misleading averages? A "bare-bones" description, such as 95 in the North, 75 in the East, etc., can certainly leave room for doubt.

However, the analyst has some alternatives to providing tabular proof in the way he summarises his results.

(a) He can show that he is aware his data contain scatter, and that he knows roughly how big it is, by mentioning that the quarterly figures deviate by an average of about 3 units. (At least he should say, or imply, that the deviations are small.)

(b) He can demonstrate that he has actually looked at the data in some detail by mentioning that the scatter is apparently irregular, and that its size does not vary much from area to area or quarter to quarter.

(c) He can show that he has sorted out the nature of the data and that he is not trying to hide anything by mentioning the two large QIII exceptions. (Reference to large exceptions—or to the *absence* of any special features—is also a quick way of implying the previous steps without having to spell them out explicitly.)

All this can then be usefully illustrated by a clear and well-digested tabular *extract* of the available data, here for *one* year as in Table 3.1, not for *both* years. This makes the summary results easier to comprehend and accept. It shows what kinds of averages are referred to, and the reader can mentally note how the quarterly readings do in fact differ from these by something like the stated 3 units (i.e. 3, −3, 6, −5 in the North; 0, 0, −1 in the East and so on).

TABLE 3.1 **The 1969 Area Results Quarter-by-Quarter**

1969	Area				
	No	Ea	We	So	Av.
Q I	98	75	50	48	68
Q II	92	75	57	42	67
Q III	101	(100)	(80)	50	(83)
Q IV	90	74	51	39	64
Average	95	75*	53*	45	67*

* Excluding QIII

Detailed data "for proof" are usually requested only when very simple analyses are presented (e.g. averages and percentages). Few readers expect to be shown the full raw data when *complex* analysis techniques have been used, but the reader should still want some assurance that the analyst himself has looked at the data. This is not provided by simply reporting all the facts. Adequate forms of summary reporting and illustration therefore become particularly important.

Tables to Dramatise

Some repetition or redundancy generally improves communication. The fact that the quarterly figures in our example varied relatively little is brought out in Table 3.2.

TABLE 3.2 **The Fit of the Theoretical Model**

(Observed and **Theoretical** Values)

		Area								
	No		Ea		We		So		Av.	
	Ob	**Th**	Ob	**Th**	Ob	**Th**	Ob	**Th**	Ob	**Th**
1969 Q I	98	**95**	75	**75**	50	**53**	48	**44**	68	**67**
Q II	92	**95**	75	**75**	57	**53**	42	**44**	67	**67**
Q III	101	**95**	100	**75**	80	**53**	50	**44**	83	**67**
Q IV	90	**95**	74	**75**	51	**53**	39	**44**	64	**67**
1970 Q I	96	**95**	74	**75**	53	**53**	46	**44**	67	**67**
Q II	94	**95**	77	**75**	49	**53**	50	**44**	68	**67**
Q III	91	**95**	72	**75**	59	**53**	42	**44**	66	**67**
Q IV	98	**95**	76	**75**	53	**53**	37	**44**	66	**67**
Average	95	**95**	75*	**75**	53*	**53**	44	**44**	67*	**67**

* Excluding QIII in 1969

Here the area averages are explicitly promoted to the status of a theoretical model and shown alongside the observed figures. This makes the general similarity of the observed quarterly figures more apparent. Such a demonstration of the agreement between observed and theoretical figures becomes even more important when the theoretical figures themselves vary, as many cases later in this book will illustrate.

Dramatisation of this kind requires a well laid-out table. It need be done only on a small illustrative basis. Graphs can also be effective here, as long as they present a single, clear-cut story-line.

What Makes a Good Table?

In Table 3.2 the reader still has to work out the differences between the observed and theoretical figures (e.g. $98 - 95 = 3$, $92 - 95 = -3$, etc.). This appears to negate the "Don't leave it to the reader" precept. The reader's task might seem easier if the differences between each pair of figures were given *explicitly*, e.g. as the "D" values 3, -3, 6, etc. in Table 3.3.

However, this table is far too complex. If the intention were to give the reader an easy view of the "difference" figures, making his eye jump three columns to compare like with like is not much help. Placing different kinds

TABLE 3.3 **A Table showing the Observed and Theoretical Results (Ob and Th) and the Differences (D) between Them**

1969	Area												Av.		
	No			Ea			We			So					
	Ob	**Th**	D	Ob	**Th**	D	Ob	**Th**	D	Ob	**Th**	D	Ob	**Th**	D
Q I	98	**95**	3	75	**75**	0	50	**53**	-3	48	**45**	3	68	**67**	1
Q II	92	**95**	-3	75	**75**	0	57	**53**	4	42	**45**	-3	67	**67**	0
Q III	101	**95**	6	(100)	**75**	*	(80)	**53**	*	50	**45**	5	(83)	**67**	*
Q IV	90	**95**	-5	74	**75**	-1	51	**53**	-2	39	**45**	-6	64	**67**	-3
Av.	95	**95**	0	75*	**75**	0	53*	**53**	0	45	**45**	0	67*	**67**	0

* Excluding QIII

of figures next to each other, and separating ones that need comparison, should generally be avoided.

The golden rule is that a table is good if in looking at it the *next* steps are easy to do. In Table 3.3 this is not so; there is nothing simple that can be done with it. In contrast, the mental arithmetic needed in reading Table 3.2 is easy because the observed and theoretical readings are adjacent and only one or two digits are involved. For example, it is simple to make a visual check of the summary statement that the differences are irregular and average about 3. Furthermore, actually working out these differences and running the eye down the North column to see that 3, −3, 6, −5, 1, −2, etc. *do* average at about 3 (as stated in the text) is a good learning process. It helps the reader to assimilate the information more deeply.

In more complex situations the differences between the observed and theoretical figures might have to be shown explicitly. Then a separate display, as in Table 3.3a, tends to be best. Here again it is easy for the reader to take the next step mentally, like seeing that the QIV values are all negative, and those in the East all small.

TABLE 3.3a **The Differences between the Observed and Theoretical Figures**

1969	Area				Av.
	No	Ea	We	So	
Q I	3	0	-3	3	1
Q II	-3	0	4	-3	0
Q III	6	(25)	(27)	5	6
Q IV	-5	-1	-2	-6	-3
Average	0	0	0	0	0*

* Average size ignoring sign = 3

3.2 Tables for the Record

Instead of presenting data to aid immediate communication, many tables set out data for possible subsequent use. Extreme examples are census reports and other official statistics where data are published which nobody has as yet attempted to analyse. The data are provided for the record, so that anybody can refer to them at some future stage.

Storage of undigested data is becoming even more common nowadays with the development of computerised data banks and information systems. The idea is to provide "all the facts at the touch of a button". But this can be overdone. Great sums are invested in elaborate methods of data-retrieval, even though no one knew what to do with the data before they were lost in the first place. (A fraction of this expenditure devoted to some *analysis* of the data would often yield results that could be stored along more old-fashioned lines, in filing-cabinets or books, or that could even simply be *remembered.*)

Data for Reanalysis and Research

Less extreme examples of data storage are cases where the data *have* already been analysed and summarised but the full details are provided in case somebody wants to reanalyse them, usually for research purposes. In this situation, the data need to be given in an Appendix at most, or copies of the data might be kept and made available to the occasional enquirer. They need not be reproduced in every report.

Data One Does Not Understand

Many published tables of data are ones the author did not understand. He is in effect saying to the reader "Here are some data. I do not understand them. Perhaps you can." This is on the whole not the way to address either one's colleagues or the public at large. If the *author* has not been able to summarise and communicate the main patterns of the data, what chance is there for anyone else to do so?

However, there are situations where this seems a legitimate approach. Table 3.3a of the quarterly deviations was given deliberately to emphasise that one could not see any pattern. The readings *appeared* irregular, but perhaps somebody else could make something of them?

Another example of undigested data given legitimately was the presentation of the summary figures:

North 95, East 75, West 53, South 44.

Here, one was saying that one does not know why these figures differ from each other. But at least they represent an effective summary of much more extensive data, and the potential pay-off is there: somebody might suggest checking the size of each area. As noted in Chapter 2, this leads to the simple result of 3 units per capita. In this case, reporting data one does not understand led to progress. However, this usually occurs only when the figures are already well-digested summary results rather than the raw, and unorganised, original observations.

3.3 Graphs

Graphs have two uses: for working-purposes and for presenting final results. In the first case, the analyst may plot the readings in the early stages of the analysis to gain a quick visual impression of whether a relationship is linear, as in Figure 3.1A, and whether the scatter is "homoscedastic" (equal scatter all along the line) or heteroscedastic, as in Figure 3.1B. Such graphs are for private use. They may be rough (drawn on the back of an envelope) and will be superseded by a proper analysis of the data, leading to summary figures or models.

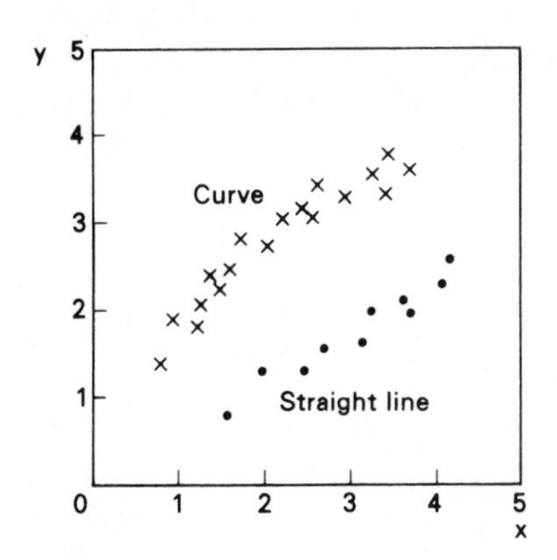

Figure 3.1A Plotting y against x

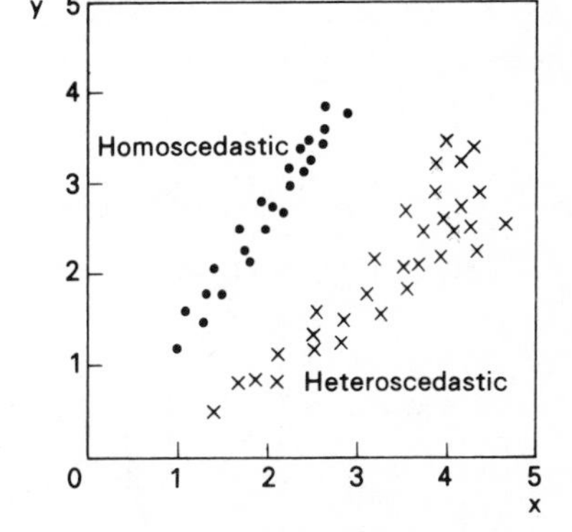

Figure 3.1B Constant or Increasing Scatter

The use of graphs when presenting final results is either to simplify or enliven them: in a word, to make the results more *graphic*. To achieve this both the limitations and the strengths of graphs need to be understood.

Figures 3.2A and 3.2B give monthly temperature and rainfall results for England and Wales from *Facts in Focus* (1972), a handbook compiled by the U.K. Central Statistical Office. It is clear that it is hotter in the summer and rains less in the spring. What else do the graphs tell us quickly? That the 1970 temperatures were mostly close to normal, but the 1970 monthly rainfall figures differed a good deal from the longer-term averages? Is *that* the kind of thing the chartist meant us to see? If so, should it be up to the *reader* to note that although the November rainfall was exceptionally heavy, this may have been more than made up for by the relatively dry months since May?

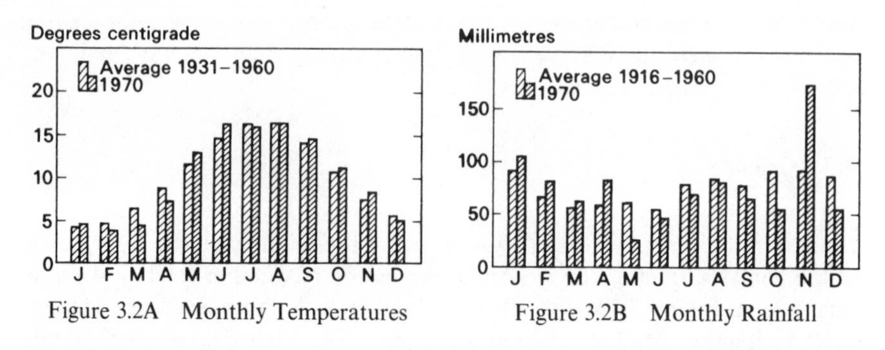

Figure 3.2A Monthly Temperatures

Figure 3.2B Monthly Rainfall

A graph must have a well-digested message. If the analyst does not know what the figures are saying, then this will be the message he communicates to the reader.

Two further examples are Figures 3.3A and 3.3B, giving the age-breakdown of men in the U.K. over 4 decades and the growth in public expenditure in the U.K. since 1953. It is hard to read off the detailed figures. Is the

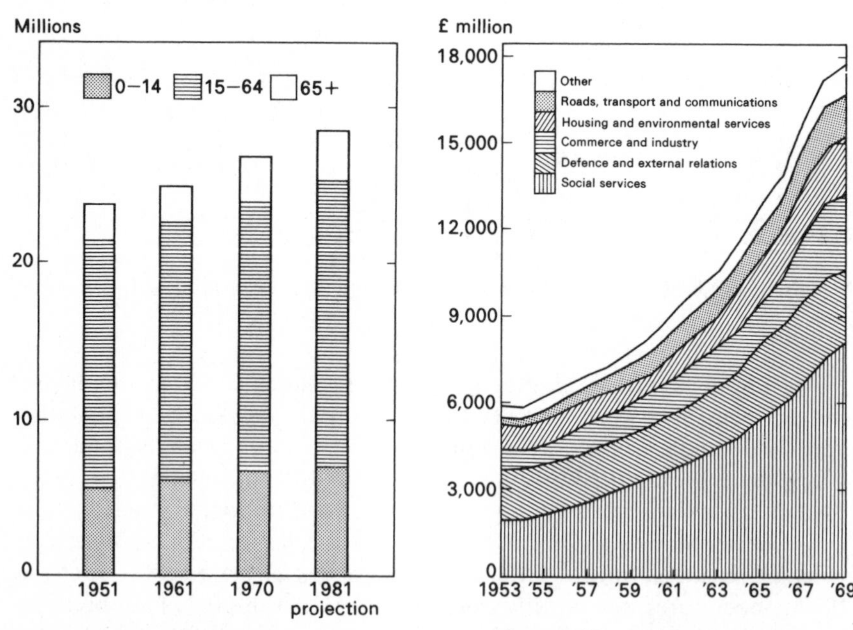

Figure 3.3A The Age-Distribution of Males

Figure 3.3B Public Expenditure

proportion of old men less than it used to be? Which of the areas of public expenditure have grown more than proportionally, and which less? These graphs illustrate the difficulty of taking in *quantitative* information from a graph. (In the forthcoming revised edition of *Facts in Focus*, Figure 3.3A has been dropped completely and Figure 3.3B replaced by a table plus a chart showing the detailed position in just one particular year.)

In general, graphs can demonstrate a *qualitative* result—e.g. "Public Expenditure has gone up", or "Most people are aged 15 to 64", or "It is hotter in the summer". But graphs cannot communicate *quantitative* results (although it is this quantitative detail which is generally quite laborious to plot initially). Even with pictorial devices, such as the population data in Figure 3.4A from the *Pocket Data Book: USA 1971* or the "pie-chart" in Figure 3.4B of the Incidence of Poverty in the U.S., we do not get our information by counting little men or trying to assess the size of the angles. Instead we look at the *numbers*.

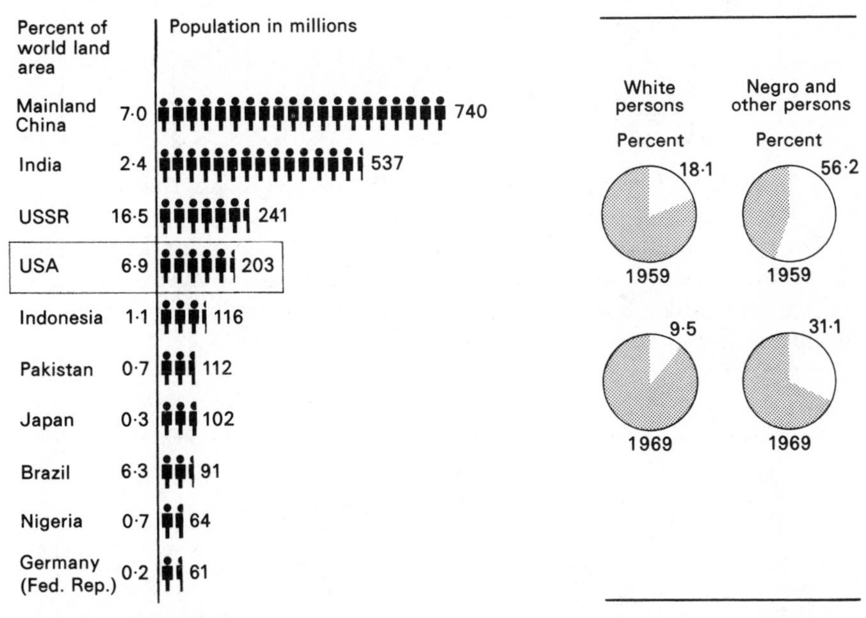

Figure 3.4A The Ten Largest Countries

Figure 3.4B The Incidence of Poverty

Graphs are generally useless for detailed analytic purposes, since it is impossible to manipulate any of the readings. In contrast, with a table it is easy to form averages, to take differences, etc. More important still, graphs seldom present a succinct and memorable summary of the results, except at a very broad qualitative level, like "It is hotter in the summer".

The Strengths of Graphs

The kind of story a graph can tell well is that something is *constant*. This is a qualitative property that is easy to take in visually. Thus Figure 3.5A, showing that B is bigger than A, is not a very good graph because it does not show clearly how *much* bigger B is (twice as big, *more* than twice, or *less* than twice?). Figure 3.5B is much more effective because it is clear to the eye that the increase from A to B is about the same as that from B to C, i.e. more or less constant.

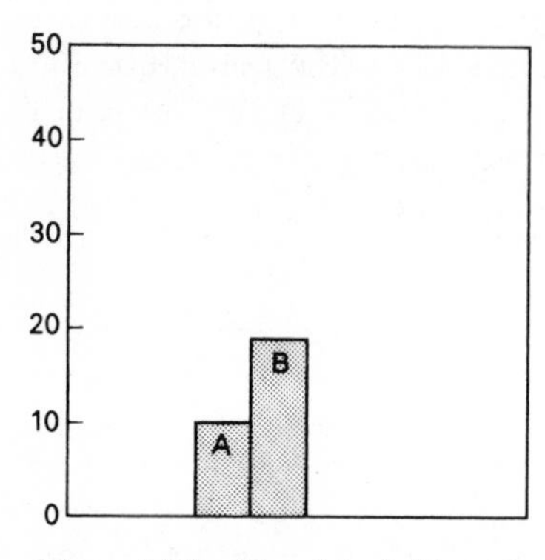

Figure 3.5A How Much Bigger?

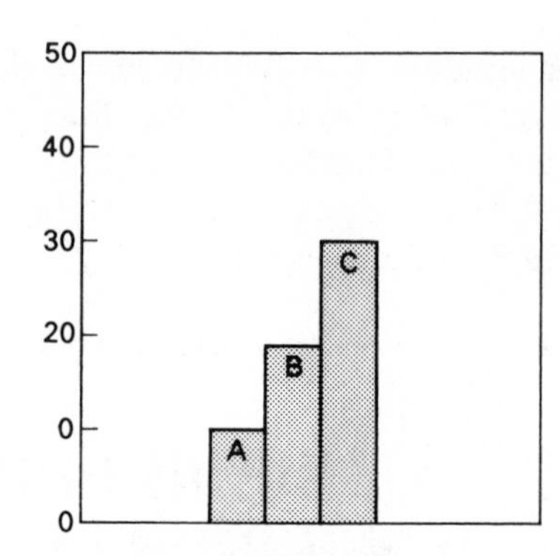

Figure 3.5B Equal Differences

The typical example of a good visual story is a straight line, as in Figure 3.6A, where the increase in y for any unit increase in x is always the same. By the same token a simple *curve* is also easy to appreciate visually as *not* being a straight line.

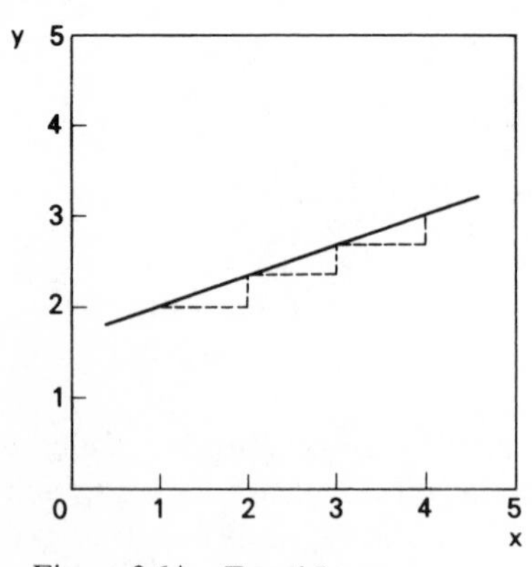

Figure 3.6A Equal Increases

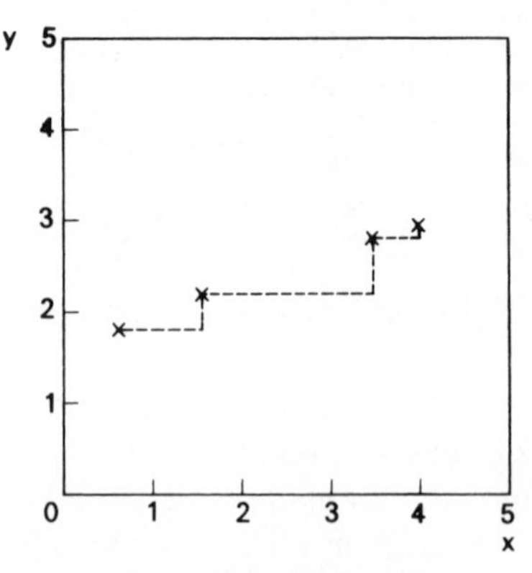

Figure 3.6B Proportional Increases

To be really effective a graph has to show that one variable is more or less constant *despite another variable varying a great deal.* The precise amount of variation involved is unimportant—a graph cannot easily communicate that anyway. Figure 3.6B illustrates how this holds for a straight line. It is visually easy to see that for any two points the ratio of the difference in y to x is the same, irrespective of how far apart the points are.

It follows that the form of a graph should generally depend on the numerical pattern of the data to be represented, and not on the nature or meaning of the variables as such. For example, in Figure 3.7A, the number of families with high incomes has remained roughly the same over the 30-year period, but the way it is charted does not make this particularly clear. Putting the high income group at the *bottom* of each column, as in Figure 3.7B, makes the pattern obvious.

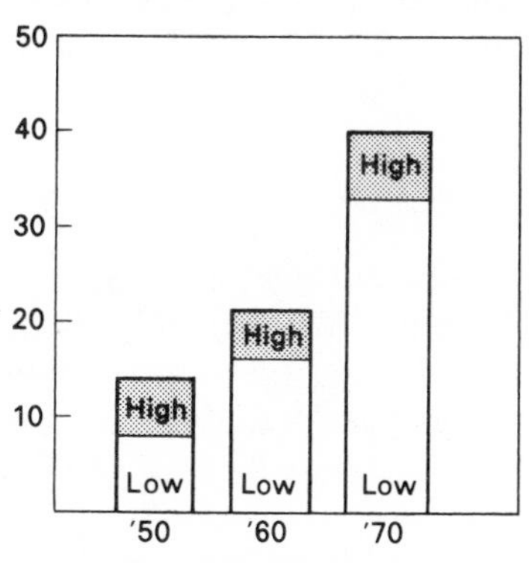

Figure 3.7A High and Low Incomes

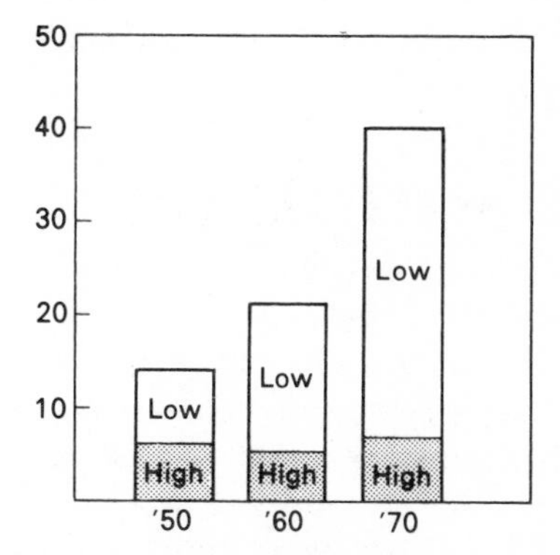

Figure 3.7B Constant Numbers with High Incomes

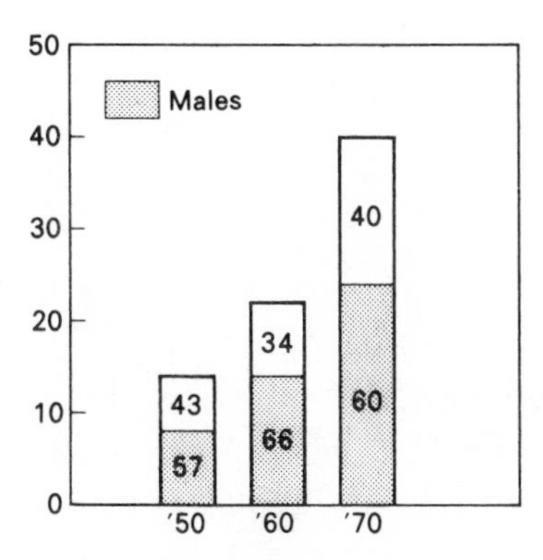

Figure 3.8A Males and Females

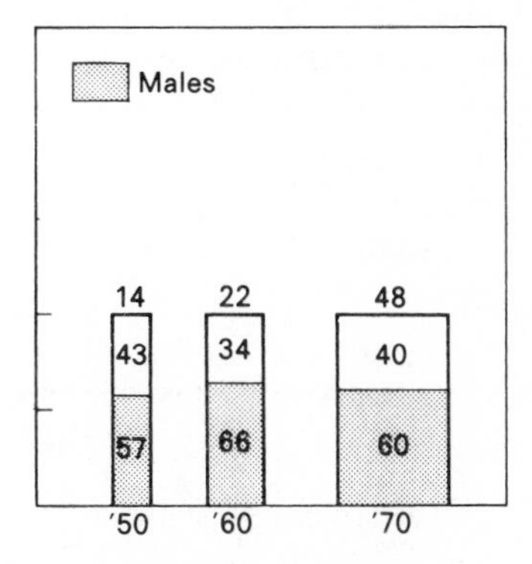

Figure 3.8B Approximately Constant Proportions

In Figure 3.8A it is not the number of men but their *proportion* that has remained fairly constant, at about 60% over the three decades. But neither this nor the degree of variation about 60% is altogether clear from the figure. In contrast, Figure 3.8B makes real use of the *graphic* tool by showing the approximately constant proportion of males as a line that remains more or less at the same level, despite the large increase in the total numbers. Figure 3.8B would tell its story *without* any of the numbers being written in, but these help to make the result more memorable: i.e. that despite the three-fold increase in population from 14 to 48, the proportion of males has stayed at about 60%.

Once we have a clear story, it is no longer so obvious that we need a graph to demonstrate or dramatise it. A few words may do ("about 60%"). If it is not clear what a graph is meant to say, probably more analysis of the data is needed. It is often said that graphs are easier for people to understand than tables of numbers, but this comparison is usually between simple graphs and undigested tables.

3.4 Summary

Undigested tables of numbers leave the work to the reader but have a place when the only aim is to provide records for possible subsequent analysis. Otherwise, tables should be presented only to support or illustrate summary statements that are made about the data. A table needs then to tell a clear story, and the next steps in looking at the numbers visually should be easy for the reader to do.

Graphs also need a simple story-line to communicate. They can be used to highlight a *qualitative* feature of the data. In particular, they excel at showing that one aspect of the data is more or less constant although another varies a great deal.

The use of tables and graphs for the analyst's own working-purposes is distinct from their use in communicating the results.

CHAPTER 3 EXERCISES

Exercise 3A. The Function of Tables

Go through the tables in this book and establish their function.

Discussion.

The tables will be found generally to be either:

records of data for analysis
intermediate working-tables
tables to illustrate results.

The crucial test for the latter purpose is that the tables could be omitted with only slight adjustments to the wording of the text. Their role of aiding communication usually relies on deliberate redundancy.

Exercise 3B. Data for the Record

Table 3.4 reproduces Table 1 from the Bureau of the Census's *Pocket Data Book: USA 1971*, a typical table of detailed records. Discuss the nature and role of such a table.

TABLE 3.4 **U.S. Population and Area from 1970**

	Resident Population			Area (1,000 sq. mi.)	
Year	Number (1,000)	Percent increase over prior year shown	Per square mile of land area	Land	Water
1790	3,929	(X)	4.5	865	24
1800	5,308	35.1	6.1	865	24
1810	7,240	36.4	4.3	1,682	34
1820	9,638	33.1	5.5	1,749	39
1830	12,866	33.5	7.4	1,749	39
1840	17,069	32.7	9.8	1,749	39
1850	23,192	35.9	7.9	2,940	53
1860	31,443	35.6	10.6	2,970	53
1870	39,818	26.6	13.4	2,970	53
1880	50,156	26.0	16.9	2,970	53
1890	62,948	25.5	21.2	2,970	53
1900	75,995	20.7	25.6	2,970	53
1910	91,972	21.0	31.0	2,970	53
1920	105,711	14.9	35.6	2,969	53
1930	122,775	16.1	41.2	2,977	45
1940	131,669	7.2	44.2	2,977	45
1950	151,326	14.5	42.6	3,552	63
1960	179,323	18.5	50.5	3,541	74
1970	203,185	13.3	57.4	3,541	74

X Not applicable.

Discussion.

The primary purpose of official statistics is to record facts accurately. But not much is gained by recording data to six digits or the like. While it can be argued that very precise figures ought to be available somewhere, a general guide like a *Pocket Data Book* does not seem the right place. The aim here is to create some *understanding* of the data and generally to help the user.

This is recognised in the table by giving two columns of *derived* statistics, the percentage increase and the population per square mile. Each quantity could easily be calculated from the other figures recorded but has been given explicitly to aid the reader. But these derived figures are also given in too much detail to see the pattern easily.

Table 3.4a gives the figures in rounded form. We now see that since the founding of the United States,

the population has increased 50-fold (from about 4 million to 200 million),

the area has increased about 4-fold (from about 900,000 square miles to 3.5 million),

the density per square mile has increased 12-fold (from about 5 to almost 60).

The population increased by about 35% every 10 years until 1860, then by about 25% until the end of the 19th century, and by about 15% this century, with a low of 7% over the 1930's.

TABLE 3.4a The U.S. Population and Area Data to 1 or 2 Significant Digits

	Resident Population			Area (1,000 sq. mile)	
Year	Millions	% Increase	Per sq. mile	Land	Water
1790	4	-	5	860	20
1800	5	35	6	860	20
10	7	36	4	1,700	30
20	10	35	6	1,700	40
30	13	34	7	1,700	40
40	17	33	10	1,700	40
50	23	36	8	2,900	50
60	31	36	11	3,000	50
70	40	27	13	3,000	50
80	50	26	17	3,000	50
90	63	26	21	3,000	50
1900	76	21	26	3,000	50
10	92	21	31	3,000	50
20	110	15	36	3,000	50
30	120	16	41	3,000	40
40	130	7	44	3,000	40
50	150	15	43	3,500	60
60	180	18	51	3,500	70
70	200	13	57	3,500	70

These simple overall results are the kind it is important to have well embedded in one's mind when looking at comparable figures for individual states, regions or cities, or for other countries, or at data for other related variables.

If may also be noted that the two columns of derived statistics are now hardly necessary. The main recorded figures are easy enough to use for rough mental arithmetic. For example, anyone can see that between 1890 and 1900, say, the population had increased by about 20% (i.e. 76 − 63 = 13, which is a fifth of 65), and that the density had increased to 76 million persons over 3 million square miles, which is about 25 per square mile.

Of course, something has been lost in the rounding, and the question is whether it matters. Perhaps rounding the water area to just *one* significant digit was too much. Nonetheless, it is still noticeable from Table 3.4a that between 1920 and 1930 somebody lost about 10,000 square miles of

U.S. water (or 8,000 to be more exact, from Table 3.4). The fact that the landmass of the States *increased* at the same time by 8,000 square miles can admittedly no longer be discerned in the rounded figures. One's guess is that these changes were due to a minor redefinition of land and water; but it is doubtful whether that is what one should be learning from Table 1 of the U.S. *Pocket Data Book*.

Exercise 3C. An Information System

In dealing with unemployment or balance of payment results, sales or production data, etc., the latest figures are usually examined with great care and the following type of table may be circulated or published.

	Latest Quarter	Previous Quarter	Quarter Year Ago
North	92	98	96
East	72	76	74
West	57	53	53
South	48	37	46

Suppose that these figures report the same kind of data that we discussed in Chapters 1 and 2. What is the value of providing a table of such up-to-date information?

Discussion.

The interpretation of such data is usually left to the reader, although some commentary might be given, e.g. that the latest quarter's results in the North and East are down and those in the West and South are up.

This makes no explicit use of the earlier analysis of previous results, which showed that quarterly figures varied irregularly by about ± 3 around the area averages of 95, 75, 53 and 44. It follows from these results that the latest quarterly figures are well within these usual limits of variation. Since one has never been able to interpret such variations in the past (not even with hindsight well after the event) the new table of figures is literally meaningless. All one need report is that the figures are normal.

Exercise 3D. A Graphical Representation

Later on in this book (in Exercise 10G of Chapter 10) the figures in Table 3.5 are discussed. Without going into the fuller meaning of the data here, develop a graphical representation.

TABLE 3.5 Attitudinal Data from Chapter 10

	% of the Population Using	"Right Taste"		"Convenient"	
		Users of Stated Brand	Non-Users of the Brand	Users of Stated Brand	Non-Users of the Brand
Brand A	50	67	6	19	3
Brand C	30	62	4	55	48
Brand B	10	69	5	17	2
Brand D	5	60	3	17	2

Discussion.

To represent these data graphically, we do not necessarily have to understand their meaning, but we must be able to see the patterns in them.

We can note certain regularities, e.g. that the figures in each "Users of Stated Brand" column are generally similar to each other and much higher than those in the Non-Users column, despite the fact that the numbers of users (and non-users) varies markedly among the brands.

Figure 3.9 brings out these results graphically. It shows which figures are similar and makes a dramatic exception stand out (like "Convenient" for both users and non-users of Brand B), without letting the differing incidence of users distract attention.

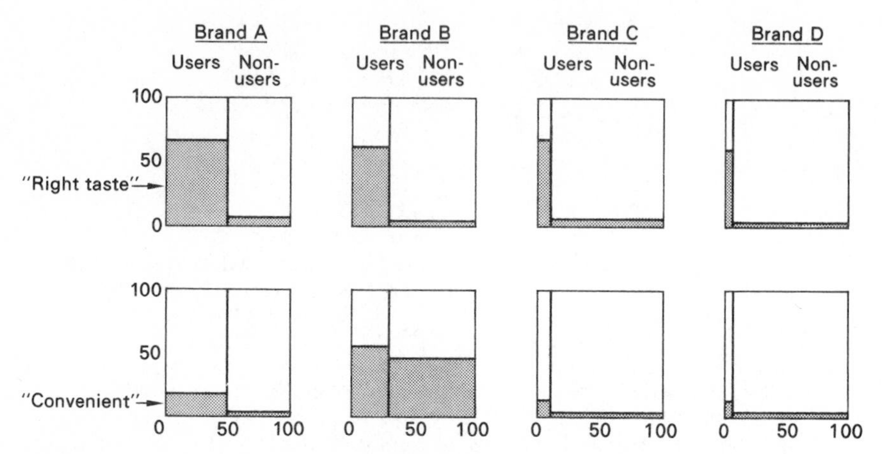

Figure 3.9 Approximately Constant Proportions of Users and Non-Users of a Brand Holding an Attitude, with Exceptions

Exercise 3E. Misleading Graphs

It is commonly alleged that graphs can be used to mislead by not showing the zero-points on the vertical scale. Thus a very small increase, as in Figure 3.10A, appears very large in Figure 3.10B.

Discuss this and some other misleading graphs.

Discussion.

Like most other things, graphs *can* be used to mislead, but anybody who does not look at the scales is likely to misunderstand *any* graph. It would be pointless to plot the data as in Figure 3.10A if the readings never varied by more than a few units from 80.

Changing the units on the *horizontal* scale can also influence one's view of the data, as shown in Figures 3.11A and 3.11B. In Figure 3.11A there appear to be some important increases and decreases, but in Figure 3.11B these same movements are merely irregular errors.

The danger with graphs is not so much misleading others as misleading oneself. Figure 3.12A shows the percentage of illegitimate births over the last century in England and Wales, Northern Ireland, and Scotland (from

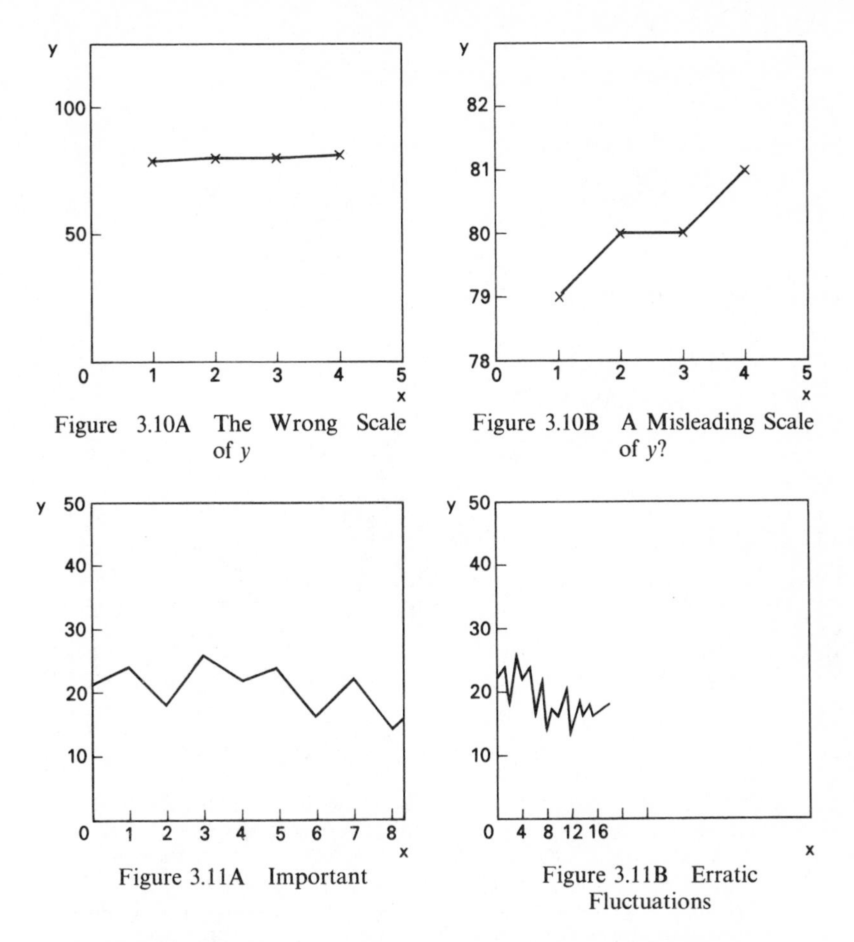

Figure 3.10A The Wrong Scale of y

Figure 3.10B A Misleading Scale of y?

Figure 3.11A Important

Figure 3.11B Erratic Fluctuations

Variations

Britain in Figures, Sillitoe, 1971). Most people would agree that the graph shows that illegitimacy rates dropped markedly from 1870 until about 1960 and then (in more permissive times?) rose again sharply.

But in England and Wales (by far the biggest of the three regions) the illegitimacy rate hardly declined at all (presumably young Scottish and Irish girls flocked to London during the '30s). Yet somehow the graph does not make this clear.

Figure 3.12B shows two financial indices over a period of 15 years. The visual impression is that the two indices go together. But this is partially misleading. Not only are the later blips in the continuous line far more extreme than those of the broken line (whereas earlier the scales of the two were more similar), but the phasing of the variations is variable. Sometimes the two lines vary together, sometimes the continuous line leads the broken one by a year, sometimes it lags by a year. Yet one does not immediately "see" this when looking at the graph (and many governments behave as if they had not noticed it either).

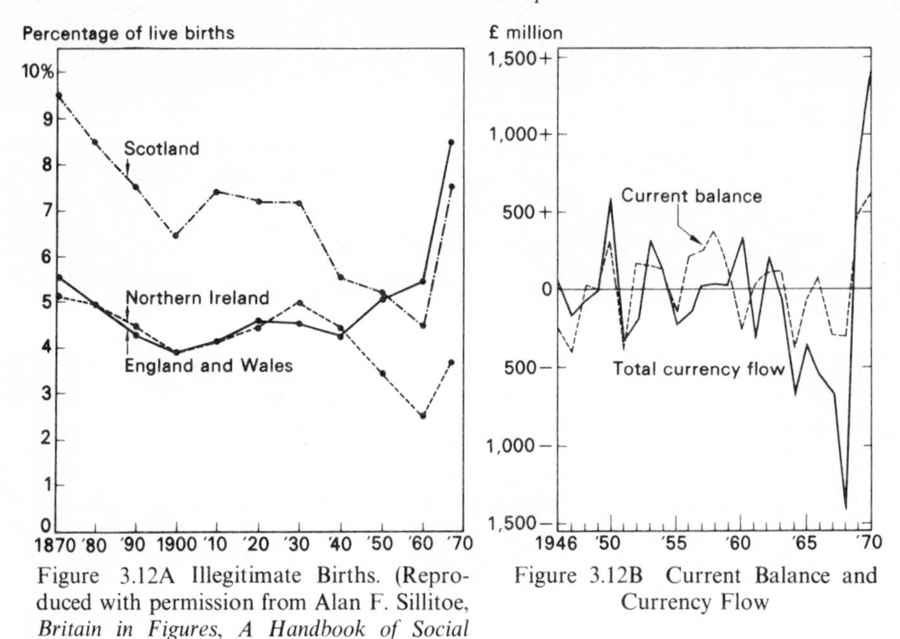

Figure 3.12A Illegitimate Births. (Reproduced with permission from Alan F. Sillitoe, *Britain in Figures, A Handbook of Social Statistics* (Pelican Original 1971). Copyright © Alan F. Sillitoe, 1971.)

Figure 3.12B Current Balance and Currency Flow

Exercise 3F. Graphs to Dramatise

Figure 3.13A shows time-series for two types of employment statistics (*Facts in Focus*, 1972). It is obvious that when unemployment is high, unfilled vacancies are low, and vice versa. Would it be advantageous to plot the unfilled vacancies as in Figure 3.13B, in effect reversing the scale and plotting a quantity like (800,000 minus Unfilled Vacancies), so that the variations go parallel to those for the figures of unemployed?

Discussion.

Plotting the data in parallel as in Figure 3.13B would help if the aim were to show the quantitative details of the data, e.g. that a particular variation in unemployment is larger or smaller, earlier or later, than the corresponding variation in unfilled vacancies. But a graph will not succeed in communicating such detail until the analyst himself has worked out precisely what the patterns are.

The original Figure 3.13A may in fact be better precisely because it stops us looking at quantitative details, as one is tempted to do in Figure 3.13B, which may obscure the main picture. It also avoids using a rather arbitrary-looking measure like (800,000 minus Unfilled Vacancies). Further, since it is so easy for the reader to work out that the two lines in Figure 3.13A are in broad agreement, this acts as a *learning* device. (The reader is pleased with himself for having worked it out.) Thus for presenting the one point, that unemployment and unfilled vacancies tend to be negatively correlated, Figure 3.13A may be more effective.

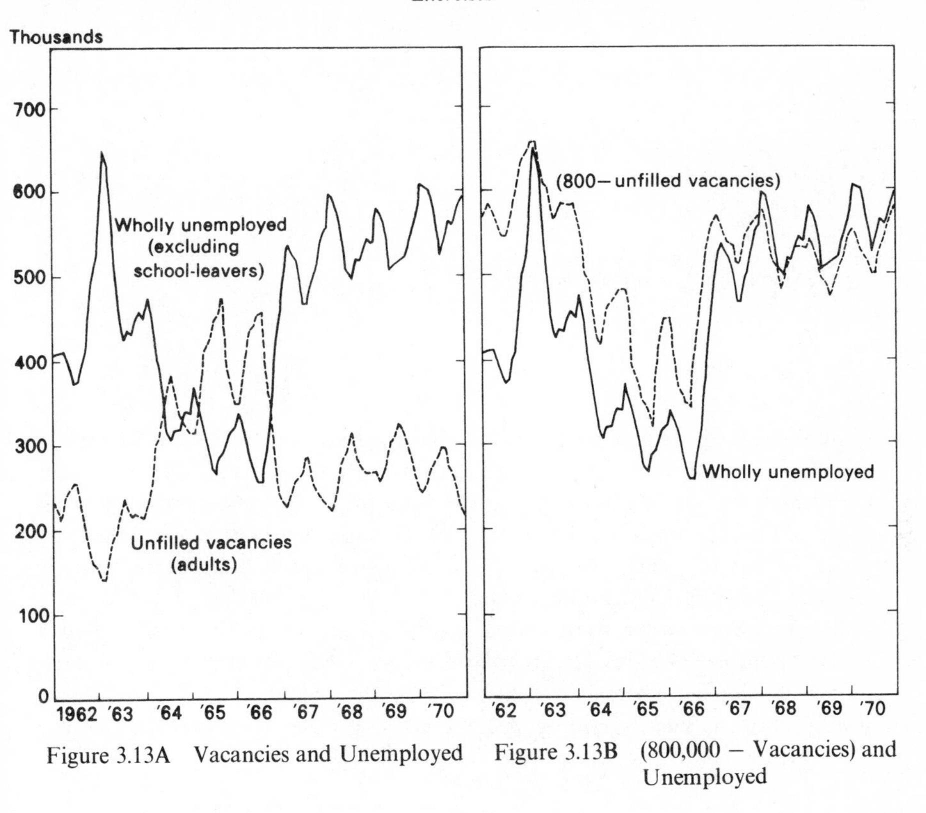

Figure 3.13A Vacancies and Unemployed

Figure 3.13B (800,000 − Vacancies) and Unemployed

CHAPTER 4

Compared With What?

A numerical result cannot be interpreted in isolation. Instead, it has to be compared with other findings. For example, a temperature of 92°F is high compared with most air temperatures in London, but low compared with body temperatures. The *implications* of any fact, whether it is "good" or "bad", and what one should do about it, also depend upon seeing how it fits into the wider scheme of things.

In this chapter we try to develop a greater feeling for the practical interpretation of an observed result. Any new finding needs to be interpreted against empirically-based norms or concepts. The crucial step is to have found usable patterns and relationships by analysing other measurements of the same type.

4.1 A Simplifying Concept

Many problems arise year after year. The answers, if only we knew them, should therefore also be similar year after year.

This changes the emphasis from collecting new data to organising and using past information. The process can be illustrated by the ordinary problem of forecasting how long it will take to drive from town A to town B (cf. Ehrenberg, 1967). Such forecasts affect many practical decisions: what form of transport to use, when to leave A to arrive in B at a certain time, whether to stay overnight, when to return, whether to go at all, etc.

The standard market research or administrative method of providing information for these decisions is to collect facts: e.g. to arrange for some investigators to drive routinely between towns A and B, C and D, etc., and regularly report their latest driving times. This would produce the kinds of large arrays of tabulated statistics which are well-known. Usually any extensive analysis of the data is precluded by the high cost of collecting it and the demand for quick, practical decisions. Since even the latest facts are already just out-of-date, it seems difficult to wait for (and pay for) a lengthy analysis.

At some stage however, a more systematic understanding of the data may evolve, or perhaps a sudden discovery is made. The data that the travelling investigators have been collecting can generally be described by dividing the distance between each pair of towns by the number 35. This figure represents a new theoretical concept called the "average speed".

Forecasting now becomes much easier and cheaper: one simply looks up the mileage between any two towns and divides by 35. Most of the forecasts are successful because the notion that average speeds are approximately constant is based on driving times obtained under a wide range of conditions. But there are errors, and some of them are large.

Using the simple law "time = distance/35", further study leads to categorisation of the forecasting errors. It develops that these are broadly of four kinds.

(a) Errors attached to the type of road; more accurate forecasts result by making simple generalisable allowances for this in the calculations (e.g. an average speed of 55 mph for motorways, adding 30 minutes for leaving or entering a conurbation, etc.).

(b) Errors related to particular driving occasions, such as weekend traffic, being affected by fog or ice, making better speed by travelling in the dead of night, etc.

(c) Errors related to the specific driver or car, such as a fast driver or a mechanical breakdown.

(d) Errors that cannot be accounted for yet.

In none of these situations would knowledge of the latest up-to-the-minute driving times be of any help.

Successful forecasting of driving times therefore requires three types of information: (i) the generalisable law that "time taken = distance/35", (ii) generalisable adjustments concerning motorways, conurbations, etc., and (iii) specific facts about distances, types of roads, weather forecasts, etc., to slot into the laws and sub-laws. No direct measurements of recent driving times are needed.

The practical man might consider it too risky to rely on a mathematical formula and correction factors; he would rather have the facts. Yet, in practice, no one would dream of running routine surveys to measure the latest just-out-of-date driving times, unless he were doing academic research. In practical life, we use generalisations and concepts rather than isolated facts.

4.2 High, Low or Normal: The Relationship Between I and U

Much decision-taking depends upon the interpretation of some specific figure or number. For this a background of organised previous knowledge and norms is needed.

Consider having to assess the future progress of Brand X, a frequently bought consumer product launched several months ago. Suppose that in a sample survey housewives are asked which brands in the product-field they intend to buy and that about 14% say they intend to buy Brand X. By itself this figure means little. Thus it does not follow automatically that 14% of housewives *will* buy Brand X. More people may say they will buy new brands than actually do or can.

A second finding in the survey is that only 3% of the sample claim to be currently using Brand X. Now we have *two* figures, as shown in Table 4.1. One obvious interpretation is that if so many more people say they intend to buy Brand X than are currently using it, future prospects for the brand must be good.

TABLE 4.1 **Intentions-to-Buy I and Brand Usage U for the New Brand X**

I:	% of Sample who Intend to Buy X	=	14%
U:	" " " Currently Using X	=	3%

But is this interpretation so obvious? Do we really know that intentions-to-buy at 14% are unusually high compared with a current usage level of 3%? Do we know that a relatively high level of expressed intentions-to-buy is generally followed by an *increase* in the purchasing level of the brand? Both these notions may seem intuitively acceptable but are in fact wrong. To interpret what the readings in Table 4.1 actually mean we need more empirically based understanding of these variables.

I is High. The average I/U ratio observed for the other brands in the product-field is about 2/1. Comparing the Brand X ratio of 14/3 with this seems to confirm the common-sense impression that the observed value of I for Brand X is high. This might also seem to imply a good subsequent sales potential.

However, this comparison has been made against the *average* value of I/U for the other brands, where the variation was from $1\frac{1}{2}/1$ to 6/1. First we have to explain this large variation for the other brands before we can understand the I/U value for Brand X and its possible sales implications.

This is where earlier research results come in. Some previous basic research (Bird *et al.*, 1966) had shown that the level of I tends to vary with the square-root of the usage score U. Thus the values of $I/\sqrt{U}$ for different brands in a product-field were approximately constant at some value K, i.e.

$$\frac{I}{\sqrt{U}} = K.$$

In other words, $I = K\sqrt{U}$, and this relationship had been found to hold within an average of 3 percentage points under a wide range of conditions, as is summarised in Table 4.2.

TABLE 4.2 The Conditions under which the Relationship $I = K\sqrt{U}$ has been Found to Hold

- For frequently-bought non-durable branded goods in over twenty different product-fields (both food and non-food).
- For large brands and for smaller brands in each product-field.
- For some American data as well as for the main British data.

- For differing demographic sub-groups of the population.
- For different points in time, ranging over five years.
- For a variety of different forms of Usage questions.
- For certain different forms of Intentions-to-Buy questions.

- For established brands, whether with steady or with varying Usage levels.
- For successful new brands (i.e. ones with increasing Usage levels), except that some 5 to 8% fewer people then expressed an Intention-to-Buy.
- For old, dying brands (i.e. ones with slowly but steadily decreasing Usage levels), except that relatively more people than normal then expressed an Intention-to-Buy.

The relationship explains why the ratio I/U varies so much between different brands. If $I/\sqrt{U}$ is constant, then I/U cannot be constant. It must be bigger for small values of U than for large ones. Perhaps this will explain the high ratio of 14/3 for Brand X.

I is Low. We can work out that $K = 11.5$ in the equation $I/\sqrt{U} = K$ for our product-field from the average ratio of I and $\sqrt{U}$ for the other brands observed in the sample survey. Accordingly, a brand with a 3% usage level should have an intention-to-buy level of $11.5\sqrt{3} = 20\%$. It follows that the observed level for Brand X is not high, as thought before, but *low*.

This new finding is, however, still not of practical use. We need to know under what conditions such low values are generally found and their implications in terms of sales potential.

I is Normal. One of the specific findings in Table 4.2 provides the explanation we need. Newly launched brands generally had I levels 5 to 8 percentage points lower than established brands in the same product-field. Since Brand X is a new brand, its observed I value of 14%, 6 points lower than the theoretical value of 20%, is therefore *normal* for a new brand.

We now know that the result observed for Brand X does not tell us anything special at all; we cannot use it to predict the brand's future sales potential. This often happens with observed data. Once we understand an area of study, we find either that most observations turn out to be normal and predictable, or that there is no discernible pattern at all. This may be unexciting but it is inevitable. Forecasting is not as easy as asking a simple question in a market research survey.

4.3 The Role of Research

What is therefore needed is basic research to build up a background of knowledge in order to understand what a particular measurement means. Its verbal or operational definition is not enough. For example, asking people their intention-to-buy does not necessarily mean what one thinks it says. This is not peculiar to attitudinal measures or even to social research in general. In physics, we place some mercury in a glass tube and expect it to measure something. This is an odd thing to do. Sometimes it measures temperature (in a thermometer), sometimes pressure (in a barometer), and so on. It all depends on the detailed circumstances and what has been found out about them through past observation and analysis. Research means not only collecting new data but also theoretical analysis of past data and the use of previous knowledge (e.g. looking up books and references to see what has already been found out about such measurements).

The analysis of data depends on establishing simple patterns and relationships. By far the simplest pattern is that something is more or less constant, i.e. unrelated to the other variables in the situation. In some cases, like that "average speed = distance travelled/time taken" is relatively constant, the resulting concept of "average speed" seems to make intuitive sense, at least with hindsight. In other cases, like that $I/\sqrt{U} = K$, the "constant" aspect of the data arises merely as an empirical finding.

In fact, the latter result raises many new questions for further research. Why does the intention-to-buy variable relate to the usage level of different brands as $I = K\sqrt{U}$? How does I relate to subsequent *changes* in U? Why does a new and subsequently successful brand tend to have relatively few people expressing an intention-to-buy, when in fact increasing numbers buy it? Why does an old and slowly dying brand have relatively many people expressing an intention-to-buy, when fewer and fewer people subsequently buy it?

The answers to such questions will be touched on in the Exercises and help us to understand better the on-going nature of research. However, in this book our main concern is not so much the general conduct of such research but the handling of the resultant data.

4.4 Summary

A fact has to be compared with other facts before it can be interpreted. We must establish whether an observed number is "high", "low" or merely "normal". Interpretive norms can be established only after the analysis of past data; this is a matter of either experience or of deliberate research. It involves the development of new concepts and the uncovering of simple patterns and relationships that generalise under a wide range of circumstances.

Simply reporting the observed facts is not enough. When a good statistician is asked "How is your wife?" he replies "Compared with whom?" Many people, however, prostitute their data by merely treating them as "better than nothing" instead of looking for any deeper relationships.

CHAPTER 4 EXERCISES

Exercise 4A. An Awful Warning

"Fifteen people died on the roads over Christmas." Is that fact worth reporting?

Discussion.

The government used to issue the number of road deaths at Christmas every year as a warning against drink. Then it was recognised that the numbers were well below normal (fewer people drive at Christmas).

Exercise 4B. Catch Them Early

A widely held view in television circles is that if a popular programme is screened early in the evening it will lead to higher audiences on that channel for the rest of the evening.

Facts quoted in support of this view are that, typically, a programme shown at 10 pm on a Monday will attract 32% of the audience watching the 7 pm programme that evening on the same channel, but only 18% of those *not* viewing the earlier programme. It therefore seems that the more people who can be induced to watch at 7 pm, the higher the audience will be at 10 pm. But is the 32% attracted from 7 pm really a high figure?

Discussion.

No, the figure of 32% is not especially high. The 10 pm Monday programme would attract 32% of the audience of the other programme even if this had been shown on a different day altogether.

Research has shown that there is a general phenomenon known as "channel loyalty", namely that even programmes shown on different days on the same channel tend to share the same audience. Only if two programmes are virtually adjacent on the same evening is there any extra overlap in their audiences (Goodhardt *et al.*, 1975).

Exercise 4C. A Successful Treatment

A group of patients is treated by a certain method. If 64% recover, is the treatment successful?

Discussion.

We are given here only a single result. The question therefore is how this compares with other treatments or no treatment? Perhaps the spontaneous recovery rate of patients with this illness was 85%. (Then we might ask whether only chronic or abnormally severe cases were given the treatment.)

Exercise 4D. "Only 70% or As Many as 70%?"

Brand M is a frequently bought consumer product. During the peak-season quarter it was purchased by only about 70% of those who had bought it in the preceding off-peak quarter, but its total sales increased by a third. Does this imply a catastrophic decrease in brand-loyalty, temporarily hidden by the seasonal increase in sales?

Discussion.

What is the normal number of repeat-buyers? Does the 70% figure for Brand M mean *only* 70% or *as many as* 70%?

The incidence of repeat-buyers of a frequently bought branded good can be closely predicted when there is no trend in the brand's sales level (e.g. Ehrenberg, 1972). The prediction for Brand M is that about 67% will buy it again under no-trend conditions. The observed level of 70% therefore seems normal and there is no evidence of any loss of brand-loyalty.

But this figure was observed when sales followed a seasonal trend. This implies that off-season buyers were not affected by the seasonal upswing! The market for this product must therefore be segmented into two types of consumers, namely year-round buyers who are quite unaffected by the seasonal trend, and peak-season-only buyers: a result which was unexpected but has since been confirmed.

Exercise 4E. Explaining the Relationship between I and U (Section 4.2)

Why does the relationship between the percentage I of consumers who express an Intention-to-Buy a brand and the percentage U using the brand follow the form $I = K\sqrt{U}$? Furthermore, why is I relatively low for a new and subsequently successful brand, yet relatively high for an old and slowly dying brand?

Discussion.

Answers to these questions obviously required further analysis or research. They can be found by turning to the reference already cited in Section 4.2 (Bird *et al.*, 1966).

Given that I varies with U, the level of expressed intentions-to-buy among users and non-users was examined. Table 4.3 gives a typical result in some detail. The percentage of consumers who expressed an intention-to-buy decreased as more time elapsed since they last used the brand. Current users are likely to say they will buy the brand again; past users are less likely to. It follows from these results that if Brand A has twice as many current users as Brand B, A has fewer *non-users* and therefore less of a contribution to its intentions score from this source. Intentions-to-buy must therefore vary *less* than proportionately to the usage level. This accounts for the non-linear relationship between I and U. (The fact that the equation uses the *square root* of U has as yet no rationale.)

TABLE 4.3 Intentions-to-Buy Amongst Current and Former Users of a Particular Brand

	Intending to buy
Current users	95%
Used in last 6 months	45%
Used more than 6 months ago	10%
Never Tried*	5%

*Possibly including some who were users long ago

The reason that I is relatively low for a new brand is that it has fewer former users than an established brand, and therefore relatively fewer people expressing an intention-to-buy, given its usage level. In contrast, an old, slowly dying brand has a long tail of former users some of whom express an intention-to-buy it.

Exercise 4F. Intentions-to-Buy and Future Usage

Do consumers' expressed intentions-to-buy branded goods predict future changes in their usage behaviour?

Discussion.

Expressed intentions-to-buy vary with current and past usage. They will therefore also successfully predict the level of future usage, as long as there is no general change in the usage pattern. (But it might be easier to assert that the levels of current and past users will not change!) The data discussed so far provide no direct evidence of how expressed intentions-to-buy relate to *changes* in future behaviour, because these had not been measured. Additional research is therefore needed.

Some results have been reported in the reference already cited. For example, it was established whether or not users or non-users of a brand in 1963 also used it in 1964, giving four groups, as shown in Table 4.4. The body of the table shows the percentage of each group who had in 1963 expressed an intention-to-buy the brand.

TABLE 4.4 The Percentage of Users and Non-Users of a Brand in 1963 and in 1964 expressing an Intention-to-Buy in 1963

% who in 1963 expressed an intention to buy the brand	Users in 1963	Non-Users in 1963
Users in 1964	75%	27%
Non-Users in 1964	77%	15%

The results show that these 1963 intentions were hardly related to people's subsequent actions in 1964. Over 70% of users in 1963 said they intended to buy, and this turned out to be the case, irrespective of whether they actually bought the brand in 1964 or not—75% and 77%. In contrast, far fewer of the *non-users* in 1963 said they intended to buy, and this also showed relatively little difference by whether or not they did so in 1964—27% and 15%. (The lower percentage among non-users in *both* years is probably due to this group including people who never used *any* brand in the product-class in question.)

The conclusion is that expressed intentions-to-buy do not predict future changes in behaviour, even though they firmly reflect current and past behaviour.

Exercise 4G. Information for Decision-making

How does practical decision-making depend on information?

Discussion.

A classic illustration is the case of the two shoe-manufacturers many years ago, who each sent a salesman to Africa to explore the possible market for boots. One cabled back, "Splendid market for boots—nobody wears any!" The other, "No market for boots—nobody wears any!"

The new information is obviously relevant, but does not necessarily tell one what to do. Nonetheless, it can still be valuable. Firstly, if it *is* decided to go into Africa, it is helpful to know the market conditions there; i.e. the information helps to *execute* the decision once it is made. Secondly, the new information can be combined with *other* information to guide the decision-maker's judgment. A shoe-manufacturer whose past expertise has been producing highly competitive versions of well-established lines might be thought to have a harder task in Africa than one who has specialised in getting new fashions adopted.

Even so, there could be no clear prediction that the latter manufacturer would actually succeed in Africa. To make a valid prediction one would need many past cases where entering a new market *had* succeeded under more or less similar conditions. But if such empirical evidence were available, the situation would no longer be regarded as a major decision-problem since the answer would be obvious.

PART II: LAWLIKE RELATIONSHIPS

The analysis of data is basically a matter of relating one variable to another. In effect we have already been doing this in Chapters 1 and 2 where we related the readings in one row to those in another, and in the last chapter where we related distance travelled to time taken, or consumers' expressed intentions-to-buy Brand X to whether they actually used it or not.

A relationship becomes *lawlike* when different sets of data are summarised or modelled by the same quantitative equation. Its status depends upon the range of empirical conditions under which it holds. This is discussed in Chapter 5. The *uses* of lawlike relationships are discussed in Chapter 6.

One particular use of a lawlike relationship is in analysing further data. However, sometimes we have to start without any previous result and in Chapter 7 we discuss the problem of fitting an equation to data for the first time. Often one fits a linear equation as a start. But many relationships are not that simple and Chapter 8 discusses the basic steps in dealing with non-linear relationships. Chapter 9 introduces problems where *many* variables have to be interrelated.

In the early stages, fitting a relationship is usually very empirical, a matter of reducing extensive data not merely to summary figures as in Part I, but to an *equation*. Chapter 10 emphasises the emergence of *theory* with increasing knowledge of one's subject-matter.

CHAPTER 5

Descriptive Relationships

A relationship between two observed variables can often be represented by a simple mathematical equation. The primary criterion of such an equation is that it be *descriptive*. It has to summarise adequately the observed values taken by each of the variables.

The usefulness of a relationship lies in the range of conditions under which it holds. This notion of empirical generalisation is fundamental to science in general and to the study of relationships in particular.

5.1 Linear Relationships

The simplest relationship between two variables x and y is a straight-line equation of the form

$$y = ax + b.$$

Here a and b represent two numerical coefficients that stay constant for a particular relationship, such as

$$y = 0.5x + 2.$$

This equation says that for any given value of x, y is equal to 2 plus half the value of x. Figure 5.1 illustrates this equation on a graph.

The quantity a is usually called the "slope-coefficient" because it measures the slope of the line when the equation is plotted on a graph. It shows that y varies by a units for every unit difference in x. The fundamental property of a linear equation is that the slope stays the same everywhere along the line.

The quantity b in the linear equation is often called the "intercept-coefficient" because it shows where the line intercepts the vertical (y) axis when $x = 0$. (This value may be well outside the range of the data and does not need to represent a physically relevant condition.) In our example, $b = 2$. A larger value of b would give a line parallel to that in Figure 5.1, but higher up.

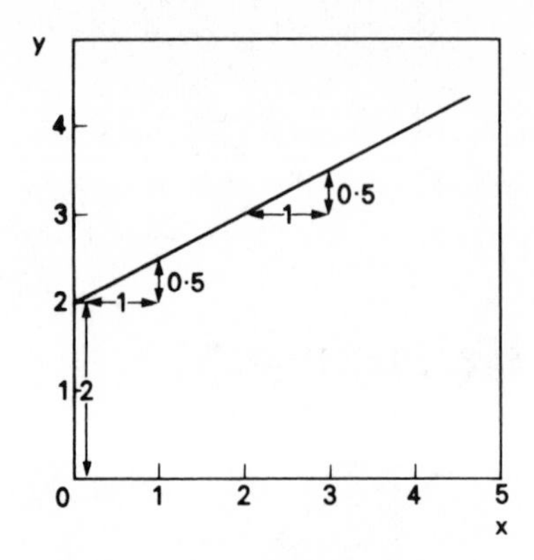

Figure 5.1 Equal Increases in y of 0.5 for a Unit Difference in x

When $b = 0$, the line goes through the origin, as shown in Figure 5.2A for $y = 0.5x$. Then y is *directly proportional* to x. When the coefficient $a = 0$, the equation $y = ax + b$ becomes $y = b$. This is illustrated in Figure 5.2B for $y = 2$. It means that the variables are independent of each other; i.e. y is *not* related to x and takes the same constant value of 2 irrespective of the value of x. This is the simplest possible result.

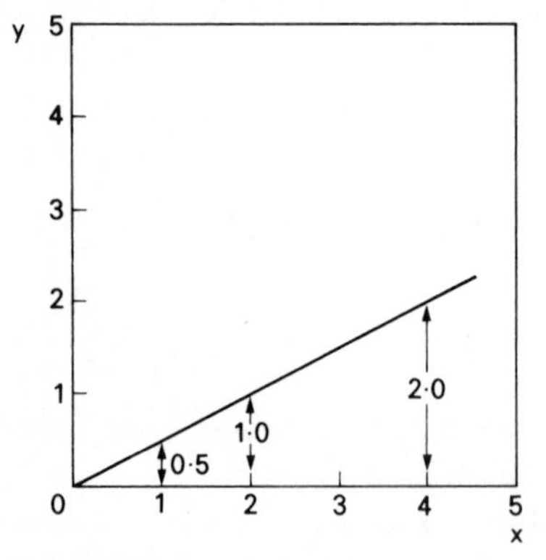

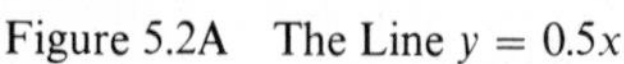
Figure 5.2A The Line $y = 0.5x$

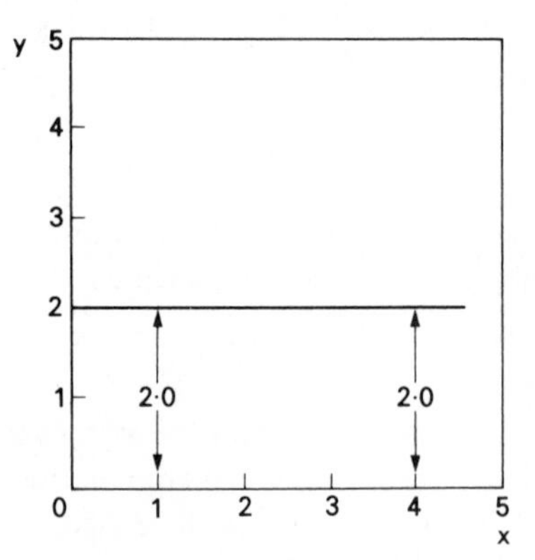

Figure 5.2B The Line $y = 2$

5.2 Deviations and Scatter

When a mathematical equation is used to describe a relationship between two variables x and y, the observed values do not generally lie exactly on the line. The line therefore is only an approximation to the observed data, which lie scattered around it. For this reason an equation like $y = ax + b$ should strictly speaking be written as

$$y \doteqdot ax + b,$$

where the symbol $\doteqdot$ means *approximately* equal and is perhaps the most important concept in applied mathematics. The equation could also be written

$$y = ax + b \pm c.$$

Here the "plus or minus" symbol $\pm$ indicates that some observed readings have positive and others negative deviations from the theoretical line $y = ax + b$, and that these deviations have an average size of c units.

The existence of such deviations does not necessarily invalidate the relationship. Nor does the *size* of the scatter interfere with one's perceptual recognition of a relationship, as is illustrated by Figures 5.3A and B.

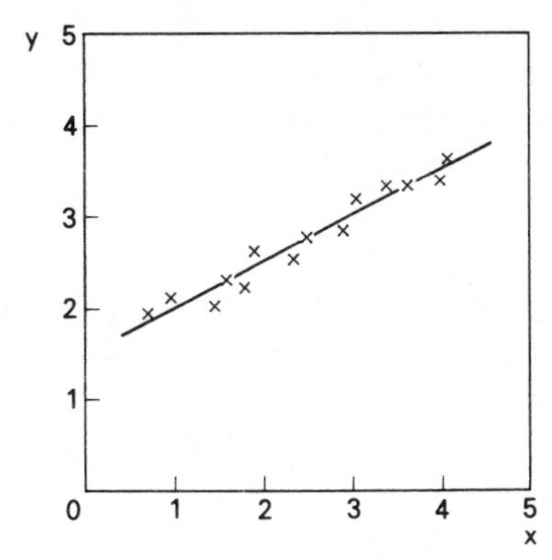

Figure 5.3A Small Scatter

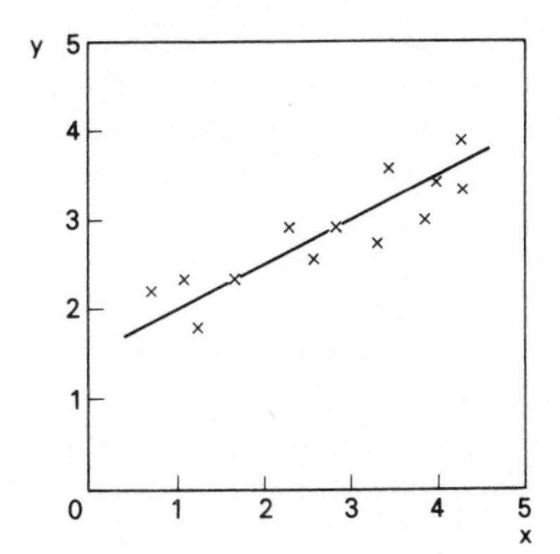

Figure 5.3B Large Scatter

The existence of scatter is generally taken for granted. Often it is not even explicitly described. Such an omission is open to purist criticism, but does not raise major practical problems. People are mainly concerned with knowing the *systematic* relationship between variables, e.g. that y varies as $0.5x + 2$, and not with the precise levels of scatter or error attached to it.

The crucial criterion is that the scatter should be *irregular*, i.e. show no systematic pattern, as in Figures 5.3A and B. Only then can it be "summarised away" in simple statistical terms, as being irregular (i.e. individually unpredictable) and of such and such an average size.

When the observed readings deviate *systematically* from the theoretical equation or line fitted, the deviations are more complicated to describe. This is the case irrespective of the size of the scatter. Figures 5.4A and B give examples of both relatively small and larger regular scatter. In Figure 5.4B we would have to say that when x is about $\frac{1}{2}$, y is almost $\frac{1}{2}$ a unit above the

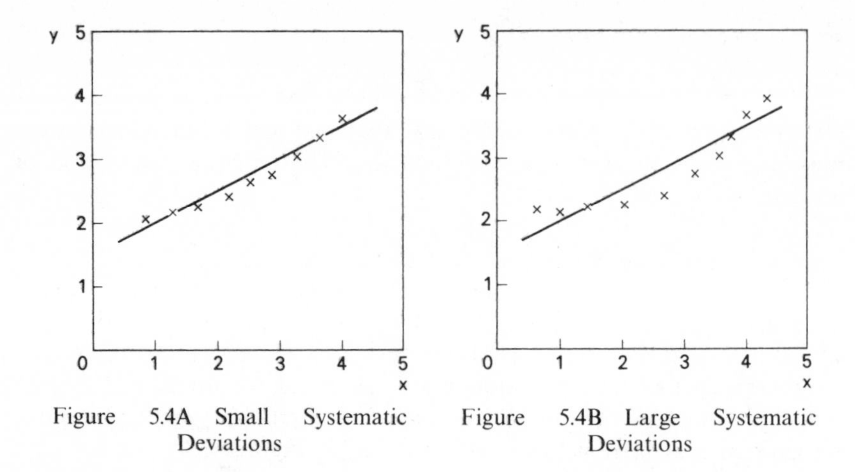

Figure 5.4A Small Systematic Deviations

Figure 5.4B Large Systematic Deviations

line, when $x = 1.5$, y is on the line, when x increases to 2.5, y lies increasingly below the line, etc. The description of the data has become very complicated.

In the initial stages of data analysis, a systematic pattern in the deviations means that the wrong descriptive relationship has been fitted, even if the deviations are small. For example, although the deviations in Figure 5.3B are larger than those in Figure 5.4A, in the first case the scatter is irregular and in the second it is systematic. Usually it is more convenient to model such data with a suitable *curve* that has irregular deviations rather than with a straight line that has systematic deviations. Only in more advanced work are systematic deviations sometimes acceptable as deliberate oversimplifications, as long as their nature is understood.

5.3 Non-linear Relationships

Many scientific relationships and laws are highly non-linear in their original units of measurement. Boyle's Law in physics is a typical example. It says that the relationship between the pressure P and the volume V of a body of gas is

$$P = \frac{C}{V},$$

where C is a constant that depends on the amount and type of gas, the temperature, etc. This law states that for any given body of gas, the greater the volume, the less the pressure, but in such a way that the product PV remains approximately constant at C, or

$$PV = C.$$

The equation is non-linear because every unit change in V produces a *different* change in P. Figure 5.5A illustrates this relationship when C takes the value 10. The relationship between P and V is *different* everywhere along the line.

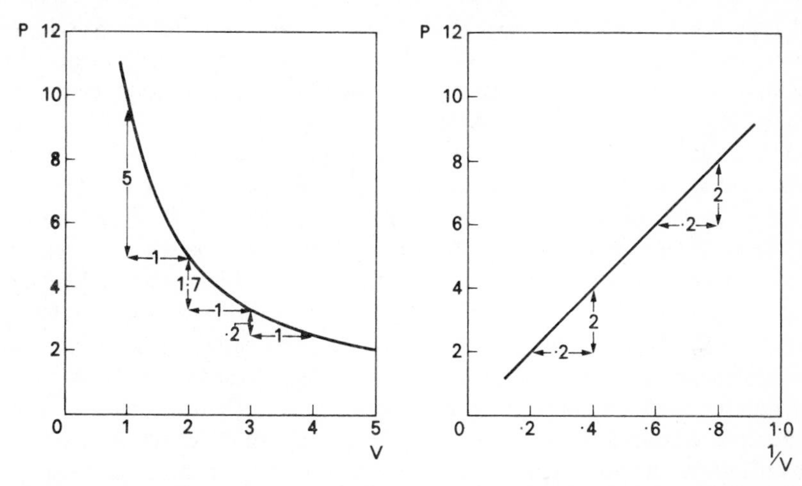

Figure 5.5A Boyle's Law: $P = C/V$ Figure 5.5B Plotting P against $1/V$

Many curved relationships can be described with a linear equation by changing the form of one of the variables. For example, with Boyle's Law we can "transform" the volume variable to $1/V$ (the reciprocal of V) and then analyse the relationship

$$P = C\left(\frac{1}{V}\right).$$

This is the same as writing the linear equation $y = ax$, where y stands for P and x stands for $1/V$. Figure 5.5B shows that this transformed relationship produces a straight line where P varies by 2 units for every 0.2 difference in $1/V$—i.e. $P = 10(1/V)$.

Transformations like this are not always possible, but they are worth doing because it is easier to fit a linear equation to transformed variables than to fit a curve to data in its original form.

5.4 The Status of Lawlike Relationships

A common comparison is between laws in physics and social science. Physical laws often appear to be exact, well-embedded in theory, and capable of giving simple, accurate predictions. By contrast, relationships in the social and biological sciences seem merely to be statistical associations without any background of well-established theory, and capable of leading only to very uncertain and erratic predictions.

Such comparisons usually pit a well-established law from physics against some recent result from sociology, economics, or biology that may be based on only one or two isolated studies. But a relationship like Boyle's Law is based on thousands of experiments. It took decades before it was accepted as an empirical generalisation and its theoretical explanation came long after that.

Even today, the laws of physics and chemistry are not exact. They are oversimplifications. For example, Boyle's Law only holds for "perfect" gases (defined as hypothetical substances for which Boyle's Law holds); Newtonian mechanics has balls sliding down planes without friction; Galileo's weights drop without air resistance; the atomic weights of hydrogen and oxygen are said to be 1 and 16 (instead of the more exact values of 1.00797 and 15.9994); etc. The justification for the simpler results is that they are *so* much simpler to use, and that they approximate the data closely enough for many practical purposes.

Thus lawlike relationships are not necessarily 100% exact. Nor do they *initially* have to form part of an explanatory theory or have an immediate practical use. It is usual for explanation, understanding and practical application to follow subsequently, as part of the historical development of a subject.

But this process cannot start until a relationship has first been discovered and described. We cannot explain *why* pressure varies with volume unless we know *how* it does so, e.g. as $P \doteqdot C/V$.

Such a descriptive relationship has to be a *generalisable* one. If for the next set of data the quite different relationship $P = -V$ were found instead, we would have to explain not only why pressure varies inversely with volume but also why it does so quite differently from one case to another. In contrast, explanation becomes relatively easy if the *same* relationship is found to hold in a variety of different circumstances.

5.5 Empirical Generalisation

The crucial step is establishing that the same quantitative relationship holds for different sets of data and different conditions of observation. Then the relationship becomes practical and useful.

In effect, this is what the laws of physics are: empirical regularities that have been laboriously isolated for a certain range of conditions of observation and that are equally well-known *not* to hold under other conditions. For example, Boyle's Law $PV \doteqdot C$ has been found to hold for different gases and mixtures of gas, different amounts of gas, different kinds of apparatus, different experimenters, different times and places, when pressure is *increasing* and when pressure is *decreasing*, etc. But this was established only because of a massive amount of empirical observation that something like it in fact happened here and there, this year and last year, in the morning and at night, etc.

Empirical observation also showed that the relationship $PV = C$ does *not* hold when the temperature changes, when there is a chemical reaction, when there is a leak in the apparatus, when there is physical absorption or condensation of the gas, or when we tried to prove the law at school. Each time the law did not hold extensive empirical study was needed to generalise the relevant conditions. Was a particular deviation due to the specific gas being examined? A trace of water vapour that condensed? A careless laboratory assistant? Could the deviation be repeated, or was the situation not even well enough understood to do that?

The direct meaning and usefulness of Boyle's Law is therefore bound up both in our knowledge of the conditions under which it holds and the conditions under which it does *not* hold. The same process of empirical generalisation applies to any lawlike relationship. A recent, less-developed case is the study of how children's heights vary with their weights. The form of such a relationship might itself vary with a whole host of other factors like race, nationality, socio-economic class, sex, nutrition, age, point-in-time, etc. Yet it has been found that the same equation

$$\log w = .02h + .76$$

between the logarithm of children's weights w (in lbs) and their heights h (in inches) holds *despite* these differences, as listed in Table 5.1.

TABLE 5.1 Summary of Conditions Under Which the Height/Weight Relationship log w = .02h + .76 Has So Far Been Found to Hold

(Lovell 1972, Kpedekpo 1970, 1971, Ehrenberg 1968)

Race:	White, Black, Chinese (in the W. Indies).
Countries:	U.K., Ghana, Katanga, West Indies, France, Canada.
Time:	1880 - 1970 approximately.
Age:	2 - 18 years.
Sex:	Male (2 - 18) and Female (2 - 13).
Socio-Economic Class:	Various in U.K., France and Canada.

The relationship also holds despite other less explicit differences in the conditions of observation, such as measurements being made by different observers and differences in the average size of families, housing conditions, intelligence levels, etc. No matter what differences there were, we know they did not matter because the same relationship has been found to hold.

The conditions covered in Table 5.1 look somewhat haphazard, like "Chinese children living in the West Indies". This is because the investigation is at an early stage and no comprehensive or systematic checks of conditions have yet been made. Nevertheless, the range of conditions is already so wide that one may already begin to refer to the equation $\log w = .02h + 0.76$ as a "lawlike relationship". One would now be surprised if the relationship did not hold for the next set of data from white or black children.

This does not mean that the relationship holds universally. There is always a variety of conditions under which a scientific law does *not* hold. Often this is so obvious that we automatically exclude such conditions even when thinking about the law. For example, the relationship between children's heights and weights obviously will not hold if they are measured when sitting or lying down, or for children who are seriously undernourished —we already *know* they will be light for their height. Less extreme exceptions are that teenage girls are heavy for their height compared with boys (which fits in with general experience of girls), and that babies may be relatively light for their length (babies being measured in the prone position).

5.6 Other Things NOT Equal

An empirical generalisation tells us how the values of the variables relate together despite variation in other factors. For example, the relationship $\log w = 0.02h + 0.76$ between heights and weights holds despite differences in children's races, nationalities, sex, ages, social classes, point-in-time, etc. Similarly, in Part I we saw that the quarterly readings in the North were about 95 even though Quarter I was Winter, with ice and snow and long dark nights, Quarter II was Spring with April showers, apple blossoms and lambs gambolling, and so on.

The popular saying "other things being equal" therefore does not necessarily mean that all these other things have themselves to be equal. It only means that their *effects* have to be equal.

5.7 Summary

A relationship between two variables usually can be represented by a mathematical equation. The prime purpose of this equation is to be descriptive: to summarise in a convenient form the observed values taken by one variable for any given value of the other variable.

The simplest form of relationship is the straight line equation $y = ax + b$, which says that y varies with x at a constant rate. In practice most empirical relationships are non-linear in their original units of measurement. However, one variable can often be "transformed" to reduce a curved relationship to a linear form.

Observed readings are usually scattered about a theoretical equation or fitted line instead of lying exactly on it. Deviations that have no regular pattern of their own are easy to summarise statistically as being irregular and of a stated average size.

To be lawlike, a relationship has to be based on many different sets of data and hence generalise to a wide range of different conditions of observation. This does not imply that the relationship holds universally, but only that it holds within the stated range of conditions and that the exceptions can also be generalised.

CHAPTER 5 EXERCISES

Exercise 5A. What is a Variable?

A common dialogue about variables runs as follows:

Teacher: "Suppose that x is a variable, i.e. a quantity which can take any value."

Student: "Yes, I think I understand that."

Teacher: "Let $x = 20$."

Student: "But you just said x was a variable!"

At this stage, many students are lost for good. But is there really a contradiction in what has been said?

Discussion.

In algebra, symbols like x are used to represent any value that a particular quantity can take—usually within certain stated or implied limits, e.g. that x is always a positive integer (as in counting 0, 1, 2, 3, 4, . . .), or that x varies only from 0 to 1 (e.g. a proportion, .3, .6, .1, etc.).

If x is a variable, it can therefore take various possible values. Saying that $x = 20$ is a way of looking at one of these values. It presents no contradiction.

Although we shall be using almost no complex notation in this book, it is worth briefly illustrating some common elaborations. For example, more detailed notation is sometimes used to distinguish x as a general variable from x as a particular value. We might use the symbol x' (usually called "x dash") for a particular value and say that $x' = 20$. Even here, x' still represents a variable quantity: it could be *any* particular value of x. But for the moment we are saying that we are considering the case where x' takes the value of 20.

Introducing such additional notation is especially useful when we wish to say more about some particular value of x. Suppose we want to consider two values of x which are not equal. We can write this as $x' \neq x''$, where the symbol $\neq$ means "not equal", and x'' is another particular value of x (called "x double dash"). We could not write this without some notation distinguishing the two values of x.

Or we might want to consider all those pairs of values of x which differ by 5 units. These we could denote by the equation $x'' = x' + 5$. Again we are referring to all possible values of the variable x which satisfy the condition laid down by the equation.

The values x' and x'' in the two equations $x'' \neq x'$ and $x'' = x' + 5$ do not necessarily refer to the same possible values of x. For example, the values

$x' = 20$ and $x'' = 21$ would satisfy the first equation but not the second. If we wanted to differentiate the two cases symbolically, we would have to introduce a further difference in notation, like $x' \neq x''$ and $x_2 = x_1 + 5$.

Using such different symbols can become cumbersome. When no real confusion can arise it is therefore common to use the simple symbol x to stand for both the variable itself and some specific value of it. While mathematics is generally a very precise subject, the symbolism is often used very loosely to keep some flexibility and simplicity.

Exercise 5B. Algebraic Relationships

In the linear equation $y = \mathrm{a}x + \mathrm{b}$, x and y are two variables and a and b are two constants that can take any values. Why are a and b called "constants" and not considered variables as well?

Discussion.

When we are speaking about linear equations *in general* we denote them by $y = \mathrm{a}x + \mathrm{b}$. But whenever we speak of a particular linear equation, the coefficients a and b automatically take numerical values that remain constant for that specific equation. On the other hand, x and y are always variables in all linear equations.

For example, we can speak of the linear equations

$$y = 0.5x + 2,$$

$$y = \quad 2x + 6.$$

In each equation a and b take specific constant values and x and y are variables.

Exercise 5C. Points on a Straight Line

On a graph of y and x, the values (x, y) refer to the point which is x units measured along the x-axis and y units along the y-axis. If (x_1, y_1) and (x_2, y_2) are two points that lie on the straight line $y = ax + b$, what can we say about the relationships between the four values x_1, x_2, y_1 and y_2?

Discussion.

Since the point (x_1, y_1) lies on the straight line, it must satisfy the equation

$$y_1 = ax_1 + b.$$

Similarly, for (x_2, y_2) we must have

$$y_2 = ax_2 + b.$$

Subtracting one equation from the other, we have

$$\begin{aligned} y_1 - y_2 &= ax_1 - ax_2 \\ &= a(x_1 - x_2). \end{aligned}$$

This says that for any two points that lie on the straight line, the difference between the two y-values $(y_1 - y_2)$ is always a times the difference between the two corresponding x-values $(x_1 - x_2)$.

It follows that two points determine a particular straight line. Thus from the above equation we have that

$$\frac{y_1 - y_2}{x_1 - x_2} = a.$$

This allows us to calculate the slope, a, of that line. For example, for the two points (2, 3) and (4, 4) we have

$$a = \frac{3 - 4}{2 - 4} = \frac{-1}{-2} = 0.5.$$

The slope-coefficient $a = 0.5$ will be the same for any two points on that line, which is the fundamental property of the straight line.

The intercept-coefficient b can be calculated by noting that the point (2, 3) is supposed to lie on the line. Thus

$$3 = 0.5(2) + b$$

and so $b = 2$. The linear equation is therefore

$$y = 0.5x + 2.$$

(The same result could be obtained using any point, e.g. (4, 4), on that line.)

This is the normal way of determining the numerical values of the coefficients a and b, given two points which lie on the straight line.

Exercise 5D. A Change in Units

If $d = 0.5t + 2$ is an equation between distance d in feet and time t in seconds, what is the corresponding equation using miles and hours?

Discussion.

There are 5,280 feet in one mile and 3,600 seconds in one hour. Thus the distance D in miles is $D = d/5{,}280$ and the time T in hours is $T = t/3{,}600$. (D and T must be numerically smaller than d and t because they are measured in larger units.) Substituting for d and t in the equation $d = 0.5t + 2$, we therefore have

$$5{,}280D = 0.5(3{,}600T) + 2.$$

This gives

$$O = .34T + .00038.$$

Exercise 5E. *x* in terms of *y*

If $y = 0.5x + 2$, what is x in terms of y?

Discussion.

Isolating the x-term gives $0.5x = y - 2$. Dividing the equation by 0.5 (or multiplying by 2), we have

$$x = 2y - 4.$$

In general, if $y = ax + b$, then $x = (y - b)/a$.

Exercise 5F. Scatter about the Equation

If the observed reading (x', y') does not lie exactly on the theoretical line $y = ax + b$, what is the deviation?

Discussion.
Given the line $y = ax + b$ and the value x', the theoretical y-value on the line is $ax' + b$. Hence the difference between the theoretical and observed values of y' is

$$y' - (ax' + b), \quad \text{or} \quad y' - ax' - b.$$

If the point is (4, 3) and the line is $y = 0.5x + 2$, the deviation measured on the y-scale is

$$3 - 0.5(4) - 2 = -1,$$

as shown in Figure 5.6.

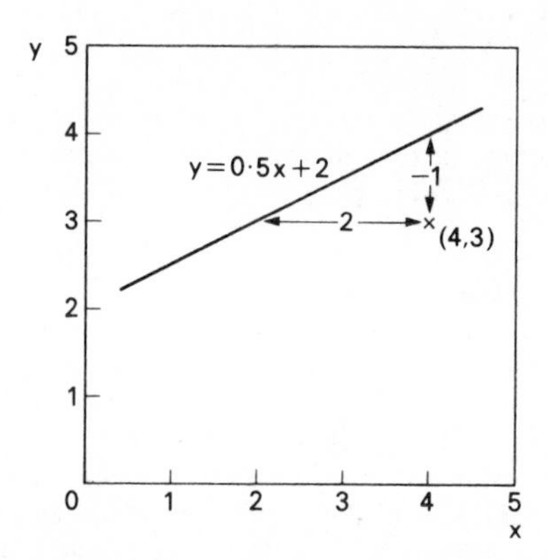

Figure 5.6 Deviations in the y and x Directions from $y = 0.5x + 2$

Measured on the x-scale, the deviation between the observed and theoretical values of x' is

$$x' - (y' - b)/a = x' - (y' - 2)/0.5.$$

For the point (4, 3) this is $4 - (3 - 2)/0.5 = 2$, as is also shown in Figure 5.6.
In other words, the point (4, 3) is 1 y-unit too low or 2 x-units too high, compared with the line.

CHAPTER 6

Using a Given Relationship

Lawlike relationships can have many practical uses. The illustrations in this chapter are based largely on the relationship between the height and weight of children referred to in Section 5.5. This case is typical of what one meets in practice, since it is non-linear, new, and incomplete.

6.1 Summarising the Available Data

The most direct function of an empirically-based relationship is to have summarised the data on which it is based. For example, the height/weight relationship, $\log w \doteqdot .02h + .76$, summarises the heights and weights of certain British boys aged 5 in 1880, of West Indian Chinese boys of various ages in the 1960's, of French girls aged $8\frac{1}{2}$ in the 1950's, and so on. Once one has this relationship one need not refer again to the raw data. That is a major achievement.

6.2 Prediction and Extrapolation

A relationship like $\log w = .02h + .76$ can also be used to predict the value of one variable from the known value of the other. In effect one is then asserting that the relationship with its associated scatter will hold again for a new set of children. That is what one expects to happen if the new data fall within the range of conditions already covered in the previous analyses.

However, one may sometimes need to extrapolate to new conditions, outside the range previously covered; e.g. for some very different racial or ethnic group, for children living under unusual circumstances or suffering from some illness, for those who lived 300 or more years ago, for another species, or whatever.

The difference between these two uses turns on the extent to which the new data lie within the range of conditions already covered. The distinction between a well-founded prediction and an extrapolatory guess can be seen by considering the actual outcome in each case: i.e. either success or failure.

With a prediction, success would be just what one had to expect. All the previous experience has been that $\log w = .02h + .76$ holds under such conditions and now it has done so again; a dull but comfortable outcome. In contrast, *failure* of such a well-founded prediction signifies a discrepancy with all the previous knowledge, something new and potentially interesting which must be studied further and explained.

But with an extrapolatory guess, it is *success* that represents an exciting discovery, a further generalisation of the relationship $\log w = .02h + .76$ to quite new conditions. *Failure* merely means that the analyst is not very good at making such guesses; there is no actual discrepancy because nothing was previously known about the relationship under these conditions (e.g. for the heights and weights of chimpanzees).

6.3 Understanding and Theory

A well-established relationship also allows us to reach a better understanding of the phenomena in question. It might be thought that a purely descriptive generalisation like $\log w = .02h + .76$ only shows *how* height and weight are related, but does not tell us *why*. But this is not entirely true. Consider all the factors which might affect the way height varies with weight: race, nationality, sex, social class, age, time, observers, etc. We now know that these factors generally do *not* affect the relationship. We also now know that some variables *do* affect the height/weight pattern, such as puberty in girls. Clearly we are beginning to understand something about the system.

Nonetheless, the relationship is still a limited result. For example, it hardly links up with other kinds of findings. The height/weight relationship is in fact lacking in "theory". This is essentially a matter of time because the relationship is still relatively new. A generalisable result has first to be *established* before it can be incorporated into any broader system of equations or theory. Low-level empirical generalisations are the essential building-bricks of more advanced theory; examples of such extensions are discussed later, particularly in Chapters 9 and 10.

6.4 Technological Applications

A technological use of the height/weight relationship is to assess whether an individual child is above average weight, as a step toward diagnosing possible obesity.

Suppose a particular child is 51 inches tall and weighs 63 lbs: which from a table of logarithms is about 1.80 in log lb units. These values do not satisfy the equation $\log w = .02h + .76$ exactly. For a value of $h = 51$, the equation gives $.02h \times 51 + .76 = 1.78$. This differs by .02 log lbs from the observed weight.

The child is therefore "overweight" in the sense that the equation gives the *average* result and the child is heavier for its height than average. None of the sources quoted in Table 5.3 reported individual deviations from the equation, but from some data supplied by Dr. E. M. B. Clements (1954), it appears that the average scatter of individual children might be about .04 log lbs. A child who is .02 log lbs above average therefore is not *abnormally* overweight. There are many children of that height who are even heavier (i.e. who differ even more from the equation).

It may seem difficult to think of many other technological applications for the height/weight relationship. But this is usual for a *single* relationship in its early days. For example, by itself the law of gravity does not tell us how to build an aeroplane that will fly or to adjust a pendulum clock to give the correct time. The law is just one of many that engineers use in these cases.

6.5 Decision-making

In using data for *decision-making*, such as whether to introduce free milk at school for children from less prosperous backgrounds, or deciding whether to believe that black children are of a different shape from white ones, it is best to have first summarised and understood the available data. Decision-making and data-analysis are separate processes.

6.6 The Analysis of Further Data

Another use of a relationship is in dealing with further data. This makes the new analysis very simple. One merely checks whether the previous result holds again. For example, in Section 6.4 we very easily concluded that the child's weight was well within the usual limits.

Use of previous results helps keep the analysis simple even when the new data are extensive. Table 6.1 gives an extract of the average heights and weights of about 5,000 children in Ghana (Kpedekpo, 1970). They are classified by sex, yearly age-groups, and race, with the black children further sub-classified as living under rural, urban non-privileged and urban privileged conditions (i.e. attending élite schools).

The biggest variation in Table 6.1 is between age-groups. We therefore start by analysing a particular type of child at different ages and in Table 6.2 compare the expatriate boys with the earlier relationship. The fit is clearly good. The corresponding analyses for the expatriate girls and the privileged black children show that they also follow this relationship. The deviations are within the same average limits of $\pm$.01 log lb that were reported for such age-groups in earlier studies (Ehrenberg, 1968). The slightly larger

TABLE 6.1 Average Heights and Weights of Groups of Children in Ghana

(From Kpedekpo 1970)

HEIGHT (in inches)	Age in Years						
	5	6	7	8	9	10	etc.
BOYS							
Rural	*	45	47	49	51	.	.
Urban	42	45	47	48	50	.	.
Urban privileged	45	48	49	51	54	.	.
Expatriate (white)	45	46	49	52	54	.	.
GIRLS							
Rural	44	45	47	50	51	.	.
Urban	*	46	48	50	51	.	.
Urban privileged	45	47	48	51	54	.	.
Expatriate (white)	45	46	49	52	53	.	.

WEIGHT (in lbs.)	Age in Years						
	5	6	7	8	9	10	etc.
BOYS							
Rural	*	41	45	51	55	.	.
Urban	35	41	44	48	54	.	.
Urban privileged	46	52	58	62	69	.	.
Expatriate (white)	46	50	59	66	69	.	.
GIRLS							
Rural	41	41	45	52	54	.	.
Urban	*	46	50	52	57	.	.
Urban privileged	46	50	59	64	73	.	.
Expatriate (white)	44	49	57	66	67	.	.

* No data

TABLE 6.2 Expatriate Anglo-American Boys in Ghana and the Prior Relationship log w = .020h + .76

Expatriate Boys in Ghana	Age							Av.
	5	6	7	8	9	10	etc.	(5 to 9)
Av. height: h	45	46	49	52	54	.	.	49
Av. weight: log w	1.65	1.70	1.76	1.82	1.83	.	.	1.75
.020h + .76	1.66	1.68	1.74	1.80	1.84	.	.	1.74
log w - (.020h + .76)	-.01	.02	.02	.02	-.01	-	.	.01*

*Average size ignoring sign = .02

deviations in the table at .02 log lbs do not generalise (e.g. to older Ghanaian boys, girls, etc.).

But analyses of the black non-privileged urban and rural children show consistent negative deviations from the relationship, as typified in Table 6.3. They are consistently lighter by about .04 log lbs for any given height than white or "privileged" black children. (The apparent trend in the deviations in Table 6.3 does not generalise.) Here use of the prior relationship makes it clear what *new* relationship will describe these discrepant data: something like $\log w = .02h + .72$.

TABLE 6.3 **Non-Privileged Urban Ghanaian Boys and the Prior Relationship log w = .020h + .76**

Non-privileged Ghanaian Boys	Age 5	6	7	8	9	10	etc.	Av. (5 to 9)
Av. height: h	42	45	47	48	50	.	.	48.0
Av. weight: log w	1.54	1.61	1.64	1.68	1.73	.	.	1.64
.020h + .76	1.60	1.66	1.70	1.72	1.76	.	.	1.69
log w - .020h - .76	-.06	-.05	-.06	-.04	-.03	.	.	-.05

These analyses have clearly been very easy to do, regardless of whether they showed agreement or disagreement, because we could use the prior relationship.

6.7 The Meaning of Failure

Using a prior relationship to analyse new data therefore works even if the new data do not agree with the earlier relationship. This is illustrated in Figure 6.1, where the previously established line clearly does not hold for the new data. Such a result may spell failure to what one hoped to find, but technically it is very simple: there is no generalisable relationship and one variable cannot be predicted from the other, at least not until the discrepancy is itself explained. But this may not be possible and such unresolved failure is worth illustrating.

Some years ago, a series of experiments was mounted to see whether a certain chemical measure could be used to predict the "eating quality" of white fish (cod, haddock, etc.). The practical relevance was that while the sensory assessment method was known to be highly reliable, it was not very acceptable to trawler captains for price-setting purposes, nor was it easy to operate routinely at 6 am each morning at the quay-side. An "objective"

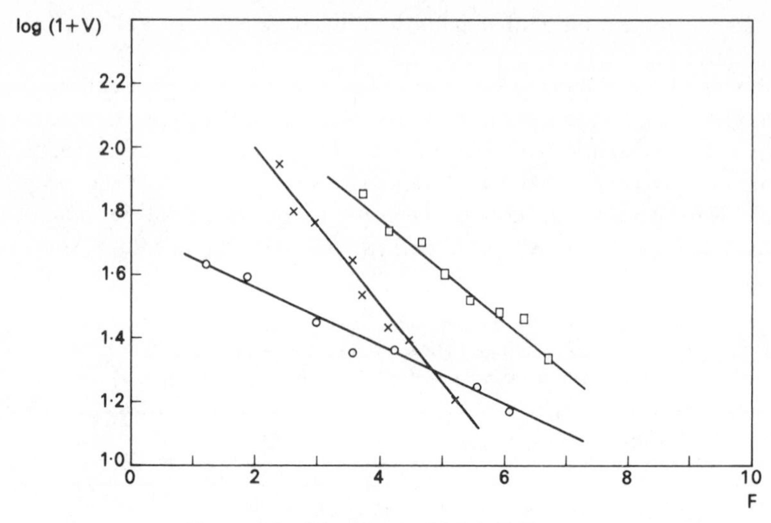

Figure 6.1 New Data which is Different

chemical measure of eating quality was therefore desirable. The two variables were

(i) V, the amount of volatile bases per milligram of fish muscle (i.e. chemical compounds such as ammonia and various amines);

(ii) F, the flavour of the fish, a direct indicator of its "eating quality" as measured on a 10-point scale by a highly trained laboratory taste panel.

In some pilot studies in 1953, the relationship $F = -6.2 \log (1 + V) + 15.0$ had been found to hold between F and V for batches of fish stored in ice for various periods (Shewan and Ehrenberg, 1955). It therefore seemed that one variable could be used to predict the other. However, the relationship had not yet been established for the range of conditions that existed in practice: fish from a variety of fishing-grounds, caught at different seasons of the year, handled and stored in different ways on board trawlers, etc. As a first follow-up, further measurements were made for several hundred batches of fish caught at different seasons of the year in 1954 and 1955 (Shewan and Ehrenberg, 1957).

Each new experiment gave well-fitting relationships of the form $F = a \log (1 + V) + b$, but the coefficients a and b differed markedly each time, and therefore also from the initial equation. For example, in one case the relationship was something like $F = -12 \log (1 + V) + 20$, in another $F = -4 \log (1 + V) + 10$, in a third different again, and so on, as illustrated graphically in Figure 6.2.

No explanations for these discrepancies were found. Possible factors studied included the different seasons of the year, the size and sexual maturity

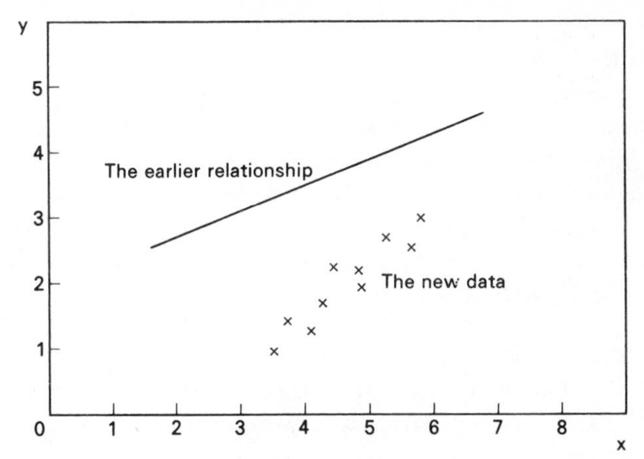

Figure 6.2 Irreconcilable Equations between the Volatile Bases Content V and the Flavour F of Cod from Three Experiments

of the fish, differences in the nature and size of the initial bacterial load of the fish, variations in the chemical composition of the fish muscle, differences in handling or storage methods, and the influence of errors of measurement (which were small and well-understood for both measuring techniques). But whatever additional factors were involved in the relationship between F and V, they remained completely unknown. Neither the initial relationship nor any of the subsequent results were reproducible. The chemical measure might appear more "objective", but the analysis showed that it objectively failed to measure eating quality in a reproducible way.

6.8 The Purpose of Analysis

The purpose of an analysis does not affect how it is done. For example, a result used for prediction is of the same form as one used to analyse new data. In all cases one has to describe how y varies with x within the observed range of conditions. One's purpose, however, influences *other* things, such as whether to do the analysis at all, how much effort to put into it, which kinds of deviations to follow up, what range of conditions to cover, how much accuracy or precision to try for, and so on.

In the white fish study for instance, it might be enough for some purposes to know that flavour and volatile bases content are always inversely related, the more volatile bases in the fish, the lower its flavour-score, even if the numerical details vary from case to case. But for quality-control and price-setting purposes the unpredictable variations in the numerical coefficients are too large. Far too many fish of acceptable flavour would be wrongly given a low price (suitable for fish-meal, say) and too much spoilt fish awarded a high price.

Again, for many purposes the pressure and volume of a gas can be taken to follow Boyle's Law, $PV \doteqdot C$. But where gases are under high pressures, as in oil refineries, this law is far too inaccurate.

More generally, the size of the observed deviations from a law is often irrelevant as long as they tend to be "small enough". But in making decisions about individual cases, as in medical diagnoses, detailed understanding of the variation between individuals may be essential.

Since analytic results are much easier to use if they are relatively simple, we generally strive for results that oversimplify and approximate the data rather than fit them exactly. The nature of one's specific problem determines how close the degree of approximation needs to be.

6.9 A Common Misuse

Relationships are often misused by applying them unthinkingly outside the range of conditions for which they have been established. If the equation $y = 5x + 10$ fits certain data, it is often assumed that if x is changed by a certain amount, then y should correspondingly change by 5 times that amount. But this is not necessarily true.

The initial data may not have referred to *changes* in x and y, and the relationship therefore cannot tell us anything directly about such changes. For example, the earlier height/weight relationship describes how the heights and weights of many different kinds of children are related. It does not say that any child will grow taller if one increases his weight by feeding him a lot, or that he gets shorter whenever he loses weight. Even for two variables like weight and *girth*, where we know from everyday experience that short-

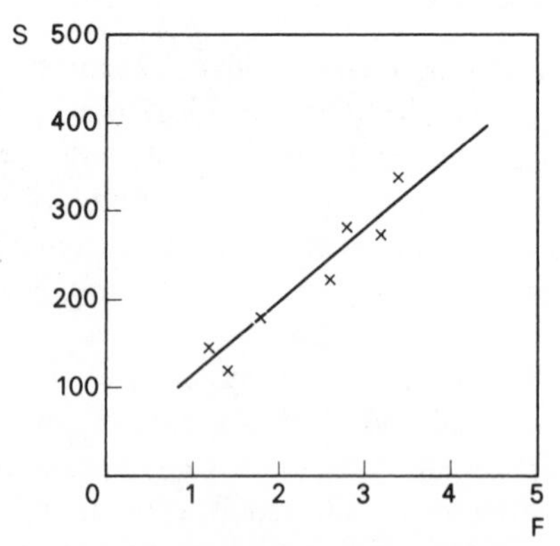

Figure 6.3 Floor-Space F and Sales-Level S of Some Different Retail Stores

term ups-and-downs do tend to correlate, it would be wrong to assume that these short-term changes necessarily follow the same quantitative pattern as those which occur as children grow.

As another example, suppose there is a close relationship, $S \doteq 80F + 41$, between the sales-level S (in thousand of pounds sterling) and the floor-space F (in thousands of square feet) of a number of retail shops, as illustrated in Figure 6.3. It does not follow from this that increasing the floor-space of a shop will lead to a corresponding increase in sales.

Variations among different shops and changes within a particular shop are two different things. For example, we know that there can be seasonal increases in sales with no change in floor-space, as shown in Figure 6.4A. This is quite unlike the relationship $S = 80F + 41$. Or all the shops could be rearranged to increase their effective floor-space by about 25%, but this might have no effect on sales, as Figure 6.4B illustrates.

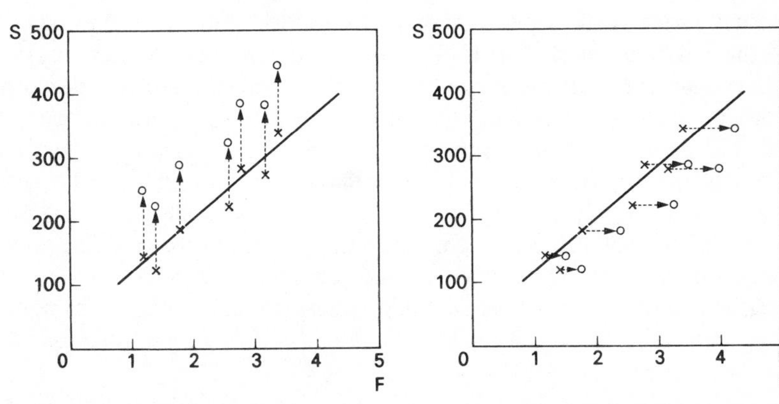

Figure 6.4A A General Increase in Sales

Figure 6.4B A General Increase in Floor-Space

Many relationships do not tell us directly what we want to know for practical decision-making purposes. The decisions may be concerned with some deliberate man-induced change occurring over time, the given relationship with existing, static and cross-sectional differences. (Much of economic analysis, for example, ignores this distinction.) Nevertheless, a static cross-sectional relationship tells us about certain constraints in the system which we may be trying to change. The equation $S = 80F + 41$ shows how S and F are related for the different shops within the conditions covered by the data. It does not say what will happen to one variable if one deliberately changes the other variable, but it does tell us about the context in which any change would take place.

Suppose for instance that the floor-space of a particular shop is doubled (e.g. rebuilding, or taking over the next-door premises), but sales have

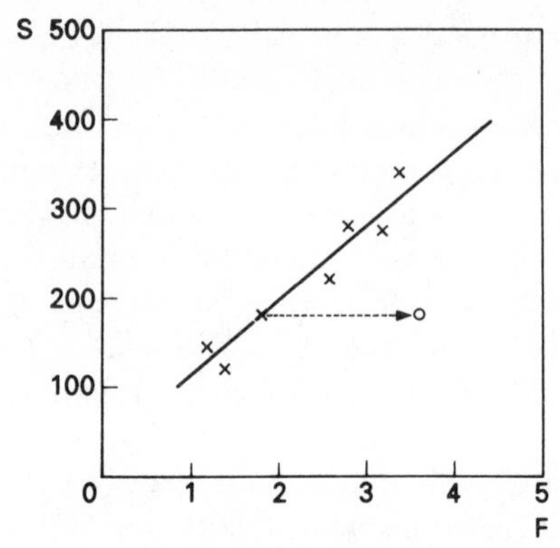

Figure 6.5 Doubling the Floor-Space of a Particular Shop

remained virtually the same, as shown in Figure 6.5. Clearly we have a clash with the equation $S = 80F + 41$. The new result lies well outside the previous range of scatter for different shops. The result is exceptional; we would not in the past have observed the relationship $S = 80F + 41$ with its relatively small scatter, if there had been many such cases.

That is the apparent rationale behind any supposition that the cross-sectional relationship $S = 80F + 41$ *ought* to predict change, i.e. that if F increases, S must also. But such a supposition ignores that the exceptional situation depicted in Figure 6.5 need not occur for long. Thus

(i) Something may have been done to increase sales of the shop *after* the increase in floor-space (e.g. publicity), thus bringing it into line again with the norm $S \doteqdot 80F + 41$.

(ii) Floor-space may have been reduced or used for other purposes after a while because sales did not warrant the extra space.

(iii) The shop may have gone bankrupt or been closed because it was losing money, so that it would disappear completely from the system modelled by the equation.

(iv) The shop may have continued with its abnormally large floor area with the extra costs covered by higher retail margins. Other such cases might have been deliberately excluded in the past from the initial analysis because they were atypical of the type of shop being studied!

(v) The shop may remain as the exception to prove the rule, the one case where a manager doubled his floor-space and did not increase his sales: an "awful warning" so well-known that it stopped repetition.

(vi) Maybe no such cases have ever been observed, anyway!

Any attempt to interpret the cross-sectional relationship $S = 80F + 41$, as telling us that sales will increase if we increase the floor-space ignores all such possibilities. It also ignores the possibility that if a causal chain exists, it may be in the *reverse* direction, with increases in floor-area being decided on after sales *have* increased (or are "certain" to do so because a new housing development is nearing completion).

6.10 Simple Prediction Again

The complexities in the last section arose because of extreme extrapolations beyond the range of conditions covered by the observed relationship, $S = 80F + 41$. The equation only approximated how sales and floor-space varied together among the different shops in Figure 6.3. It might therefore be thought that the relationship could be used more easily in its own context, e.g. to predict that the sales of an additional shop with floor-space F' should be about $(80F' + 41)$. But there are some difficulties even in that. For example, the sales of the new shop will obviously be nothing like that unless someone remembers to put in stock, hire staff, and unlock the doors at the right times. More generally, the equation $S = 80F + 41$ may be relevant only if the shop is managed in the same style as the original shops.

The prediction for a particular shop will therefore be firmer if the original equation held for shops managed in various *different* ways. Then we would know that certain variations in management style do not affect the relationship between S and F, and it does not matter precisely how the new shop is to be run, as long as it is within the range already covered. Similarly, we need to know the type of locations of the seven shops, what kinds of shops they were (supermarkets, corner-shops etc.), the ages of the premises, and so on. In particular, if the previous data referred to *established* shops and the additional shop is altogether new and not based on a previous business in the same location, presumably it will take some time for its sales to build to a "normal" level.

Thus even the simple predictive use of an equation like $S = 80F + 41$ is not really that simple. It does not depend so much on statistical matters, such as the number of readings on which the relationship is based, or how closely the equation fits these readings. It depends on two quite different factors: the *range* of conditions under which a generalisation has been established, and whether the additional shop lies *within* that range of conditions.

6.11 The Need for Research

The need throughout is for empirical generalisation. This shows that trying to establish lawlike relationships is not for the amateur. One is unlikely to have much success by merely collecting some data and a statistical technique and applying one to the other. The problems are not necessarily difficult, but they are laborious. Developing a well-based empirical law requires a great many different sets of data and much analytic effort and perseverance, mostly at a rather tedious level of detail.

For example, current doubts about the validity and practical applicability of the height/weight relationship rest only marginally on the nature of the analyses carried out so far. The real worry is whether the same relationship will hold under other kinds of conditions that might be met in the future. This worry can be reduced only by extending the range of conditions that have already been examined in the past. So far the results are based on four studies only, the earliest published in 1968. The authors were primarily concerned with data to which they had easy access. Although each case covered an extensive range of different sets of data no *comprehensive* studies have yet been undertaken. Thus the empirical basis of the relationship is still relatively weak. Typically, more cross-checking with other data is needed.

It follows that the potential user of a relationship—the engineer, doctor, economist, administrator or scientist—can seldom have instant answers to his problem, let alone expect to obtain valid do-it-yourself results from scratch. Instead, investment in research is needed that will provide well-established results for subsequent use. This takes time. The skill in managing the necessary research and development work lies in starting the right research at the right time.

6.12 Summary

The most immediate practical use of a lawlike relationship is that it reduces bulky data to a succinct summary or model. We can then predict that the relationship will hold for other cases within the range of conditions already covered. We can also *extrapolate* outside these conditions by making an informed guess.

Other uses of lawlike relationships lie in technological applications, in leading to a better understanding of the phenomena in question, and in providing the basis for more ambitious theories and explanations.

Lawlike relationships are misused when they are applied unthinkingly to conditions not covered by the previous data. Typically, the relationship $\log w = .02h + .76$ tells us how the height and weight of different children are related. It does not say how a *change* in the weight of a particular child will be related to a *change* in his height. The two situations are different and the relationships between the variables cannot be assumed to be the same. This is a question for *empirical* investigation and analysis.

A previously established relationship also greatly facilitates the analysis of new data. One need merely check the new data against the already available equation. This is easy to do. If the result is successful, it leads to a wider generalisation of the relationship. But if the previous relationship and the new data fail to agree, this means the relationship does not generalise, and then one has no basis for prediction or for technological applications. This result is also worth knowing.

CHAPTER 6 EXERCISES

Exercise 6A. The Meaning of a Relationship

If the sales-volume S and floor-area F of different shops are related by the equation $S = 80F + 41$, does this mean that shops will *shrink* if sales drop?

Discussion.

This kind of example is a popular way of warning against the misinterpretation of equations by applying them to conditions for which they have not been validated. Unfortunately this lesson is then often ignored in less obviously unreasonable situations. For example, it is still assumed that increasing floor-area will increase sales. The point is that of course it *might* do so, but the "cross-sectional" relationship $S = 80F + 41$ as such does not provide any direct evidence that it would.

Similarly, decreasing sales *do* often cause shops to shrink, but not because of the equation $S = 80F + 41$. The reductions in floor-space are not even in line with it numerically. A drop in sales below a certain level may cause a shop to be closed completely, and we may regard a closed shop as effectively having no floor-space. But the result is quite unlike $S = 80F + 41$.

This also raises the practical question of how one's variables are operationally defined. For example, how does "floor-space" differentiate between selling space, storage space, space for staff amenities, etc? Before attempting any deep explanation or extrapolation of a relationship, it is wise to clarify what the variables mean.

Exercise 6B. Less Simple Data

The analysis in Section 6.6 of the height and weight data for children in Ghana was easy to do because we chose to look at a particular type of child (e.g. expatriate white boys, or non-privileged Ghanaian girls) across different age-groups. In each case this gave consistent results, either consistent agreement as in Table 6.2, or consistent disagreement as in Table 6.3. This was not accidental. We knew age was the biggest single factor and structured the analysis accordingly. But sometimes this prior knowledge can be misleading, and one does not always have it.

As an alternative approach to the data in Table 6.1, analyse each age-group separately. An example for the 9-year-olds is given in Table 6.4, where the prior equation, $\log w = .02h + .76$ is also fitted.

Discussion.

Table 6.4 for the 9-year-olds shows mixed results. Expatriate white and privileged black children are in line with the previous result

TABLE 6.4 Heights and Weights of 9-Year-Old Children in Ghana and the Prior Relationship log w = .02h + .76

(From Table 6.1)

9-Year-Olds	Sex and Living Conditions								
	Boys				Girls				
	Rur	Urb	Priv	Exp	Rur	Urb	Priv	Exp	Av.
h	51	50	54	54	51	51	54	53	52
log w	1.74	1.73	1.84	1.85	1.73	1.76	1.86	1.83	1.79
.02h + .76	1.78	1.76	1.84	1.84	1.78	1.78	1.84	1.82	1.80
Difference	-.04	-.03	.00	.00	-.05	-.02	.02	.01	-.01

$\log w = .02h + .76 \pm .01$. But rural and urban black boys and rural girls are substantially lighter for their height, by about .04 log lbs.

The result for the urban (non-privileged) girls is somewhat unclear. The deviation of $-.02$ is no more than the occasional large deviation found for the basic relationship. These girls could therefore be in line with this relationship. But they might also fit in with the other rural and urban 9-year-olds.

However, the urban (non-privileged) girls of *other* ages are generally .04 log lbs too light, as are all the urban boys and rural children. Thus the urban 9-year-old girls fit in with *them*.

Another query is for the *privileged* black 9-year-old girls. Other ages in this category are on average within .01 log lb of the basic equation. Thus the somewhat large deviation of $+.02$ for this group in Table 6.4 seems to be merely the occasional larger deviation from zero.

This analysis is only a slightly clumsier way of reaching the same conclusion as in Section 6.6. But it is typical of the slower "teasing-out" of results that tends to be required in practice. One's first attempt to select sub-groupings of new data is not always the simplest to use.

Exercise 6C. The Fit for Sub-groups and Aggregates

If a relationship holds for a certain set of data, will it also hold for any sub-group?

Discussion.

If there are a number of groups of readings and a relationship holds for each of them, it must also hold for the aggregate or combined set of data. (Table 6.2 in Section 6.6 gives an example.)

But the opposite is not necessarily true. If there is scatter about an equation, any particular sub-group may consist of individual readings which are biased in one direction. This could "average out" when considering the *total* set of readings. Table 6.4 in the previous Exercise gives an example. The fit for *all* the 9-year-olds appears good, but the fit for the rural children is certainly not.

Exercise 6D. $y = ax + b$ or $x = y/a - b/a$?

In Chapter 4 we used the equation $I = K\sqrt{U}$ between intention-to-buy I and usage U. Could this also be written as

$$I^2 = K^2U, \quad \text{or}$$

$$U = I^2/K^2, \quad \text{or}$$

$$U = LI^2, \quad \text{where } L = I/K^2?$$

Discussion.

The three equations are mathematically identical. The formulation $U = I^2/K^2$ would be used to determine the value of U corresponding to a given value of I, for example when predicting U from I.

But for the general reporting of results, the form $I = K\sqrt{U}$ is preferable because the scatter of the data is easier to summarise in this form. The deviations $(I - K\sqrt{U})$ have a constant average size of about ± 3 percentage points all along the line: i.e. they are "homoscedastic". It follows that the deviations $(I^2 - K^2U)$ of the observed values about the line $I^2 = K^2U$ will *not* be constant: they will differ for low and high values of U. Similarly, the scatter for the formulation $U = I^2/K^2$ will not be constant. We illustrate this for the equation $I = 10\sqrt{U} \pm 3$, where $K = 10$.

Consider first a relatively low value of U, say $U = 4$. For this, I takes the value $10\sqrt{4} = 20$. The average limits of scatter of I will be from 17 to 23. In the formulation $U = I^2/K^2$, these limits correspond to a scatter of U-values from $17^2/100 = 2.9$ to $23^2/100 = 5.3$, an average of about 1.2 units about the theoretical value $U = 4$.

For a much higher value of U, say 36, $I = 10\sqrt{36} = 60$. The average limits of I will again be ± 3 from 57 to 63. However, in the $U = I^2/K^2$ formulation, these limits of I correspond to average limits of U from 32 to 40, a scatter of 4 units about the value 36. The scatter of U is nearly three times as large when $I = 60$ as when $I = 20$.

Therefore, $I = K\sqrt{U}$ is the simpler formulation to use because no matter where we are on the line, its average scatter of $(I - K\sqrt{U})$ can be denoted as ± 3. (If needed, the size of the scatter of U values can still be estimated with this formulation, as just demonstrated.)

Exercise 6E. Prediction and Decision-making

An equation $y = ax + b$ has been found to fit previous data within limits of $\pm c$. The prediction of y in future data will be that y is distributed about $(ax + b)$ within average limits of c. Is this of any use for decision-making, where a single value of y is usually needed?

Discussion.

Prediction and decision-making are separate processes. For example, suppose that we have to decide on the maximum load L which a certain bridge can take when it is made of girders of a certain thickness T.

First we have to arrive at a relationship $L = aT + b \pm c$ between maximum safe loads L and thickness of girders T. From this we predict that L is distributed about $aT + b$ within average limits $\pm c$.

Then we have to make a decision about the maximum load to be allowed. One possibility is to choose the average value $(aT + b)$, but this is unlikely when dealing with *safety*. Another might be to use the engineer's adage of "multiply by three for safety" and hence use $(aT + b)/3$ as a maximum permissible load. Another might be to restrict the load by some multiple of the average scatter c, e.g. $(aT + b) - 5c$, being about the highest load-level at which no bridge with girders T has ever collapsed, but this restriction might be too costly.

The decision here clearly involves considerations of risks and costs. These are separate from the task of predicting at about what load the bridge would actually collapse.

CHAPTER 7

Deriving a New Relationship

In this chapter we discuss fitting a linear equation to measurements of two variables. The situation considered is where there are two or more sets of readings. The problem of having to fit such an equation arises primarily with data that are being handled for the first time.

7.1 The First-time Problem

Treating data as if they were being analysed for the first time is fairly common, but strictly speaking such a first-time situation can happen only once. After that, there must be a previous result: e.g. that $y \doteqdot 5x + 10$, with which the new data can be compared, as discussed in Chapter 6.

At times one may not know about any earlier results, or they may not be in a usable form. Sometimes it is easier to fit a first-time equation to the new data and compare it with the prior results. Or the previous equation may not fit and we want to find another equation to summarise the new data. These are proper reasons for having to fit an equation.

In other cases we should use the previous results in the analysis. But people are often afraid to use earlier findings because the conditions were different (the earlier results came from Mexico, or were pre-war, or whatever) and are thought not to be comparable. This view is wrong. One cannot know whether or not the relationships in different sets of data are the same without comparing them. In general, the aim is to study a relationship under a variety of different conditions. If the same result emerges, it is the more powerful just because the conditions were so different. If the results are different, that also is highly informative: the relationship was different last year, or in Mexico, etc.

7.2 A Degree of Prior Knowledge

To illustrate the fitting of a new relationship we again use the data on the heights and weights of children. But we suppose now that we do not know of any relationship fitted to previous data. Nonetheless, a good deal of other background knowledge already exists.

For example, everyone knows something about children's heights; that children grow taller as they get older, boys tend to be taller than girls, some races are shorter than others, tall parents tend to have tall children. We know that height does not vary hourly or daily, that individual children vary a good deal from each other (some girls are taller than some boys), and that despite this there are statistical regularities, so that boys *on average* are taller than girls. We also know that age is generally the dominant factor in accounting for children's height. Any comparison of different children (boys with girls for example) has to be done *at the same age* if it is to be meaningful.

There is similar knowledge about children's weights. In addition, weight can vary marginally in the short-term (e.g. by eating a 2 lb meal) and sometimes fairly markedly in the medium-term (people losing or gaining weight).

Something is also known about the relationship between the two variables. The two measures do not always go together, taller children usually weigh more than shorter ones, but there are also tall "thin" children and short "fat" ones. We also know that height and weight both increase as children get older: there is no doubt that as children grow, height and weight are correlated.

But at this stage we do not know the quantitative form of the relationship. Nor do we have much theoretical insight into it, or know which other factors (such as sex, race, age, country, etc.) influence it. These are the questions that we have to start answering in our "first-time" analysis.

7.3 The Design of the Study

Our background knowledge helps in choosing what kind of data to analyse. These may be selected from available sources or be newly collected. With children's heights and weights many measurements already exist. Simple findings can be established if a purposeful approach to the available data is adopted.

One obvious aim in any study is to select children who differ from each other in their heights and weights, so that there is some variation to analyse. Since we know that certain races tend to be taller than others (e.g. Caucasian versus Chinese), we can study how racial differences in height relate to those in weight. But since we know that age is the largest factor affecting children's heights and weights, an efficient design for a first study is a comparison of children of different age-groups.

Following the growth pattern of a group of children over a period of years (a so-called "longitudinal" study) is, however, expensive and technically difficult. We can therefore either turn to another factor to study first, or analyse children of different ages measured at the same point-in-time (a "cross-sectional" study). This is very simple. Most available height and weight data for children are already grouped by yearly ages (or in school-classes of children of *similar* ages). An example for three age-groups is given in Table 7.1.

TABLE 7.1 The Average Heights and Weights of Ghanaian Privileged Boys Aged 11, 12 and 13

Ghanaian Boys	Age (in years)		
	11	12	13
Average height (ins.)	57	58	59
Average weight (lbs.)	83	86	88

It is accidental that age, the main design variable used here, is itself quantitative. This will not affect anything we do in the analysis during this chapter. In principle, the data could equally well consist of groupings of children who differ *qualitatively*, e.g. by race, sex, religion, school-grade, socio-economic condition, colour of eyes, and so on.

Having used our prior knowledge to control the data selection, we are now not faced with an unstructured set of height and weight readings, but with data which are ordered into several sub-groups. And as expected, the older children are both taller and heavier. All we have to do is describe the relationship.

Each reading in Table 7.1 is the average of a group of about 50 children of the stated age. This produces more regular statistical patterns, since individual children generally differ from each other considerably. In other kinds of studies one may have to make do with a single reading in each sub-group (as with the time-series data analysed in Chapters 1 and 2). This can affect the scatter or precision of the results, but not the principles of the analysis.

7.4 A First Working-solution

A glance at Table 7.1 shows that average height increases by an inch per year, and weight by 2 or 3 lbs. The increases in height and weight are therefore in roughly similar ratios from year to year, so that the relationship is more or less linear, as is also shown by plotting a graph like Figure 7.1. As a simplifying approximation it is therefore worth fitting a linear equation of the form

$$w = ah + b$$

to summarise the data, rather than look for some kind of curve at this stage.

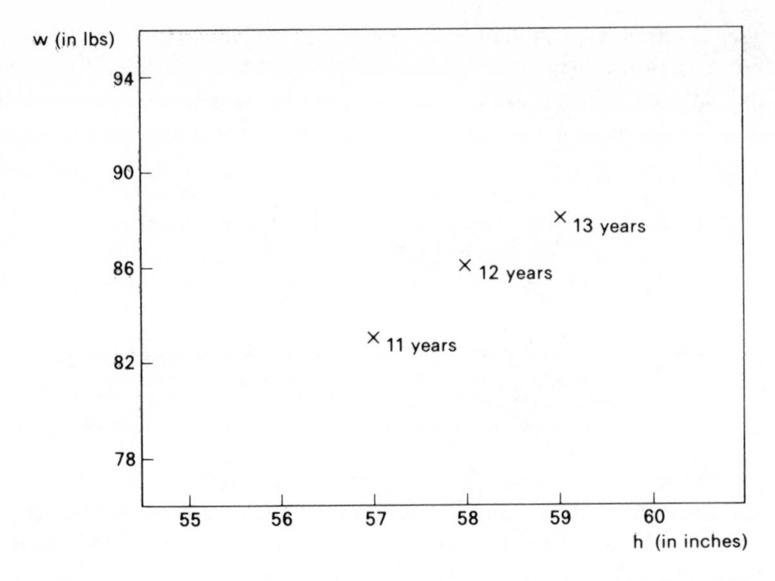

Figure 7.1 The Average Heights and Weights of 11-, 12- and 13-Year-Old Ghanaian Boys

But since the three points do not lie *exactly* on a straight line, any linear equation will be a deliberate oversimplification to provide a simple, approximate summary of the data. No one straight line will reflect the data perfectly, and there can be no unique answer. The equation that we choose will therefore be no more than a tentative initial working-solution.

One way of determining a value for the slope-coefficient a in such a working-solution is by the formula

$$a = \frac{y_2 - y_1}{x_2 - x_1},$$

where (x_1, y_1) and (x_2, y_2) are the two extreme pairs of mean values in the data. For the data in Table 7.1, they are the readings for the 11- and 13-year-olds: i.e. 57 inches and 83 lbs, and 59 inches and 88 lbs. This gives

$$a = \frac{88 - 83}{59 - 57} = \frac{5}{2} = 2.50,$$

working to three significant figures at this stage. So far the equation therefore reads

$$w = 2.50h + b.$$

To determine a value of the coefficient b, we put the line through the overall averages of the readings, which are 58.0 inches and 85.7 lbs. This gives

$$85.7 = 2.50 \times 58.0 + b,$$

$$b = 85.7 - (2.50 \times 58.0) = -59.3.$$

The resulting equation is

$$w = 2.50h - 59.3.$$

For a given value of h, our theoretical estimate of w is therefore $(2.50h - 59.3)$, as shown in Table 7.2.

TABLE 7.2 The Theoretical Estimates (2.50h - 59.3)

Ghanaian Boys	Age 11	12	13	Av.
Av. height: h	57.0	58.0	59.0	58.0
Av. weight: w	83.0	86.0	88.0	85.7
2.50h - 59.3	83.2	85.7	88.2	85.7

The differences between our theoretical estimates and the observed readings of w show how well the initial linear solution fits the data and are given in Table 7.2a. (With a range of only 5 lbs in the average weights it is helpful to calculate the theoretical values to one place of decimals. Hence the observed data are also shown to one place in the table.)

TABLE 7.2a The Deviations from the Working-Solution w = 2.50h - 59.3

Ghanaian Boys	Age 11	12	13	Av.
Deviations: w - (2.50h - 59.3)	-.2	.3	-.2	.0*

* Average size ignoring sign = 0.2 lbs

The important feature of these deviations is that they are irregular in sign, + − +. Their average size ignoring the sign, called the *mean deviation*, is about 0.2 lb. This is small compared to the 5 lb total range of weights in the data, so the equation provides a fairly good fit.

The rationale behind this method of deriving an initial working-solution is that any reasonable equation must go more or less through the overall means of the data, and that fitting the slope by the extreme values will tend to leave irregular deviations, if the data are in fact more or less linear. As a result the data are easy to summarise along the lines discussed earlier, namely that w tends to vary as $(2.50h - 59.3)$, with deviations which are irregular and of a certain average size (here 0.2 lb).

Sometimes the extreme values used to determine the slope-coefficient are not as clear as in this case. Figures 7.2A and B give two examples. In such situations some more *ad hoc* approach can be adopted to derive a first working-solution, like excluding an odd value or grouping several extreme readings together to determine the slope-coefficient. Of necessity the fit will not be close and the solution will be particularly tentative.

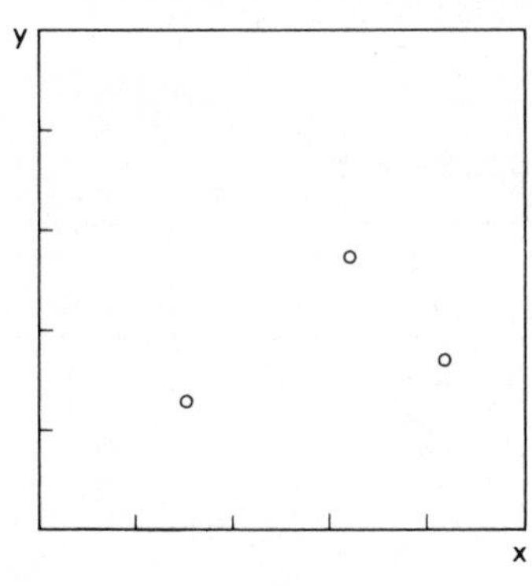

Figure 7.2A Which Extreme Values?

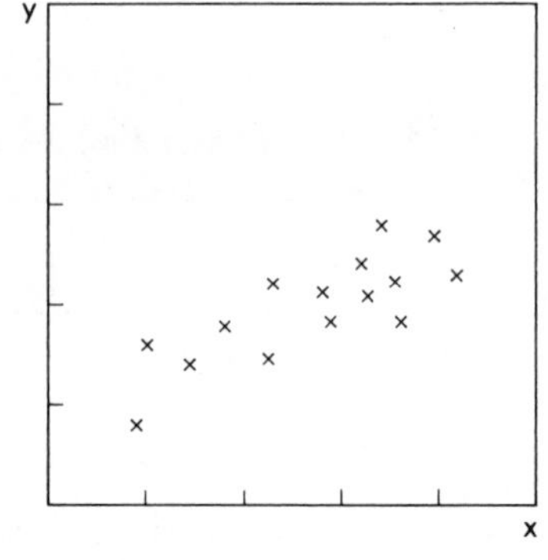

Figure 7.2B Another Example

7.5 The Existence of Alternative Solutions

Since the initial equation above did not fit the data perfectly, there will be other equations which might be just as good or even better. The differences in the fit of such alternative working-solutions are usually slight. It is therefore difficult to choose one as being clearly the best. But this also means that descriptively it is not important which equation one uses.

One alternative is the equation $w = 3h - 88.3$, which is obtained by trying a slope of 3 and again putting the line through the overall means of the data. "Trying a slope of 3" may not seem a very rigorous method of analysis, but the method of derivation matters far less than the results. The two equations in fact give very similar theoretical values of w if we insert the values of h for the 11-, 12- and 13-year-old boys:

	h	= 57	58	59
Original:	$2.5h - 59.3$	= 88.2	85.7	88.2
New:	$3.0h - 88.3$	= 82.7	85.7	88.7

The new equation should therefore also fit the observed data fairly well, and Table 7.3 shows that it does. There is one sizable deviation of −0.7 lb,

TABLE 7.3 The Fit of the Alternative Working-Solution w = 3.0h - 88.3

Ghanaian Boys	Age 11	12	13	Av.
Av. height: h	57.0	58.0	59.0	58.0
Av. weight: w	83.0	86.0	88.0	85.7
3.0h - 88.3	82.7	85.7	88.7	85.7
w - 3.0h + 88.3	.3	.3	-.7	0.0*

* Average size ignoring sign = 0.4

and the mean deviation is 0.4 lbs. Compared with the 5 lb range of weight, this is only fractionally bigger than the mean deviation of 0.2 lbs for the original equation. The two equations look fairly different, however, both in their slope-coefficients and their intercept-constants. Nonetheless, they give very similar results. Indeed, they differ hardly more from each other than either does from the data. But this is only over a limited range of variation. Outside this range the equations will differ increasingly from each other.

Figure 7.3 illustrates the point with four different equations: the two straight lines just discussed, and two *curves*, A and B. The four equations fit

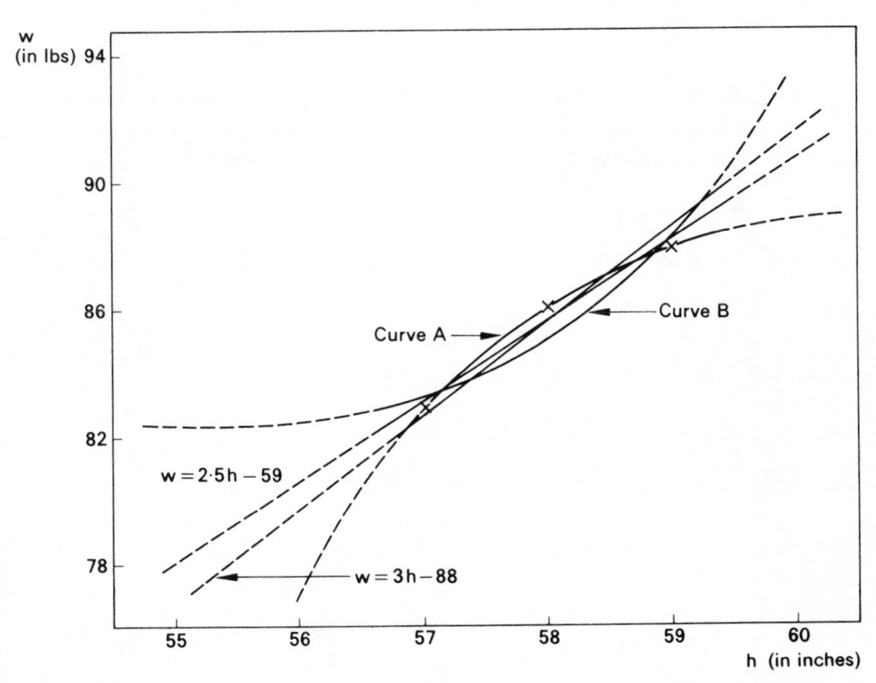

Figure 7.3 Alternative Working-Solutions for the 11- to 13-Year-Olds

the given data almost equally well, but differ markedly from each other *outside* the range covered by the 11- to 13-year olds. It follows that the apparently arbitrary choice of a working-solution for the given data can be narrowed by seeing which equation, if any, also holds for boys outside the initial range of variation.

We therefore now turn to a wider range of data, namely for the 5- to 13-year old privileged Ghanaian boys referred to in the previous chapter. Table 7.4 sets out the data. The weights here vary by 40 lbs, so that we can drop the third significant figure in the coefficients and generally work to the nearest whole pound, keeping an extra decimal place only in the overall averages for working purposes.

TABLE 7.4 The Failure of the 11- to 13-Year Olds' Equation $w = 2.5h - 59$ to Fit for Younger Boys

Ghanaian Boys	Age									Av.
	5	6	7	8	9	10	11	12	13	
Av. height: h	45	48	49	51	54	55	57	58	59	52.9
Av. weight: w	46	52	58	62	70	75	83	86	88	68.9
2.5h - 59	54	61	63	69	76	78	83	86	88	73.1
w - 2.5h + 59	-8	-9	-5	-7	-6	-3	0	0	0	-4.2*

* Average size (5-10 years) ignoring sign = 6 lbs

Table 7.4 shows immediately that our initial working-solution $w = 2.5h - 59$ does not fit the younger boys at all. It gives large and consistently negative deviations for boys under 11 years. Figure 7.4A also illustrates this. It follows

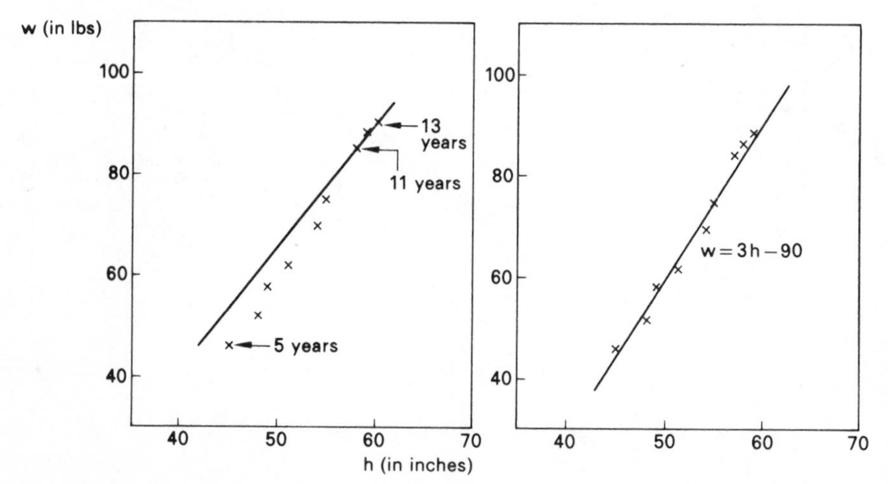

Figure 7.4A The Failure of $w = 2.5h - 59$ Figure 7.4B A New Working-Solution

that our earlier worry about precisely which equation to fit to the 11- to 13-year olds has been superseded by a much bigger problem.

However, the new situation is not hopeless: a strong relationship between height and weight clearly exists and can no doubt be described by an adjusted linear solution with different coefficients.

7.6 A New Working-solution

To find an adjusted linear working-solution we can use the same procedure as before, but this time we apply it to the wider range of data in Table 7.4.

The slope-coefficient is again determined from the two extreme pairs of readings, i.e. we divide the range of the weights by the range of the heights to get $a = (88 - 46)/(59 - 45) = 3.0$. The intercept-coefficient is calculated by requiring that $w = 3h - b$ should hold for the overall averages of all the data, so that $b = 68.9 - (3 \times 52.9) = -89.8$, or -90 to two significant figures. The new working-solution is therefore

$$w = 3h - 90.$$

Table 7.5 shows the fit of this equation. The individual deviations appear fairly irregular and their average size is 1.3 lbs, which is small compared with the 42 lbs range of average weight in the data.

TABLE 7.5 The New Working-Solution w = 3h - 90 Fitted to the Data for the 5- to 13-Year-Olds

Ghanaian Boys	Age									
	5	6	7	8	9	10	11	12	13	Av.
Av. height: h	45	48	49	51	54	55	57	58	59	52.9
Av. weight: w	46	52	58	62	70	75	83	86	88	68.9
3h - 90	45	54	57	63	72	75	81	84	87	68.7
w - 3h + 90	1	-2	1	-1	-2	0	2	2	1	.2*

* Average size ignoring sign = 1.3 lbs

The new equation, $w = 3h - 90$, therefore gives a good fit to the data as a whole. But it does not fit the 11- to 13-year-old boys as well as the earlier working-solution did. The deviations for these ages are now 1 or 2 lbs, compared with only about 0.2 lb before. Furthermore, the deviations are all positive, all three points lying *above* the fitted line. The choice facing us is between an equation which gives a close fit to a limited range of data and one which covers a much wider range of readings but less precisely, as

Figures 7.4A and B illustrate. The answer is that it is descriptively and conceptually much simpler to deal with one equation plus some irregular scatter for all the data, rather than with various *different* equations with smaller scatter for different parts of the data. This determines the approach described here.

7.7 Alternative Working-solution

On closer scrutiny the deviations in Table 7.5 seem to have a slight systematic + − + pattern. They are positive for two of the three youngest age-groups, negative or zero for the middle three age-groups, and positive again for the three oldest age-groups. This may seem like reading too much into just nine readings, but at this early stage in the analysis we are only noting possibilities. A *curved* working-solution might therefore fit better in giving less of a pattern, especially for the 11- to 13-year olds.

This analysis brings us back again to the problem of choosing among different working-solutions. For example, curve C in Figure 7.5A gives less regular deviations than the working-solution $w = 3h - 90$. Almost the same effect can, however, also be achieved with another *linear* working-solution, such as

$$w = 3.2h - 100,$$

shown in Figure 7.5B. The equation was derived by applying the usual fitting-procedure to *groupings* of the more extreme average heights and weights, namely the 5- to 7-year olds and the 11- to 13-year olds. This reduces

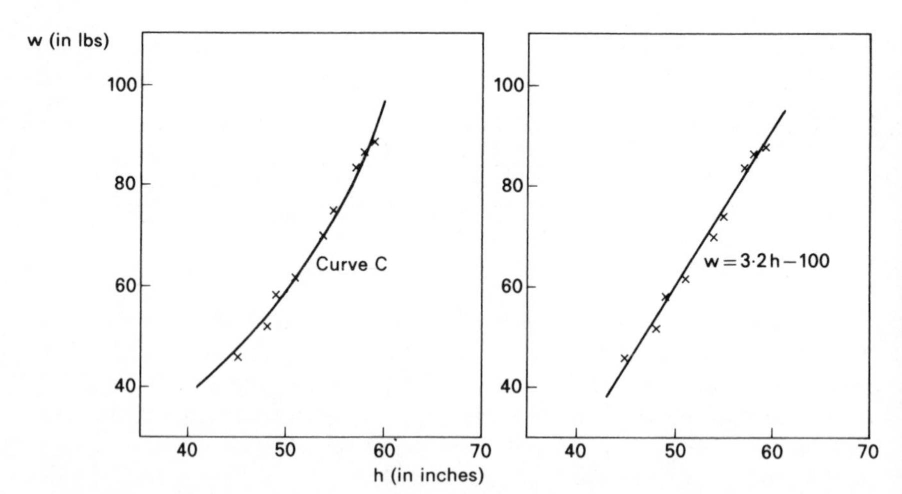

Figure 7.5A A Possible Curve

Figure 7.5B An Adjusted Line

the reliance on some particular extreme value which may be a little abnormal. (Figure 7.4B for example suggests that the reading for the 5-year olds influenced the earlier equation too much. Without it we would have fitted a steeper line, more like the new equation $w = 3.2h - 100$.)

The mean deviations of the two new alternative equations in Figures 7.5A and B are about 1.2 or 1.3 lbs, which is virtually the same as for the earlier equation $w = 3h - 90$ in Figure 7.4B. The important difference is that the deviations have a less regular pattern, in particular those for the three oldest age-groups.

There is therefore once more a certain variety of equations which fit the data "reasonably" well. We still have a problem of choice. But the degree of uncertainty has been greatly decreased by extending the range of variation covered. Figure 7.6 shows that three of the earliest working-solutions from Figure 7.3, $w = 2.5h - 59$ and Curves A and B, are no longer tenable at all, and that the fourth, $w = 3h - 88$, is also ruled out by its small but consistent bias.

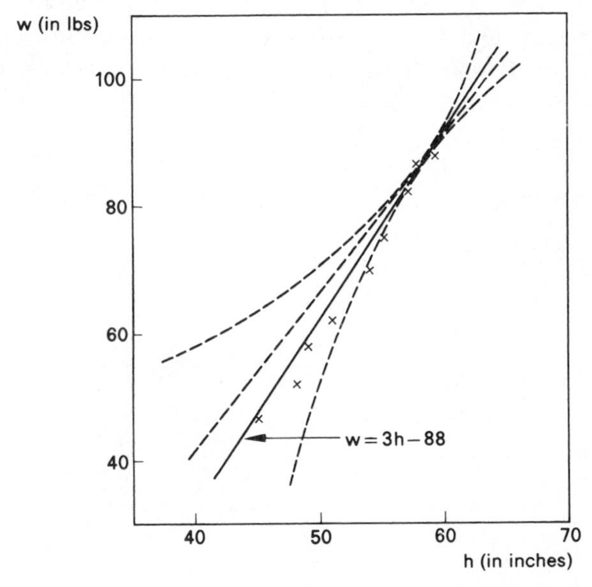

Figure 7.6 The Initial Working-Solutions of Figure 7.3

The ambiguity which now remains will be reduced by analysing yet more data, coupled in due course with the growth of theoretical understanding. The question is which model, if any, will describe not only the heights and weights of these particular 5- to 13-year-old Ghanaian boys, but other data sets as well? The analysis of further data will be discussed in the next chapter, together with the first stages of dealing with *curved* relationships.

But at this stage, with only nine pairs of readings, there is no need to worry unduly about precisely which of the possible equations to select. If the further data show quite a different height/weight relationship (with all the new data for example lying well to the left of the line in Figure 7.5B), then it no longer matters which particular equation was fitted. And if the new data more or less agree with the present readings, they will narrow the choice of equations.

7.8 The Scatter of Individual Readings

The analysis of a relationship essentially has two parts. One describes the systematic variation of the mean values between different groups of readings, as we have been doing. The other describes the scatter of the *individual* readings in each group, if there is more than one reading in each. This is a separate part of the analysis. It generally does not affect the question of which equation to fit.

Figure 7.7 gives a notional picture of the distribution of individual children's heights and weights. The size and nature of this scatter cannot affect the line fitted to the mean values (except with very small samples, when the observed means may be affected by sampling variation).

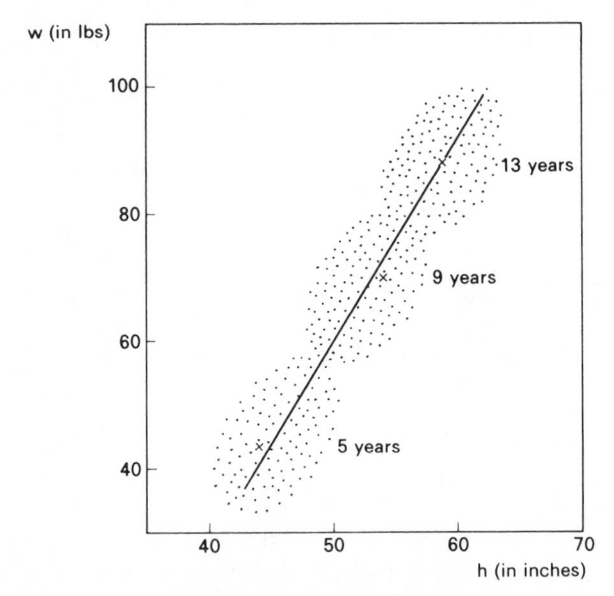

Figure 7.7 The Distribution of Readings for Individual Boys (A notional representation)

We therefore first fit an equation like $w = 3.2h - 100$ (together with any discrepancies in approximating to non-linear data). Then we describe the scatter of the individual readings by summarising the average size of the deviations ($w - 3.2 + 100$) for individual boys.

From the information on Birmingham boys supplied privately by Dr. E. M. B. Clements (1954), as mentioned earlier, we can estimate the average size of such individual deviations to be about 8 lbs, as a broad average over all age-groups. This provides a better "feel" for the data, and is necessary knowledge for certain practical applications of the results (as was illustrated in Section 6.4).

However, such information is often not reported. For example, none of the published sources of the height and weight data referred to earlier (see Table 5.1) gave information on the scatter of the individual readings. As mentioned in Section 5.2, such an omission is open to purist criticism. But, in practice, people are mostly concerned with the *systematic* variation in their data, how w varies with h in different sets of data, and this is represented by the equation fitted to the age-group means.

7.9 Summary

In this chapter we have discussed fitting a linear equation to two variables. This occurs mainly when there is no previous information about the relationship. But even when handling data for the first time, one usually has some background information about the variables in question. For example, it may already be known where the larger variations occur. One can therefore choose to analyse groups of data that differ from each other systematically.

Even for data which are more or less linear, the means of different groups of readings usually do not lie *exactly* on a straight line. Various alternative approximations are therefore possible in fitting a straight line. An initial linear working-solution can be obtained by calculating the slope-coefficient from the ratio of the differences between the highest and lowest mean values of the two variables, and the intercept-coefficient by making the line go through the overall averages. This procedure generally leads to residual deviations which are irregular and hence easy to summarise simply in terms of their average size.

Because the different sets of data do not lie exactly on a straight line, no linear equation can fit perfectly. The equation fitted is therefore one of various possible working-solutions. These all give a similar fit to the data and therefore all tell effectively the same story. In general, two solutions which fit the data with irregular deviations will differ no more from each other than each does from the observed data.

Initial working-solutions are adjusted in subsequent work, mainly to accommodate additional data. This occurs especially if the new data lie outside the initial range of variation. The degree of uncertainty involved in choosing among different possible equations is therefore reduced. The aim of the analysis lies not so much in finding an equation that fits best for the initial set of readings, but one that can generalise to a wider range of data.

CHAPTER 7 EXERCISES

Exercise 7A. Problems with Extreme Values

Fit an initial working-solution to the following five pairs of readings:

x:	15	10	20	18	12
y:	7	3	8	11	1

Discussion.

This example typifies a fairly common problem when fitting an equation by the method in Section 7.4, namely that it is difficult to determine the extreme values in the data. Ordering the readings by the size of the x variables gives

x:	10	12	15	18	20
y:	3	1	7	11	8

Using the two extremes we get a slope-coefficient of $(8 - 3)/(20 - 10) = .5$ and the equation

$$y = .5x - 1.5.$$

But if we order the reading by the y variable, we have

x:	12	10	15	20	18
y:	1	3	7	8	11

Using the two extremes here gives a slope-coefficient of $(11 - 1)/(18 - 12) =$ 1.7 and the equation

$$y = 1.7x - 19.5.$$

These two equations look very different.

The problem arises because although there is a general tendency in the data for high x to go with high y and low x with low y, this is not the case for the two *lowest* or *highest* pairs of values. The lowest x-value is not the lowest y-value, even though both are *low*.

In a case like this a better alternative working-solution can be fitted by combining the two lowest pairs and the two highest pairs of values, as suggested in Section 7.4. Fitting a slope-coefficient to these grouped means gives $(9.5 - 2)/(19 \times 11) \doteqdot 1$ and the equation

$$y = x - 9.$$

This is clearly a compromise between the first two equations.

The reader can check that each equation fits the y-values with a mean deviation of about 2 units. There is therefore not a great deal of difference among them in their average fit. But the "compromise" solution, $y = x - 9$, is more attractive because it depends less on an isolated and rather erratic extreme value and because it has the more evenly sized and irregular deviations.

The main conclusion is that if the scatter in the data is relatively large, then there will be a considerable range of equations that can reasonably be fitted. Additional data will generally reduce the ambiguity, as already stressed in the main text of this chapter.

Exercise 7B. Subjective or Objective?

Are the analysis procedures described here very subjective?

Discussion.

No. If the analyst describes what he has done, any experienced person will arrive at effectively the same result by applying the same procedure to the same data. This is the criterion of objectivity in analysis. Wat is more, if the result generalizes, anyone will be able to arrive at it by following the same procedure with *other* data. This is the criterion of objectivity in *science*.

Objectivity should not be confused with the absence of choice. For a method to be "objective" does not mean that any fool must always get the same answer as a highly experienced analyst. It is not necessary for there to be only one possible method or one possible result. A foolproof method of being objective would be always to fit the equation $y = 2x + 3$. Everyone would then very objectively get the same answer for any data but generally be wrong.

Exercise 7C. Alternative Working-solutions

How can one claim that two alternative "working-solutions" like

$$w = 2.5h - 59.3$$

$$w = 3.0h - 88.3$$

in Section 7.5 say more or less the same thing about the data? According to one, w varies 20% more with h than according to the other. And when h is 0, one says that w is 59 lbs and the other 88 lbs. Such differences are not negligible.

Discussion.

The two equations do not purport to say at precisely what rate w varies with h, since this was not altogether clear from the given three pairs of readings. That is why different working-solutions are possible. Nor do such equations purport to reflect the values of w for values like $h = 0$ which are way outside the observed range.

Instead, the equations merely aim to summarise the observed sets of data. Within the observed range of variation they generally differ from each other no more than each differs from the data. (The example in Exercise 7A illustrated an extreme type of case where some working-solutions can differ markedly.)

Once data covering a more extensive range of variation have been successfully analysed, the coefficients of the equation may bear more direct interpretation. Thus for the 5- to 13-year olds analysed in Section 7.6, it would appear that weight varies by something like 3.0 to 3.2 lbs for every inch increase in height, and working-solutions like $w = 3h - 90$ and $w = 3.2h - 100$ summarise this.

But the "intercept-coefficients" of -90 or -100 still have no direct meaning. They do not say what w would be for $h = 0$, since no such data have been observed.

Exercise 7D. Why the Problem?

Why is there this problem of different possible working-solutions?

Discussion.

Because one is forcing a linear equation on to non-linear data. Approximation is often said to be an art, and within the small limits of choice illustrated here this is to some extent true.

Exercise 7E. The Intercept-coefficient

Why do the equations fitted to the heights and weights not go through the point (0, 0), since initially children have virtually no height or weight?

Discussion.

The data being analysed do not cover babies, let alone pre-natal conditions.

Some attention can, however, be paid to such external knowledge: for example to differentiate between two otherwise equally possible working-solutions. But it would be wrong to force the empirical equation too much, since at this stage we do not know what mathematical form the height/weight relationship should take below 5 years.

Exercise 7F. The Wrong Working-solution

Suppose that the data in Exercise 7A have been summarised by the working-solution $y = x - 9 \pm 2$, in the range of $x = 10$ to $x = 20$. How does this assist in analysing the following additional data?

x:	12	14	27	30	42
y:	3	6	9	14	18

Discussion.

The earlier result does not fit the higher values here. For example, for $x = 42$, the predicted value is $42 - 9 = 33$, instead of the observed value 18.

Fitting a new working-solution to the new data gives $y = .5x - 2.5$. We must now decide whether this will also fit the previous data, without having direct recourse to them. All we know about the earlier data is that they were fitted by $y = x - 9 \pm 2$, within the range of $x = 10$ to $x = 20$.

We therefore compare the two equations in this range:

x	10	15	20
$.5x - 2.5$	2.5	5.0	7.5
$x - 9$	1.0	6.0	11.0
Difference	1.5	−1.0	−3.5

The average difference is about 2, which is the same as the reported fit of the earlier equation to the data. The *maximum* discrepancy is -3.5, which is probably no larger than the largest discrepancy for the earlier data. This implies that the equation $y = .5x - 2.5$ will fit the original data roughly as well as the first working-solution, $y = x - 9$, did.

This conclusion has been reached merely from a *summary* of the earlier data, without having seen the detailed readings. The initial working-solution proved "wrong" in the light of new data, but it still performed its primary function of first adequately summarising the original data and then leading to a better working-solution for both that *and* the new data.

Exercise 7G. Different Sets of Data

The analysis in this chapter depends on having more than one set of data, and these must have different means. (Otherwise no equation can be fitted by the procedure that has been described.)

What happens if we either have only one set of readings, or if our different sets of readings differ little (if at all) in their means?

Discussion.

Having only one set of readings cannot happen often, since any worthwhile study has to be repeated. If all the different sets of data have the same means, the observer has not managed to exercise any effective control over his variables: there may be variation in x and y, and this may correlate, but it is effectively uncontrolled or "error" variation. The observational basis for establishing an empirical generalisation does not exist.

The observer needs to see if he can differentiate his data more effectively by using better selection criteria (e.g. older boys versus younger boys rather than size of family, say).

A set of height and weight readings for a group of n children consists of a *single* set of readings if we know nothing about each boy other than his height and weight. But if for each separate child we have a description of its age, sex, race, place of residence, number of siblings, etc., and if the children differ from each other in some or all of these respects, then we have n different sets of data, each consisting of a single child. (Much observational data consists of 1 reading per set.)

Exercise 7H. Height as a Function of Weight

In Section 7.6 we fitted the working-solution $w = 3h - 90$, which expresses weight in terms of height for the 5- to 13-year-old Ghanaian boys (Table 7.5). What would we get if we used the same technique to fit an equation of the form $h = pw + q$, expressing height in terms of weight?

Discussion.

From the data in Table 7.5, the slope-coefficient in the equation $h = pw + q$ will be $(59 - 45)/(88 - 46) = 14/42 = .33$. The intercept-coefficient will be $52.9 - .33 \times 68.9 = 30$. This gives the equation

$$h = .33w + 30.$$

Dividing through by .33 (or multiplying by 3) gives

$$3h = w + 90,$$

which is $w = 3h - 90$, as before.

As long as the same method of fitting and the same "extreme points" are used, the same equation is obtained whichever way round it is written.

Exercise 7I. Height and Age, and Weight and Age

In Table 7.5 age is a third quantitative variable. What are the relationships between height and age and between weight and age?

Discussion.

Fitting a working-solution to the data in Table 7.5 for height h and age A gives $h = 1.75A + 37$. This has a mean deviation of about $\pm.5$ inches, and the deviations appear irregular.

For weight and age, the working-solution is $w = 5.25A + 22$, with a mean deviation of ± 1.6 lbs. But here the deviations show quite a marked pattern; negative for low ages, positive for 9 to 12 years, and negative for 13 years. This suggests that a curve might fit better than a straight line, as is discussed in Exercise 8F.

Exercise 7J. The Velocity of a Falling Body

The velocity of a falling body was measured at intervals of one second with the following results

Time t (in sec):	0	1	2	3	4	5
Velocity v (in ft/sec):	10	45	70	120	130	160

What relationship would you fit?

Discussion.

It is well-known in physics that a body falling freely near the surface of the earth is subject to a virtually constant acceleration (an increase in its velocity) of about 32 feet per second every second, i.e. v varies as $32t$. Given that the observed velocity at time $t = 0$ was 10 ft/sec, the velocity at any time t should therefore be

$$v = 10 + 32t.$$

This theoretical result gives a good fit to the data. It is a typical case where a "first-time" method of fitting an equation is unnecessary.

Exercise 7K. Fitting to Individual Readings or to Means

Why are working-solutions fitted to the mean values of the different sets of data and not to the individual readings of x and y?

Discussion.

Suppose there are two or more sets of data, such as the different age-groups in our numerical example. Any line fitted to the first set has to go through its mean values. If the same line is to hold for the second set of data, it also has to go through the means of *that* set. These two points determine the line. The individual readings do not affect the line that is fitted.

Exercise 7I. The Analysis of the Individual Readings

How can one analyse the scatter of the individual readings in each set of data? For example, in Figure 7.8A would it be useful to fit a separate equation to the individual readings in each age-group?

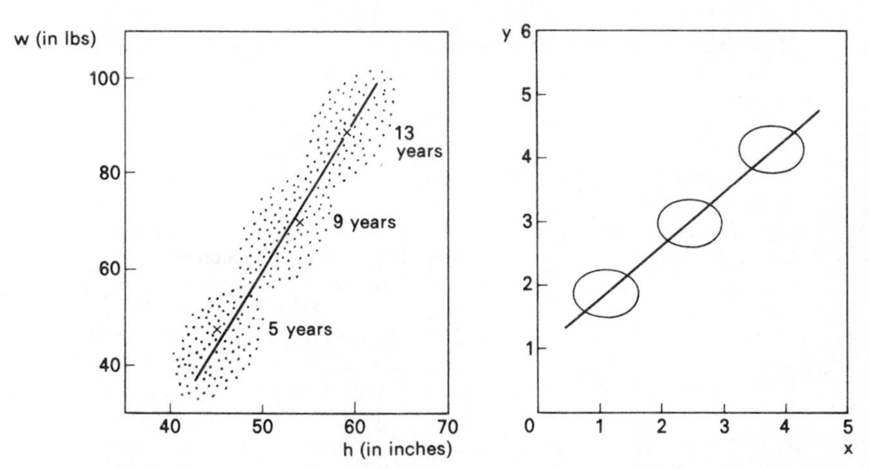

Figure 7.8A Individual Boys (Figure 7.7) Figure 7.8B A Different Form of Scatter

Discussion.

If equations are fitted to the separate age-groups in Figure 7.8A which differ from the equation already fitted to the group means, they will also differ from each other. We would therefore generally finish up with n unique equations for n sets of data. It is not clear what use they could be.

However, the data in each age-group might not be "in line" with the fitted equation, the scatter of the individual readings taking a different shape, as illustrated in Figure 7.8B. This would merit further scrutiny. One possible cause for data to be in this form is simply if the errors of measurement of one variable (here x) are much greater than those of the other.

The scatter of individual readings can be broken down into two types of "error", one attached to each variable:

$$x = \text{"true value of } x\text{"} + \text{"error of } x\text{"},$$

$$y = a(\text{"true value of } x\text{"}) + b + \text{"error of } y\text{"}.$$

But even though in the past I have engaged in some theoretical research into this kind of model (e.g. Ehrenberg, 1950, 1951), I have found virtually no call for it in practical data analysis.

In terms of our example, the main practical questions in a more detailed analysis of individual deviations seem to be (a) whether the individual boy's deviations from the relationship are consistent over time, i.e. does a boy remain overweight if he was overweight earlier? (b) how these deviations relate to other factors, such as the height/weight patterns of parents and of siblings, the boy's medical and nutritional history, and so on. To answer such questions more information is needed than just the boys' heights, weights, and ages. In particular, the *longitudinal* type of design mentioned in Section 7.3 would be required, giving repeated measurements of the same boys over a period of time.

Exercise 7M. Is There a Relationship?

How can we tell whether a relationship exists between two variables for some given data?

Discussion.

An equation of the form $y = ax + b$ reflects a numerical association between the observed values of x and y. A relationship exists if the average deviation of y from the line, $(y - ax - b)$, is less than the average deviation of the y-values from their overall mean $\bar{y}$, $(y - \bar{y})$. If the difference between the two types of deviation is small, the relationship is weak. (If the data are based on random samples and the number of readings is small, one must test the statistical significance of the result, as is discussed in Chapter 18: an association between x and y could exist in the sample due to chance errors in the sampling, without an association in the population sampled.)

This question of whether the two variables are related can only occur at the "first-time" stage of analysis. After that there will be prior information about what has already been found in other data. The more common question in practice is therefore not whether the relationship exists, but whether the relationship in one case is the same as in others.

Lack of correlation should not in any case be regarded as a *failure*. For example, establishing an empirical generalisation consists of showing that one's results (e.g. the values of the coefficients in the equation $w = 3h - 100$) are *not* related to other variables.

In general, nothing could be simpler, and hence more important, than to show that y does *not* vary with x. This is especially important if the analyst firmly expected that y *would* vary with x. Having waited a minute or two to overcome his discomfiture, he should be able to say "So y is *not* related to x—what can I make of that?".

Exercise 7N. Other Methods of Analysis

By reference to statistical textbooks and journals, discuss other methods of fitting a straight-line equation.

Discussion.

Several different approaches have been considered in the statistical literature. The main ones are as follows.

Regression Analysis. Here the coefficients of the equation $y = ax + b$ are determined by making the sum of all the squared deviations $(y - ax - b)^2$ as small as possible. This is discussed in Chapter 14.

Regression analysis is generally applied to a single set of readings, which is a different situation from that considered in the present chapter. The problem of fitting a regression equation to two or more distinct sets of data does not appear to be discussed in the literature. Nor is it usually claimed that regression analysis leads to generalisable results.

Regression Applied to the Group Means. A regression equation could be fitted to the mean values of different sets of readings, in effect treating them as a single set of data. But if the means lie exactly on a straight line, there is no fitting problem. If they do not, one would be fitting a linear regression equation to non-linear data, therefore the statistical requirements and advantages of regression theory would not apply. Thus regression analysis applied to the group means has no particularly attractive properties.

Regression Applied to Pooled Data. A regression equation could also be fitted to two or more sets of data by first pooling all the readings. But this would lose all the information about how the readings differ from each other (e.g. boys of different ages). The regression equation would also depend on the arbitrary numbers of readings in the different groups. Furthermore, pooling is unnecessary because an equation can be fitted to the group means, as discussed.

Functional Analysis. This method is based on the "errors in both variables" model outlined in Exercise 7L. But as described in the statistical literature functional analysis is generally discussed in terms of a single set of data. In such a case it is agreed that the method cannot actually provide a solution, as the slope-coefficient a in this model cannot be determined without some extraneous information.

The approach outlined in this chapter is essentially one of functional analysis, but applied to more than one set of readings. The extraneous information is then provided by making the equation fitted to one set of data go through the means of *another* set of data.

The Wald–Bartlett Approach. Another form of functional analysis of a single set of data is to divide the data either into two sub-groups (Wald, 1940) or three sub-groups (Bartlett, 1949). Then an equation can be fitted to the sub-group means essentially along the lines of this chapter.

However, dividing a single set of data into two or three sub-groups is an arbitrary matter if no *external* criterion is being used (such as age, race, sex, etc. in the height/weight example). The resulting equation therefore does not carry any of the connotations of an empirical generalisation and the Wald–Bartlett methods do not seem to have caught on in practice.

Instrumental Variables. This is a method discussed mainly in the literature of econometrics. It arises when there is a third variable, like *age* in our height and weight example. For example, in Exercise 7I we derived the equations $h = 1.75A + 37$ between height and age, and $w = 5.25A + 22$ between weight and age. Eliminating the age-variable A from these two equations (by writing $(h - 37)/1.75 = (w - 22)/5.25$) gives the equation $w = 3h - 89$: virtually the same height/weight working-soludion as in Section 7.6. The age-variable has therefore been "instrumental" in deriving this relationship.

This process is discussed more generally in Chapter 10, as part of the general use of theoretical arguments. The analytic approach described in this chapter has been in common with the broad idea of instrumental variables. But in the literature not much emphasis is placed on the property of empirical generalisation, that the same height/weight equations holds despite the different values of the instrumental variable.

CHAPTER 8

Non-linear Relationships

Most empirical relationships require non-linear mathematical functions to describe them. The purpose of this chapter is to start to bridge the gap between fitting initial linear working-solutions and deriving more complex curvilinear models.

Our discussion will largely centre on the height/weight relationship. As more data are considered, we will see how the initial linear working-solution fitted in Chapter 7 develops into the logarithmic relationship referred to in Chapters 5 and 6.

8.1 Systematic Deviations

We finished our analysis of the heights and weights of privileged Ghanaian boys in the last chapter with several possible solutions and had some doubts about which one to use. Such doubts are common when working with initial solutions based on limited evidence.

The possibilities included the linear equation $w = 3.2h - 100$ and Curve C, shown again in Figure 8.1. Since in both cases the deviations are not large, one would normally choose the straight line because it is simpler to use. But we can see in Figure 8.1 that a linear equation like $w = 3.2h - 100$ could not hold much outside the range of variations covered so far: it says, for example, that children standing 30 inches or less have no weight.

Therefore some kind of curve will ultimately be needed to integrate the wide range of data for children of other ages. But at this stage we have no indication of which kind of curve to fit and have to look at more data.

8.2 More Data in the Same Range

Table 8.1 gives readings taken in 1947/8 for 5- to 13-year-old boys from Birmingham, England, of the middle Social Class "3" (Healy, 1952; Clements, 1953; Ehrenberg, 1968.) These cover about the same range of variation as the Ghanaian boys.

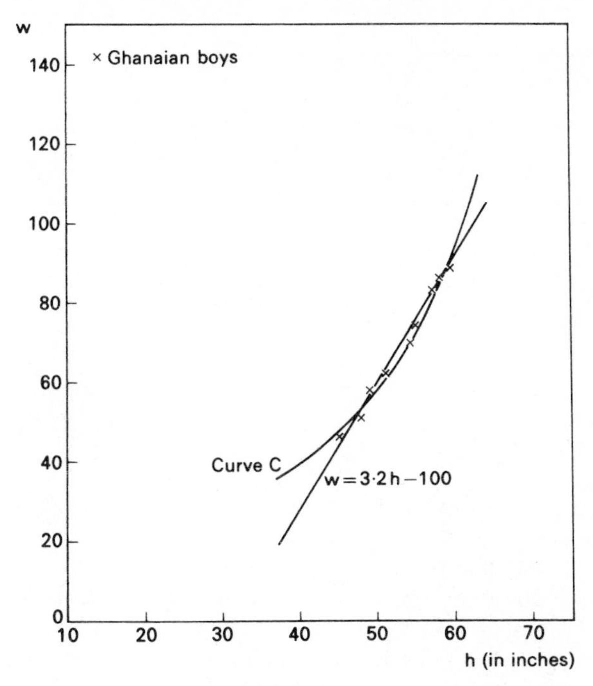

Figure 8.1 Two Working-Solutions for the Ghanaian Boys

The deviations of these data from the linear working-solution $w = 3.2h - 100$ are somewhat larger than for the privileged Ghanaian boys in the last chapter: on average 3.1 lbs compared with 1.3 lbs. They are also mostly negative and have some kind of pattern in that the two extreme values are either positive or small. This suggests that a curve would fit the Birmingham data better than any straight line, and supports the earlier suggestion in

TABLE 8.1 The Fit of the Previous Working-Solution w = 3.2h - 100 for Birmingham Boys of Social Class (3) Aged 5 to 13 Years

Birmingham Boys	Age 5	6	7	8	9	10	11	12	13	Av.
Av. height: h (ins.)	43	46	48	50	52	54	55	58	59	51.7
Av. weight: w (lbs.)	42	46	51	57	62	68	73	82	88	63.2
3.2h - 100	38	47	54	60	66	73	76	86	89	65.4
w - (3.2h - 100)	4	-1	-3	-3	-4	-5	-3	-4	-1	-2.2*

* Average size ignoring sign = 3.1 lbs

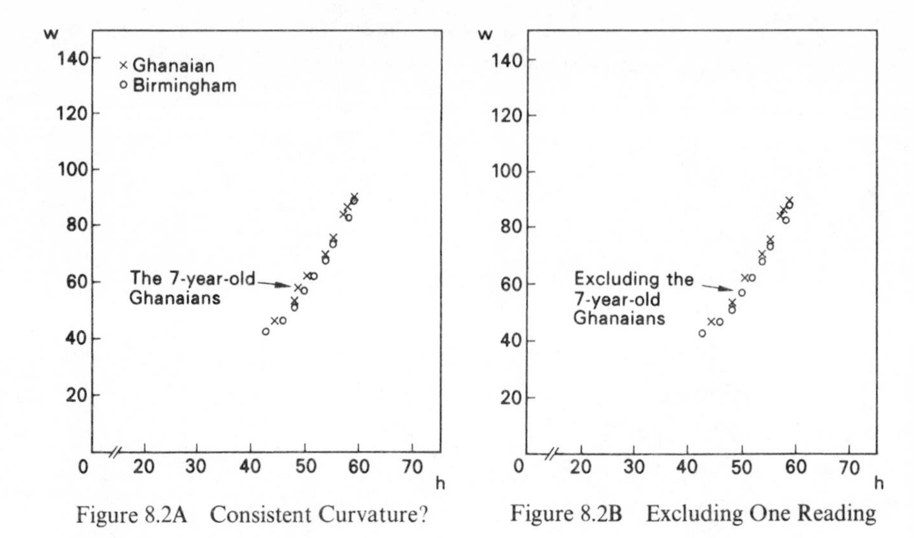

Figure 8.2A Consistent Curvature? Figure 8.2B Excluding One Reading

the Ghanaian data that the relationship generally might be curved. The next question is whether the same curve will serve both sets of data.

Figure 8.2A plots the Birmingham results on the same graph as the results for the privileged Ghanaian boys. It shows that both sets of data are fairly similar. In Table 8.2 the two sets of deviations from the equation $w = 3.2h - 100$ are compared numerically. On average the Birmingham boys are about 2 lbs lighter, but this difference is small compared with the nearly 50 lbs total range of weights. The pattern that emerges still confirms the suggestion of curvature; not a simple + − +, but a positive deviation at 5 years changing to large negative ones at 8 to 10 years, and then back on average to *small* negative ones at the higher ages.

TABLE 8.2 The Deviations from the Linear Working-Solution $w = 3.2h - 100$ for the Birmingham and the Ghanaian Boys

The Deviations ($w - 3.2h + 100$)	Age									Av.
	5	6	7	8	9	10	11	12	13	
Birmingham boys	4	-1	-3	-3	-4	-5	-3	-4	-1	-2
Ghanaian priv. boys	2	-2	1	-1	-3	-1	1	0	-1	0
Average	3	-1	-1	-2	-3	-3	-1	-2	-1	-1

Sometimes a single reading can wrongly dominate an analysis. Looking at Figure 8.2A it is apparent that the 7-year-old Ghanaian boys are relatively heavy. They have a positive deviation of 1 lb in a context of negative deviations for boys aged 6 to 8 years. If we exclude this reading, the similar curvature of the two sets of data becomes more apparent, as in Figure 8.2B.

This might seem like "subjective messing around with the evidence", but one is only noting possibilities, e.g. that the same curve might describe both sets of data. The crucial next step is finding that in fact this possibility has also been supported by a wide variety of other data, as has been summarised in Table 5.1.

This need not have happened. We could have found something like the notional data in Figure 8.3. Here the deviations from the straight line are sometimes positive and sometimes negative, even at the same point on the line. They do not fall into a simple pattern. A straight line therefore gives the simplest summary that any *single* equation in the two variables could provide.

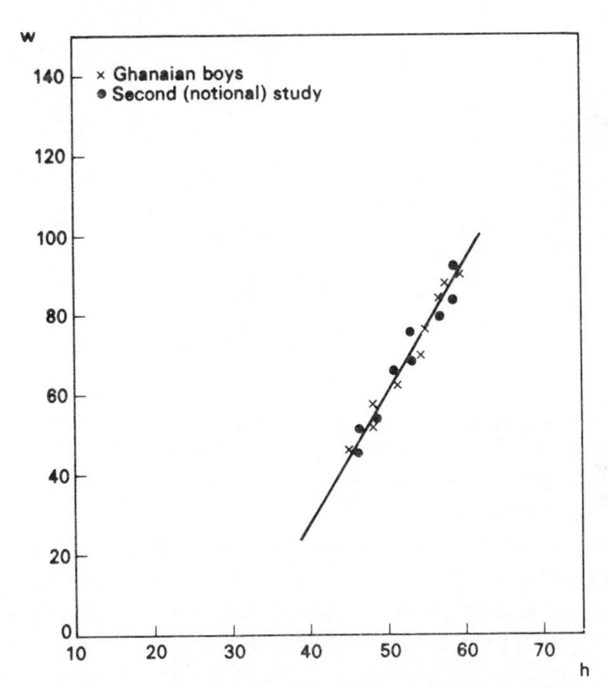

Figure 8.3 Inconsistent Deviations (Notional Data)

Another possibility is that different sets of data would not agree even about the broad nature of the relationship. This would lead to a complete failure to generalise, as was illustrated in Chapter 6.

8.3 Choosing a Curve—Transforming One Variable

Having seen that the height/weight data follow a curved relationship, we have to decide what kind of curve to fit. Rather as we saw with straight lines, a variety of different mathematical functions can closely approximate any particular set of data. This is especially true when there is a limited range of variation and only a slight degree of curvature in the data. But the alternative curves may differ greatly outside the observed range of variation, as Figure 8.4 shows. Ultimately, this will help us choose which to use.

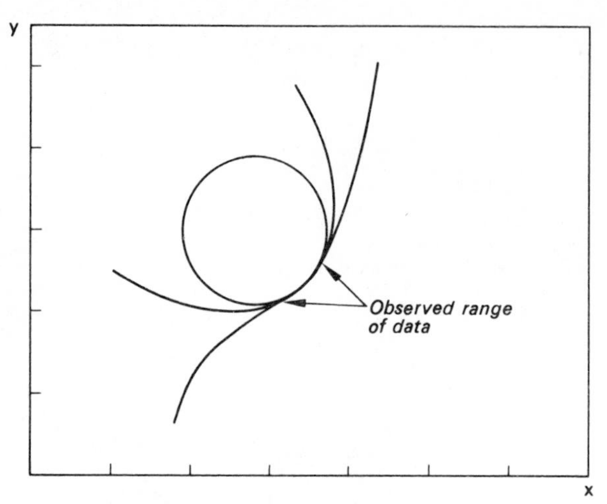

Figure 8.4 Different Curves Giving a Good Fit

We begin by selecting an initial working-solution from various possible curves, just as we did earlier with linear equations. This time the problem of choice is worse because there are so many different kinds of non-linear mathematical functions to choose from.

Sometimes a particular curve may be suggested either by an analogy or by a theoretical argument. For example, the growth pattern of a certain animal may be known, suggesting that the height/weight relationship for boys is of the same mathematical form. But in the absence of such outside knowledge, the usual procedure is to see whether the curved relationship can be dealt with by changing it into a linear equation. This is done by "transforming" the scale of measurement of one or both of the variables.

There are no clear rules for finding a suitable mathematical transformation. It is usually a matter of trial and error. The procedure depends on seeing the "shape" of the observed data and being aware of a variety of mathematical functions which might change that shape into a straight line. In practice, the analyst usually tries a small number of common mathematical functions, e.g. square roots, squares, reciprocals, exponentials, and logarithms. Other functions usually stem from a stronger theoretical basis (as illustrated in Chapter 10).

We saw in Figure 8.2 that the degree of curvature is small for the height and weight data. Transforming just one of the variables should therefore suffice. To help decide which one, we have to look for some extra evidence or suggestions. These usually exist, even if they are weak.

In our case, some evidence is provided by the separate relationships of age with height and weight. Between height and age the relationship is more or less linear, but between weight and age it is less so (see Exercise 7I). This suggests that a suitable transformation of *weight* should straighten out the relationship of weight with height and also that of weight with age, a beneficial side-effect. There is nothing mandatory about trying this approach, but at this early stage of study the suggestion is enough to make one explore it.

Next we have to choose an actual transformation for weight. Given the shape of the data in Figure 8.2, we are seeking to shorten the intervals between large weights. This will have the required effect of "pushing down" the top-end of the height/weight curve and straightening it out.

Squaring the weights would have the wrong effect since large values would be spread out more than before. Equally spaced values of 1, 2, 3, 4 when squared become 1, 4, 9, 16, with increasing gaps of 3, 5, 7. In contrast, a *square-root* transformation would have the required kind of effect. Values of 1, 2, 3, 4 would become 1.00, 1.41, 1.73, 2.00, with *decreasing* intervals of .41, .32, .27. Such a transformation would work with the height/weight data but has nothing special to recommend it.

Using the *logarithm* of weight would have the same desired effect. It also has a slightly special appeal because it fits in with certain broader notions of biological growth functions. These often take logarithmic or exponential forms because growth tends to be multiplicative rather than additive. Things often grow *proportionally* to their size, e.g. by 10% a year rather than by a fixed amount per year. The argument is not a *strong* one when applied to weight. (Why not also to height, for instance?) But historically it largely led to a logarithm transformation being tried.

8.4 The Fit of the Logarithmic Relationship

Having decided to try a logarithmic transformation of weight, we next have to write down the log values of the average weights, using a table of logarithms. The 5-year-old Birmingham boys had an average weight of 42 lbs. The number given for 42 in a log table is .62 and since it is a number in the tens, a 1 has to be added and the log is 1.62. The results for the 5- to 13-year-old Birmingham boys are set out in Table 8.3. (We have to use three digits because the first ones are always the same and hence do not contain "significant" information.)

To fit a linear equation of the form $\log w = ah + b$, we use the same procedure as before. From the extreme readings for the 5- and 13-year olds

TABLE 8.3 The Logarithmic Values of the Average Weights of the Birmingham Boys

Birmingham Boys	Age 5	6	7	8	9	10	11	12	13	Av.
Av. weight: w (lbs.)	42	46	51	57	62	68	73	82	88	63.2
log w (log lbs.)	1.62	1.66	1.71	1.76	1.79	1.83	1.86	1.91	1.94	1.79

we get the slope-coefficient

$$a = \frac{1.94 - 1.62}{59 - 43} = \frac{.32}{16} = .020.$$

Inserting the overall averages of 1.79 and 51.7 into the equation, we get the intercept-coefficient

$$b = 1.79 - .020 \times 51.7 = .76.$$

Therefore the working-solution is

$$\log w = .020h + .76.$$

Table 8.4 shows the fit of this equation, which gives a mean deviation of .01 log lb units.

TABLE 8.4 The Fit of the Working-Solution log w = .020h + .76

Birmingham Boys	Age 5	6	7	8	9	10	11	12	13	Av.
Av. height: h	43	46	48	50	52	54	55	58	59	51.7
Av. weight: log w	1.62	1.66	1.71	1.76	1.79	1.83	1.86	1.91	1.94	1.79
.020h + .76	1.62	1.68	1.72	1.76	1.80	1.84	1.86	1.92	1.94	1.79
log w - .020h - .76	.00	-.02	-.01	.00	-.01	-.01	.00	-.01	.00	-.01*

* Average size ignoring sign = .01 log lbs

The logarithmic transformation has worked in the sense of yielding a simple linear relationship between log weight and height for the Birmingham data. Figure 8.5A illustrates the relationship as a curve when plotting weight in lbs against height. Figure 8.5B shows it as a straight line when plotting *log w* against height.

The deviations, although mostly negative, show no trend. The logarithmic transformation has therefore succeeded in accounting for the curvature of the original data. The tendency for negative or zero deviations in Table 8.4 means that an equation with an intercept-coefficient of .75 would fit better.

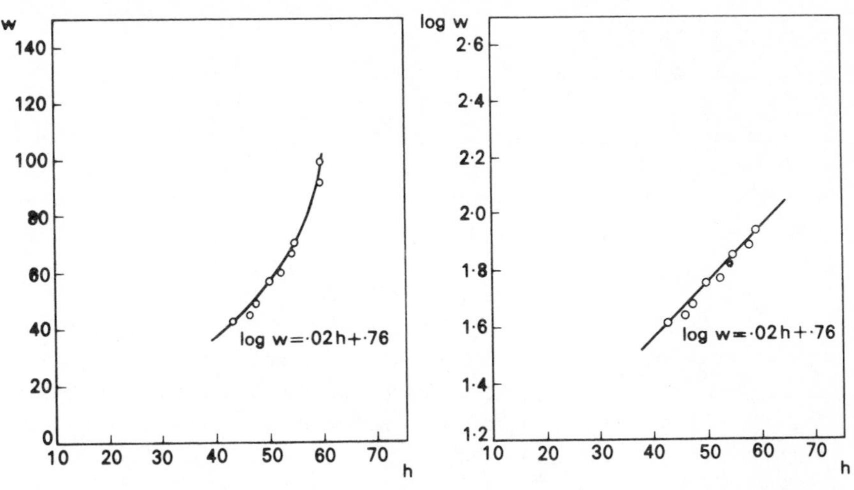

Figure 8.5A A Logarithmic Relationship

Figure 8.5B Log w against h

But the negative deviations are merely due to rounding-off. We could have eliminated the problem by working to three decimal places when calculating b. The average of the log w values is 1.787, and the average of the theoretical estimates is 1.793; these differ by $-.006$, which is $-.01$ to two places, but both numbers are 1.79 when rounded. This illustrates the rather niggling type of problem that can arise when working to only two significant figures, but which is usually worth putting up with. (The real question is which value, .75 or .76, works better with further data.)

The next step then in the analysis is to see whether the relationship $\log w = .02h + .76$ generalises to other data. We start with readings in about the same height and weight range as the Birmingham and Ghanaian boys. The results of other wider checks have already been summarised in Table 5.1. Table 8.5 gives an example for Canadian girls. The fit is good again, to within $\pm.01$.

TABLE 8.5 The Fit of the Working-Solution log w = .020h + .76 for Canadian Girls Aged 6 to 11 of Above-Average Socio-Economic Class

Ottawa Girls	Age 6	7	8	9	10	11	Av.
Av. height : h	47	48	51	52	54	56	51
Av. weight : log w	1.68	1.72	1.77	1.81	1.84	1.89	1.79
.020h + .76	1.70	1.72	1.78	1.80	1.84	1.88	1.79
log w - .020h - .76	-.02	.00	-.01	.01	.00	.01	.00*

* Average size ignoring sign = .01 log lb

8.5 Strong Curvature

The main doubt about the working-solution $\log w = .02h + .76$ is whether we have chosen the correct function to transform the weights. To subject the logarithmic function to a stringent test, we have to see whether it also holds for data outside the range of variation covered so far. Since we know that *age* is the major determinant in the values of children's height and weight, we test the log relationship with data outside the 5- to 13-year age range.

Table 8.6 gives height/weight data for Ghanaian pre-school children aged 0 to 4 years (Kpedekpo, 1971). The logarithmic relationship fits well for the 2- to 4-year olds, but not for babies aged one year or less. It is not clear whether this discrepancy is general for babies, something specific for these Ghanaian children, or a measurement bias, because this is the only data for young children and babies analysed so far.

TABLE 8.6 Pre-School Children and the Working-Solutions log w = .02h + .76 and w = 3.2h - 100

Pre-School Ghanaian Children	Age 0	1	2	3	4
Av. height: h	24	30	33	37	39
Av. weight: log w	1.16	1.32	1.42	1.48	1.54
.02h + .76	1.24	1.36	1.42	1.50	1.54
log w - .02h - .76	-.08	-.04	.00	-.02	.00
Av. weight: w	14	21	26	30	35
3.2h - 100	-25	-4	6	18	25
w - 3.2h + 100	39	25	20	12	10

Nevertheless we can see from Table 8.6 that the logarithmic relationship fits these data vastly better than the linear working-solution $w = 3.2h - 100$. (For the 5- to 13-year olds the logarithmic equation was only *fractionally* better, in the sense that it only had to account for the minor curvature in the data.)

To test the new relationship for older children, we look at data for some teenage children up to 17 years (Lovell, 1972). Table 8.7 shows that the logarithmic relationship holds well for the boys, but the deviations for the girls are all positive and about .04 log lb, much larger than average. Other data (Lovell, 1972; Kpedekpo, 1970) have confirmed the finding that older girls are relatively heavy for their height (therefore the logarithmic relationship itself cannot be blamed for these discrepancies).

TABLE 8.7 Older Boys and Girls and the Relationship log w = .02h + .76

Prosperous African Jamaicans	Age 14	15	16	17	Av.
BOYS					
Av. height : h	64	66	68	69	66.8
Av. weight : log w	2.04	2.08	2.12	2.13	2.09
.02h + .76	2.04	2.08	2.12	2.14	2.09
log w - .02h - .76	.00	.00	.00	-.01	.00
GIRLS					
Av. height : h	63	63	63	64	63.2
Av. weight : log w	2.06	2.07	2.07	2.07	2.07
.02h + .76	2.02	2.02	2.02	2.04	2.03
log w - .02h - .76	.04	.05	.05	.03	.04

We now no longer have any doubts about whether the height/weight data need to be described by a non-linear equation. Had we started our study with a wider age-range, we would at once have been faced with a strongly curved empirical relationship, like the one shown in Figure 8.6A. Then we would not even have tried a straightforward linear solution. We would probably have transformed one of the variables immediately in order to reach the type of linear situation shown in Figure 8.6B, which is easier to analyse. That way

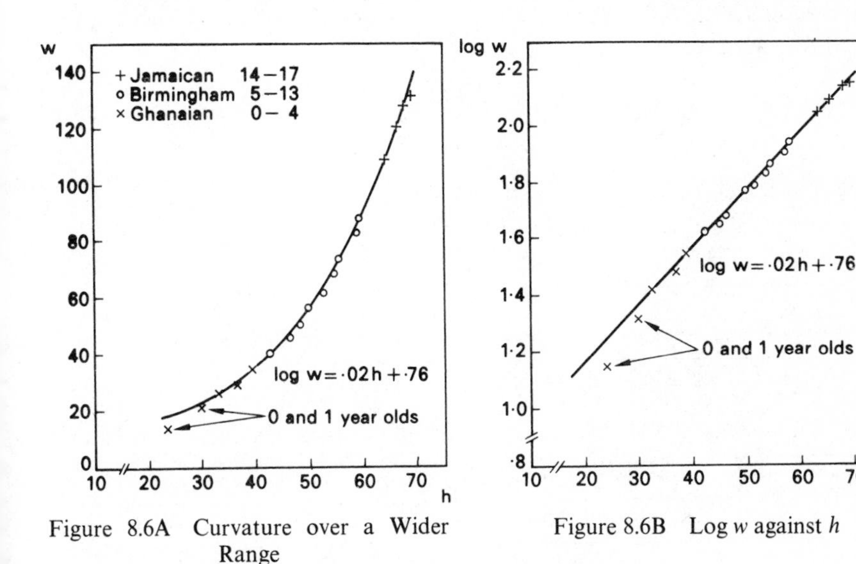

Figure 8.6A Curvature over a Wider Range

Figure 8.6B Log w against h

we could have determined more quickly whether the right kind of model had been chosen.

8.6 A Cube-root Law

The fact that the logarithmic equation fits well for boys aged 2 and over does not eliminate other possible mathematical functions. For example, a cube-root transformation can also be considered.

There is a dimensional argument in favour of using a cube-root equation. We know that the weights of different bodies vary with their volumes, and that the volume of a body varies as the product of three linear dimensions: height, width, and depth. It follows that the cube-root of volume, and hence the cube-root of weight, should be proportional to height. We can therefore try to fit an equation of the form

$$\sqrt[3]{w} = ah.$$

However, a check on the data shows that the variation is not directly proportional and that the more general equation $\sqrt[3]{w} = ah + b$ would work better. This does not mean that the theoretical argument is already being discarded, only that it is being shaped to fit the facts. This is legitimate because the deduction that height should vary directly with the cube-root of weight was not cast-iron anyway; it assumed that the shape of children was rectangular, and that this and their density would remain exactly the same as they grew. Obviously some adjustments are permissible.

Table 8.8 gives the height and weight readings for boys aged 0 to 17 (covered in Tables 8.4, 8.6 and 8.7). Fitting a straight line for $\sqrt[3]{w}$ and h in the usual manner gives

$$\sqrt[3]{w} = .060h + 1.0.$$

TABLE 8.8 The Cube-Root Relationship $\sqrt[3]{w} = .060h + 1.0$

(Ghanaian children aged 0 to 4, Birmingham boys aged 6 to 12, and Jamaican boys aged 14 to 17)

	Age													
	0	1	2	3	4	6	8	10	12	14	15	16	17	Av.
Av. height: h	24	30	33	37	39	46	50	54	58	64	66	68	69	48.9
Av. weight: w	14	21	26	30	35	46	57	68	82	110	122	131	134	67.4
$\sqrt[3]{w}$	2.4	2.8	3.0	3.1	3.3	3.6	3.8	4.1	4.3	4.8	5.0	5.1	5.1	3.9
.060h + 1.0	2.4	2.8	3.0	3.2	3.3	3.8	4.0	4.2	4.5	4.8	5.0	5.1	5.1	3.9
w - .060h - 1.0	.0	.0	.0	-.1	.0	-.2	-.2	-.1	-.2	.0	.0	.0	.0	.0*

* Average size ignoring sign = .1

This equation fits the data to within about 0.1 "cube-root lb" units. All the deviations are small. There are negative deviations in the middle of the range; but we cannot tell whether this generalises until we investigate additional data.

What matters is that the cube-root equation works almost as well as the logarithmic one for describing the height/weight relationship of children aged 2 years and over, and that it *also* gives a close fit for babies aged one year or less, as is also shown in Figure 8.7. This suggests that perhaps babies are not different from older children, but that it was a failure of the earlier theory, the logarithmic transformation, that caused the marked overestimation in Table 8.6. Further work and more data are still needed to determine this.

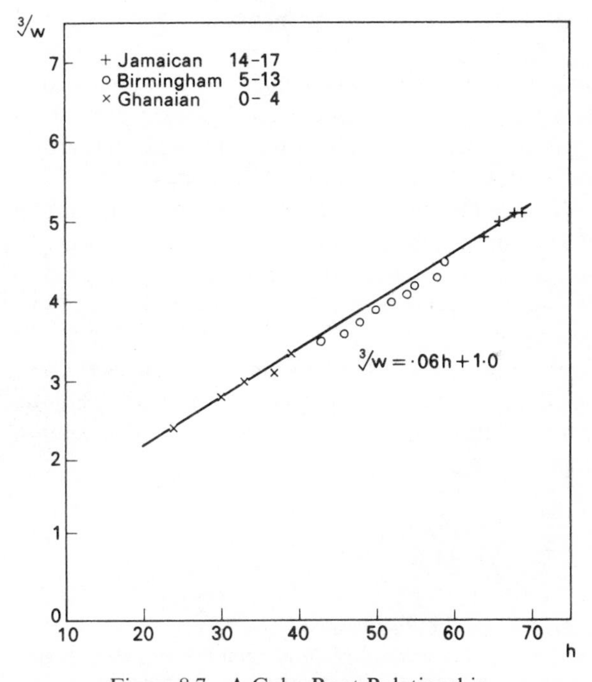

Figure 8.7 A Cube-Root Relationship

More generally, the success of the cube-root transformation so far means that our initial "theoretical deduction" has connected these results to our prior knowledge about the dimensionality of height and weight. This integrative function of theory is often far stronger than here. More examples will be discussed in Chapter 10.

8.7 Summary

When modelling a non-linear relationship, the main problem is choosing an appropriate mathematical function. Because this can be a complex operation, it is usually worth establishing first that the curved pattern generalises to other sets of data.

Usually a number of different curves or mathematical functions can approximate any particular set of data, especially when the range of variation is limited. Previous work, a theoretical argument, or an analogy can help suggest which functions to try.

Sometimes little previous information is available. Then a simple transformation of one of the variables is usually tried to lead once more to a linear type of working-solution. Unless there is some well-based theory this becomes a matter of trial and error. One has to see the shape of the observed data and be aware of a variety of mathematical functions that could meet that shape.

In our example of the heights and weights of children aged 5 to 13 years, the equation $\log w = .02h + .76$ modelled the slight curvature in the data. It also held for data that showed a much stronger curvature outside the initial range of variation, but it did not fit for babies.

While the logarithmic fit was generally good, it did not preclude the possibility of other mathematical functions. A cube-root transformation held almost as well as the logarithmic one and even gave a good fit for babies. This function also had some theoretical support in terms of the dimensions of weight and volume.

At this stage the choice between two such functions is not very important because they tell almost the same story about the available data. The nature of the functional relationship is usually clarified further as more advanced theory develops.

CHAPTER 8 EXERCISES

Exercise 8A. A Curved Relationship

Table 8.9 gives measurements of the apparent brightness of a light-source at various distances from it. Fit an equation to summarise the data.

TABLE 8.9 The Brightness of a Light-Source at Various Distances

Distance D (ft.)	1	2	3	4	5	6	7
Brightness B (lumens)	1900	530	210	120	80	62	21

Discussion.

It is obvious from the table (or from a rough working graph) that brightness B decreases as distance D increases. It is also clear that B

decreases at a much faster rate at short distances. For example, brightness at 4 ft, half the distance covered, is far less than half of the range of B-values observed.

The relationship is very highly curved, so that a transformation of at least one variable is needed. Since distance in feet is a well-known and highly controllable measure, it might seem best to leave this alone and transform brightness. Logarithms, as in Table 8.9a, would largely straighten out the relationship, but not entirely. For example, log B drops by 1.20 from 1 to 4 ft, but by only .76 from 4 to 7 ft.

TABLE 8.9a Logarithms of Brightness

Distance as D	1	2	3	4	5	6	7
Brightness as log B	3.28	2.72	2.32	2.08	1.40	1.79	1.32

To hit on a better transformation we can use the physical law that brightness and certain other measures, such as the force of gravity, vary inversely with the square of distance, i.e. as $1/D^2$ (so that we transform D after all).

Table 8.9b sets out the original brightness readings with the reciprocal of the squared distances D, multiplied for numerical convenience by 1,000.

TABLE 8.9b B and 1000/D^2

Distance as D	1	2	3	4	5	6	7	Av.
Brightness as B	1900	530	210	120	80	62	21	418
Distances as 1000/D^2	1000	250	110	62	40	28	20	216

It is obvious that B is generally about twice as large as $1{,}000/D^2$, so that the relationship is something like

$$B = \frac{2{,}000}{D^2}.$$

The main discrepancy is at 7 feet, where B is virtually equal to $1{,}000/D^2$, rather than $2{,}000/D^2$. This requires further study. Perhaps the measuring procedure used is relatively erratic at low brightness levels.

Exercise 8B. A Measurement Bias?

Table 8.10 shows that the relationship $B = 2{,}000/D^2$ does not fit data for the same light source at greater distances. Analyse the data further.

TABLE 8.10 Brightness at Greater Distances

Distance D (ft)	10	15	20	25	30	35	Av.
Brightness B (lumens)	15	9	7	7	6	6	8.3
2000/D^2	20	9	5	3.2	2.2	1.6	6.8

Discussion.

In general, the apparent brightness of a light source *does* vary inversely with the square of the distance. Since this formulation gives roughly the right pattern in Table 8.10, we shall keep to the $1/D^2$ function at this stage, but first fit a more general equation of the form

$$B = a\left(\frac{1{,}000}{D^2}\right) + b.$$

Determining a linear working-solution for B and $(1{,}000/D^2)$ in the usual manner gives the equation

$$B = \frac{1{,}000}{D^2} + 4.9.$$

(Note that Table 8.10 gives $2{,}000/D^2$, not $1{,}000/D^2$.) The new equation fits the brightness values with a mean deviation of ± 0.3 lumens.

A possible interpretation of this result is that the device for measuring brightness had a bias of about 4.9 or 5 lumens. The relationship might therefore read

$$(B - 5) = \frac{1{,}000}{D^2},$$

where $(B - 5)$ is the *real* level of brightness.

There could of course be other causes. For example, some light could be reflected from surrounding material, or other sources of light might be registered at lower brightness levels. Or it could be the *distance* measure that is biased, e.g. by being taken from the edge of the light source instead of its focus. In the latter case, the corrected form of the inverse-square law would read $B = \hat{a}/(D - \hat{b})^2$, where $\hat{a}$ and $\hat{b}$ are two coefficients to be determined.

Establishing the relationship in this form would be technically complicated because it cannot be expressed linearly by transforming B and/or D. But written the other way round, in the form $D = \hat{b} + \sqrt{\hat{a}}/\sqrt{B}$, the relationship is linear in D and $1/\sqrt{B}$.

Determining an appropriate model for these phenomena and, in particular, reconciling these data and those from Exercise 8A, will clearly take more extensive data and a reasonable degree of theoretical understanding.

Exercise 8C. A Change of Units

The relationship between children's height h and weight w is $\log w = .02h + .76$ in inches and pounds. What is it in metres and kilograms? (1 inch = 2.54 centimetres; 1 pound = 454 grams.)

Discussion.

If H is height in metres, then $100H/2.54$ or $39H$ equals h in inches. (H must be numerically smaller than h because there are fewer metres than inches in a given length.) Similarly, if W is weight in kilograms, then $1{,}000W/454 = 2.20W$ is equal to w measured in lbs.

It follows that $\log w = \log 2.20W$. This in turn equals $\log 2.20 + \log W$ because of the multiplicative property of logarithms ($\log XY = \log X +$

log Y). From log tables, log 2.20 = .342 so that

$$\log w = .342 + \log W.$$

Substituting for $\log w$ and h in the equation $\log w = .02h + .76$ it follows that

$$.342 + \log W = .02 \times 39H + .76.$$

Hence

$$\log W = .78H + .42.$$

This is virtually the equation $\log W = .8H + .4$ that was reported by Lovell (1972) and others, in metric units.

Similarly, the reader may check that changing the cube-root formulation $\sqrt[3]{w} = .060h + 1.0$ to metric units gives

$$\sqrt[3]{W} = 1.8H + .77.$$

Exercise 8D. Deviations from the Log Relationship

In Section 8.4 the mean deviation from the relationship $\log w = .020h + .76$ was $\pm .01$ log lbs. What is the equivalent deviation in ordinary pount units?

Discussion.

There is no single answer. Constant limits of $\pm .01$ log lbs correspond to differing values in pound units, depending on the absolute value of w. Thus the deviations in lbs are larger for large values of w than for small ones.

To illustrate, consider three values of $\log w$ in about the range covered by the data for the 5- to 13-year olds, namely 1.60, 1.75, and 1.90. The values of $\log w \pm .01$ are

$\log(w) + .01$:	1.61	1.76	1.91
$\log(w)$:	1.60	1.75	1.90
$\log(w) - .01$:	1.59	1.74	1.89.

Looking up the antilogarithms of 1.61, etc., the equivalent values of weight w in pound units are

w equivalent to $\log(w) + .01$:	40.7	57.5	81.3
w equivalent to $\log(w)$:	39.8	56.2	79.4
w equivalent to $\log(w) - .01$:	38.9	54.9	77.6

This shows that deviations of $\pm .01$ in log lb units are equivalent to about ± 1 lb at the 40 lb level, ± 1.3 lbs at the 56 lb level, and about ± 1.8 lbs at the 80 lb level.

When an empirical relationship between two variables x and y is non-linear and curved upwards as here, deviations from it in the original units tend to be larger for larger values of x or y. When the relationship is "linearised" by some suitable transformation, the deviations from the resultant straight line, i.e. deviations in the transformed units, tend then to be more or less homoscedastic (i.e. have equal scatter) all along the line.

This shows up here particularly in terms of the larger scatter of *individual* children about the line. Thus for the linear equation $w = 3.2h - 100$, the mean deviation of 8 lbs quoted in Section 7.8 was a broad average, there being a marked upward trend from 5 to 13 years. But in log units the scatter was virtually homoscedastic at .04 log lbs, as quoted in Section 6.4.

A related technicality is that if one first transforms the *individual* weights to logs their average will be slightly smaller than the log of the average weights in lbs that we have used. (This is related to the well-known difference between arithmetic and geometric means.) This can matter when working at high levels of precision.

Exercise 8E. The Fit in Logarithmic and Additive Units

The fit of the age-group averages to $\log w = .02h + .76$ was on average within .01 log lb units and that for the linear equation $w = 3.2h - 100$ in Section 7.7 was about 1.2 lb. Is the fit of the logarithmic equation closer?

Discussion.

Exercise 8D showed that deviations of $\pm.01$ in log lb units are equivalent to deviations of ± 1.3 lbs at the mid-range of w-values, but *smaller* than 1.3 for lower values of w and *larger* for higher values. No single "neat" comparison is therefore possible. But as a broad average, the two deviations must be roughly the same, because both equations gave a fair fit to the same data (i.e. for the 5- to 13-year olds in Table 8.3).

As a *proportion* of the weight values the two average deviations differ. From the previous Exercise we know that ± 1.3 of 60 lbs (the mid-range w-values) is about 2%, whereas $\pm.01$ of 1.8 log lbs is about 6%. But such a comparison is not very meaningful, as it depends on the rather arbitrary zero values of the two scales. A more appropriate comparison would be the size of the deviations compared to the *range of variation* on each scale. Here we have

A range of .34 in log lbs, so that $\pm.01$ log lbs is about $\pm 3\%$,

A range of 40 in lbs, so that ± 1.3 lbs is also about $\pm 3\%$.

However, the crucial difference between the two equations is not the *size* of the deviations, but the fact that the deviations for the relationship $w = 3.2h + 100$ are systematic, especially outside the range of the 5- to 13-year olds (see Table 8.6).

Exercise 8F. Children's Weight and Age

In Exercise 7I we noted that the relationship between weight and age for the 5- to 13-year-old Ghanaian boys appeared to be slightly non-linear. Examine this further.

Discussion.

One way of dealing with the weight–age curvature is a logarithmic transformation of weight. Table 8.11 shows that this also works well for 5- to 13-year-old Birmingham boys, with at most a slight pattern in the small residuals.

A stronger test for the equation would be a check outside the above range, like the data on black Jamaican teenage boys in Table 8.7. These tend to be taller at a given age than the relationship indicates. But the difference is consistent (about .05 log lb units) at various ages, so the

TABLE 8.11 The Fit of the Working-Solution log w = .04A + 1.43

Birmingham Boys	Age in Years (A)									Av.
	5	6	7	8	9	10	11	12	13	9.0
Av. weight: log w	1.62	1.66	1.71	1.79	1.79	1.83	1.86	1.91	1.94	1.79
.04A+1.43	1.63	1.67	1.71	1.75	1.79	1.83	1.87	1.91	1.95	1.79
log w - .04A - 1.43	-.01	-.01	.00	.01	.00	.00	-.01	.00	-.01	.00

logarithmic transformation has in fact succeeded in "straightening out" the relationship. More extensive study is needed to clarify how average heights differ at a given age. This will be touched on again in Chapter 9.

Exercise 8G. A Failure to Fit

Table 8.12 compares the average weights of British boys from a survey in 1880 (Clements, 1953) with the equation $\log w = .04A + 1.43$ fitted to the Birmingham boys measured in 1947. Comment on the data.

TABLE 8.12 The Deviations from the Birmingham 1947 Equation log w = .04A + 1.43 for Boys in 1880

Boys in 1880	Age in Years (A)									Av.
	5	6	7	8	9	10	11	12	13	9.0
Av. weight: log w	1.57	1.61	1.67	1.71	1.79	1.78	1.81	1.89	1.88	1.73
.04A+1.43	1.63	1.67	1.71	1.75	1.79	1.83	1.87	1.91	1.95	1.79
log w - .04A - 1.43	-.06	-.06	-.04	-.04	-.05	-.05	-.06	-.07	-.07	-.06

Discussion.

The Birmingham equation does not fit, but the deviations are consistent at about $-.06$ log lbs. The boys in 1880 therefore weighed markedly less at any given age than those in 1947. The equation

$$\log w = .04A + 1.37$$

would be a good working-solution for the earlier data.

Exercise 8H. Changing the Mathematical Function

In Section 8.6 we changed from a logarithmic to a cube-root function of weight. This is typical of the change in mathematical functions that can occurs as a subject develops. When doing this, is it necessary to refer again to the original data?

Discussion.

No. Unless more precision is required than the average deviations of $\pm.01$ log lbs, the theoretical relationship $\log w = .02h + .76$ summarises the original data sufficiently and no direct reference back is needed. (This is like the change in linear relationships discussed in Exercise 7F.)

To examine the fit of a cube-root relationship, we can reconstruct the data from the logarithmic working-solution, as shown in Table 8.13.

TABLE 8.13 **Deriving and Testing a Cube-root Relationship from the Logarithmic Relationship**

	Selected Values				Av.
Height h (in.)	20	40	60	80	50
$.02h + .76 = \log w$	1.16	1.56	1.96	2.36	1.76
w (in lbs)	14	36	91	229	92
$\sqrt[3]{w}$	2.4	3.3	4.5	6.1	4.1
$.062h + 1.0$	2.2	2.5	4.7	6.0	4.1
$\sqrt[3]{w} - .062h - 1.0$	.2	-.2	-.2	.1	0

For a systematic selection of h-values covering the relevant range, we work out $.02h + .76$. This gives the theoretical values of log w and we look up antilogarithms to find the values of w. Then we calculate $\sqrt[3]{w}$ and fit a linear equation to h and $\sqrt[3]{w}$. This gives $\sqrt[3]{w} = .062 + 1.0$. The final step is to check the fit of the new equation. (More elaborate calculations can also be made using the $\pm.01$ average limits of approximation of the logarithmic relationship.)

CHAPTER 9

Many Variables

When analysing data with three or more variables, one usually starts by looking at two variables at a time. Many of the procedures therefore remain the same as in preceding chapters.

The main difference is that by interrelating the different paired relationships in the data, one can gain a fuller interpretation and more advanced theoretical structures.

9.1 An Initial Analysis: Apple Trees

To illustrate multivariate data at the earliest stages of analysis, we consider an example provided by Dr. S. C. Pearce (for a meeting of the Biometric Society in London in 1969). The data consist of four measures of the size of apple trees:

g = girth of trunk after 4 years,
G = girth of trunk after 15 years,
e = extension growth (length of branches, etc.) after 4 years,
W = weight above ground after 15 years.

Eight trees from each of 13 different rootstocks were planted in 1918/19. (The trees were Worcester Pearmains grafted on to rootstocks raised asexually.) It is regarded as a classical experiment in elucidating the effects of different rootstocks.

To see the pattern in the results, we shall concentrate on the 13 rootstock averages for the four variables, instead of the readings for the individual trees. These averages have been arranged in Table 9.1 according to the size of G (before rounding). We chose G because girth is easier to measure than the other variables (and hence is more likely to recur in other data) and because girth *at 15 years* is probably a more stable measure than girth at 4 years. (This initial arrangement of the data is essentially arbitrary. One must check later whether any conclusions have been unduly influenced.) Even without rearranging rows and columns we can see from the table that the other three variables are quite highly correlated with G and that the

TABLE 9.1 **The Average Values for the 13 Rootstocks, in Increasing Order of Girth G at 15 Years**

	Rootstocks													
	IX	VII	VI	I	X	IV	XIII	V	II	III	XV	XVI	XII	Av.
G, in cm.	24	33	36	37	37	39	39	43	45	45	45	45	48	40
e, in m.	17	29	22	30	30	29	29	23	31	28	36	40	47	30
g, in cm.	9	10	10	11	11	11	11	11	12	11	12	12	14	11
W, in 100 lb.	4	7	7	9	8	10	11	12	13	14	13	17	19	11

relationships are roughly linear. (For example, the values for Rootstock XIII are very roughly mid-range for all four variables,) By using the normal procedure to fit working-solutions between G and each of the other variables. we obtain

$$G = e + 10,$$

$$G = 4.8g - 14,$$

$$G = 1.6W + 22.$$

The deviations from each relationship are shown in Table 9.2. They are mostly irregular and average at 3 cm, which is fairly small compared with G's range of 24 cm.

TABLE 9.2 **The Deviations between the Observed and Estimated Values of G**

	Rootstocks													
	IX	VII	VI	I	X	IV	XIII	V	II	III	XV	XVI	XII	Av.
G - e - 10	-3	-6	4	-3	-3	-2	0	10	4	7	-1	-5	-9	0
G - 4.8g + 14	-5	-1	2	-2	-2	-2	0	4	1	6	1	-1	-5	0
G - 1.6W - 22	-4	0	3	1	2	-1	-1	2	2	1	2	-4	-4	0
Av. size ign. sign	4	2	3	2	2	2	0	5	2	5	1	3	6	3

The all negative or zero deviations for Rootstocks IX, VII, XVI and XII (at the two extremes of the table) suggest that the three relationships with G might be curved. There are also exceptional deviations in the relationship with e for Rootstocks V and III, and in the relationship with g for Rootstock III. Further data is needed to pursue these various deviations.

But the three relationships provide an adequate first summary of the given data. They also imply the relationships between the other pairs of variables. For example, from the two equations $G = e + 10$ and $G = 4.8g - 14$, we can deduce

$$e + 10 = 4.8g - 14,$$

so that

$$e = 4.8g - 24.$$

Relationships between g and W and between e and W can be derived similarly.

This is one of the crucial points about the analysis of many variables. Increasing the number of variables need not greatly complicate the final results. With four variables there are six equations between pairs of variables, but one only needs three equations to summarise the data. With ten variables there are 45 equations between pairs of variables, but one only needs to describe nine of them explicitly, and so on.

A second point is that having more variables leads to stronger interpretation. There is more information on which to base one's judgement. We can see a possible general factor of *size* due to specific rootstocks; some rootstocks produced trees which tended to be bigger than others for *all* the variables measured.

A third point is that having even more data usually simplifies the results. We could reach firmer conclusions on any general patterns and on the exceptions if we had data on these trees at other ages and for additional variables (e.g. root growth, weight of apples cropped, etc.), and data for other trees. Since the results have been simple so far, it would not be difficult to integrate such additional data or to start asking ourselves what sort of additional variables we ought to try to analyse and what sort of results we would expect to find.

9.2 A More Structured Example: Consumer Attitudes

We now consider market research data on consumer attitudes as an example where a great deal of data are available, since many similar studies have already been made, and where the results all take the same general structure. It is also an example where we can start asking a few questions before going far in the analysis.

Market research surveys of consumer attitudes to frequently bought branded goods usually have a large number of variables, Consumers are asked whether they think each brand is modern, popular, sweet enough, convenient to use, good value for money, etc.: whatever seems relevant. Table 9.3 gives an extract of some typical attitudinal data for four brands of a certain product. The readings are the percentages of consumers who gave positive responses to the questions. For example, 36% of all consumers said Brand A has the "Right Taste".

The data look complex, but it is still possible to discern some order, such as the fact that Brand D scores very low for all the variables. We could now proceed as before and fit equations to pairs of variables. For example,

TABLE 9.3 **The Percentage of Consumers Giving Certain Attitudinal Responses to Brands A, B, C, and D**

	% of All Consumers Saying:					
	"Right Taste"	"Convenient"	"Good Value"	"Popular"	"Modern"	etc.
Brand A	36	11	30	49	.	.
Brand B	11	4	40	9	.	.
Brand C	21	50	18	31	.	.
Brand D	6	3	4	6	.	.
etc.	.	.	.	.	.	.

between "Popular" and "Right Taste" we could fit the linear equation

$$P = 1.4RT - 9,$$

which gives a mean deviation of about 4 percentage points. But the analysis for the other variables is not as straightforward.

In any case, is this really the best way of summarising and interpreting such data? The readings refer to consumer attitudinal responses to branded goods, for which most of us have some background knowledge, so we can start asking questions. For example, *why* are the results for Brand D all so low?

One possible explanation is that Brand D is not very good, and that few people like it and use it. Another possibility is that Brand D *is* good, but either not well-known or not widely available (because of low advertising levels or limited retail distribution). This would *also* show up in few people actually using the brand.

We have two different interpretations, but both suggest attitudinal responses might be related to the number of consumers who use each brand. (We have already seen a similar connection in Chapter 4 for consumers' intentions-to-buy.) A simple measure of brand usage is available ("Have you used the brand in the last week?") and is introduced into the analysis in

TABLE 9.4 **The Observed Attitudinal Responses and Usership Levels**

	% of Consumers Using: U	% of All Consumers Saying:			
		"Right Taste"	"Convenient"	"Good Value"	"Popular"
Brand A	50	36	11	30	49
Brand C	30	21	(50)	18	31
Brand B	10	11	4	(40)	9
Brand D	5	6	3	4	6
Average	24	18	6*	17*	24

* Averages excluding the Brand B and C exceptions

Table 9.4. Here the data are set out in decreasing order of each brand's market-share.

A pattern now becomes apparent. The attitudinal values decrease in parallel with the usership level. There are only two exceptions—for Brand C on "Convenient" and Brand B on "Good Value". (Reversing the columns and rows, as discussed in Chapter 1, would make the pattern even clearer.)

We cannot be sure of this pattern with just sixteen numbers, two of them exceptions. Again this is a case where more data would help rather than hinder. In fact, a wide range of other results exists (e.g. Collins, 1973; Bird and Ehrenberg, 1970; Chakrapani and Ehrenberg, 1974) and shows the same connection, that the level of such attitudinal responses to a brand is generally proportional to U, its user-level, but with some occasional highly dramatic exceptions.

Now we can start fitting working-solutions to the data in Table 9.4. But instead of trying to equate attitudinal variables to each other, as we would have done before, we shall look for relationships between attitudes and usership level, i.e. equations of the form

$$\text{Attitude} = aU + b.$$

The previous evidence suggests that the relationships have zero intercept-coefficients. Therefore, instead of using the highest and lowest readings to calculate the slope-coefficient, referred to by the symbol R, it is simpler to use the ratio of the average for each attitudinal variable to the average market-share, across all four brands. Thus for "Right Taste" we have $R = 18/24 = .75$ or .8, rounded. (The two large exceptions in Table 9.4 are excluded from both averages. For example, for "Convenient" we calculate the slope as $R = 6/22 = .3$.) This gives the equations:

$$\text{"Right Taste"} = .8U$$

$$\text{"Convenient"} = .3U$$

$$\text{"Good Value"} = .6U$$

$$\text{"Popular"} = 1.0U$$

Table 9.5 shows the deviations between the observed and theoretical readings RU. The fit is close, with a mean deviation of about 2 percentage points.

We therefore have a much simpler model than in the case of the apple trees. The same form of relationship with usage holds for each variable. And as can be seen from the other published results, this form holds across a wide range of product-fields and different attitudinal measures.

We could still have analysed the data directly, relating the different attitude variables to each other. The above relationships in fact imply those between the attitudinal variables themselves. But the usage variable has

TABLE 9.5 Deviations between the Observed Attitude Levels and the Predicted Values given by RU

	"Right Taste" Obs. -.8U	"Convenient" Obs. -.3U	"Good Value" Obs. -.6U	"Popular" Obs. -1.0U
Brand A	-4	-4	0	-1
Brand C	-3	(41)*	0	1
Brand B	3	1	(34)*	-1
Brand D	2	1	1	1
Average	0	0	0	0

* Excluded from the calculations

introduced much more structure. It also links the attitudinal data to the surveys' real concern, i.e. consumers' actual usage and buying behaviour.

9.3 A Simple Model: Buyer Behaviour

Multivariate data can often appear extremely complicated, but may still be reduced to a simple, well-developed model. For example, individual consumer's purchases of frequently bought branded goods are very irregular, as the data shown in Table 9.6 illustrate.

TABLE 9.6 Consumers' Purchasing of Different Brands of Breakfast Cereals in 12 Successive Weeks

(C: Corn Flakes, W: Weetabix, S: Shredded Wheat, P: Puffed Wheat)

	Purchases in Week											
	1	2	3	4	5	6	7	8	9	10	11	12
Consumer I	-	-	C	W	C	-	C	-	S	CS	-	C
Consumer II	P	-	-	-	-	-	-	-	-	P	-	-
Consumer III	-	W	W	-	W	-	W	-	W	W	-	-
Consumer IV	-	-	-	-	-	-	-	-	-	-	-	-
Consumer V	C	C	-	C	W	C	S	S	C	C	C	W
etc.	.	.	.	.	.	.	.	.	.	.	.	.

One approach to such data is to look at successive purchases, e.g. C, W, C, C, S, etc. for Consumer I. But consumers buy at different frequencies, so their purchases quickly get out of step with each other. Consumer II's second purchase follows Consumer I's *fifth*. This creates difficulties, especially when relating the data to outside events, such as seasons of the year, price changes, a new brand launch, advertising, etc.

An alternative approach is to consider specific time-periods such as successive 4-weekly ones, and to examine separately

(i) the number of consumers who buy each brand at least once in each period, and
(ii) how often on average they buy it in the period.

Experience has shown that this approach leads to simple and generalisable results. This is the crucial test in choosing an analytic approach.

To explore the relationship between the different brands, suppose that we look at the number of consumers who buy any *pair* of brands in an analysis period. Table 9.7 gives an example of such data for five breakfast cereals in a 24-week period (from Charlton *et al.*, 1972). It shows that 35% of all consumers bought *both* Corn Flakes and Weetabix at least once in the period. (Each "duplication" percentage occurs twice in the table because the percentage of consumers who bought both Weetabix and Corn Flakes is the same as the percentage who bought Corn Flakes and Weetabix.) The diagonal shows each brand's "penetration", i.e. the percentage of consumers who bought each brand at least once during the 24 weeks,

TABLE 9.7 Brand-Duplication of Purchase in 24 Weeks

(% of consumers buying any pair of brands, each at least once)

24 Weeks	and also buying				
	Corn Flakes	Weet-abix	Shred. Wheat	Sugar Puffs	Puffed Wheat
Buying					
Corn Flakes	(61)	35	25	19	11
Weetabix	35	(49)	22	17	10
Shredded Wheat	25	22	(33)	11	9
Sugar Puffs	19	17	11	(26)	8
Puffed Wheat	11	10	9	8	(16)

The five brands have been arranged in decreasing order of their penetrations. This brings out a regular pattern: the duplication figures in each column (omitting the penetrations in the diagonal) decrease in the same order. Therefore the percentage of consumers who bought two brands varies with the penetration of each brand. More people bought Weetabix than bought Puffed Wheat (49% versus 16%), and more *Corn Flakes* buyers also bought Weetabix than Puffed Wheat (35% versus 11%).

Now we have to establish the quantitative details to support this theory. One possibility is that the two penetration levels act independently of each other. If 49% of the population buy Weetabix, then perhaps about 49% of the *Corn Flakes* buyers also buy Weetabix. This gives 49% out of 61%, or about 30% who should buy both. It is close to the observed value of 35%, but a little below it.

Checking our model against the observed data in Table 9.7 shows that the Corn Flakes/Weetabix result generalises. The theoretical estimates are close to the observed values, but consistently lower. The deviations tend to be larger for the more popular brands, and this suggests adjusting the initial working-solution with a constant multiplier. Thus if b_X and b_Y represent the penetrations of Brands X and Y respectively, and b_{XY} is the percentage of consumers buying *both* X and Y, the adjusted model reads

$$b_{XY} = Db_X b_Y/100,$$

where D is the average value of $(b_{XY}/b_X b_Y)$ across all pairs of brands. For the data in Table 9.7, D is about 1.3. (The divisor of 100 in the equation is needed because we are working in percentages, whereas in *proportions* the equation would read $b_{XY} = 1.3b_X b_Y$.) Table 9.8 shows the fit of this model, the mean deviation being only 1.5 percentage points.

TABLE 9.8 The Fit of the Model $b_{XY} = 1.3b_X b_Y$

(Observed and Theoretical Values O and T)

24 Weeks	Corn Flakes O	Corn Flakes T	Weetabix O	Weetabix T	Shred. Wheat O	Shred. Wheat T	Sugar Puffs O	Sugar Puffs T	Puffed Wheat O	Puffed Wheat T
Corn Flakes	-	-	35	**39**	25	**26**	19	**21**	11	**13**
Weetabix	35	**39**	-	-	22	**21**	17	**17**	10	**10**
Shredded Wheat	25	**26**	22	**21**	-	-	11	**11**	9	**7**
Sugar Puffs	19	**21**	17	**17**	11	**11**	-	-	8	**5**
Puffed Wheat	11	**13**	10	**10**	9	**7**	8	**5**	-	-
Average	23	**25**	21	**22**	17	**16**	14	**15**	10	**9**

Other mathematical formulations could also give a reasonable fit to this one set of readings. The importance of the expression $b_{XY} = Db_X b_Y$ is that the same model holds for a wide variety of other time-periods and product-fields (e.g. Ehrenberg and Goodhardt, 1968, 1969; Ehrenberg, 1972). The same type of model also holds for other choice phenomena, such as industrial contracts (Ehrenberg, 1974b) and viewers' selection of television programmes (e.g. Goodhardt, 1966; Ehrenberg and Twyman, 1966; Goodhardt *et al.*, 1975). The theoretical model's slight overstatement of the observed results involving Corn Flakes in Table 9.8 is also a general effect. It is a marginal and still unresolved failure of the theoretical model for very large brands, not anything specific either to the present data or to Corn Flakes as such.

Despite the apparent complexity of the initial data, the main result is a simple one. The number of people who buy different pairs of brands can be represented, to a first degree of approximation, by a model which merely depends on the penetration levels of the brands and which has only a single numerical coefficient, D. The reason for this simple kind of result is an important one, and this we now discuss.

9.4 The Other Variables

Various examples in this book have shown how it is possible to establish a simple relationship between a pair of variables whilst ignoring other variables (e.g. the height/weight equation holds despite variations in age, sex, race, etc.). But in the last section this looked more blatant. It seems wrong to try to build a model for the number of consumers who buy brands X and Y while ignoring variables like pack-size, amount bought, price paid, retail outlet, household size, income, the weather, the amount of advertising, and whether they buy yet *other* brands.

The prime justification for leaving out all these variables is simply that the model *works*: the analyses lead to simple results which generalise.

But now we can also examine separately these other excluded aspects, For example, the preceding analysis ignored how *often* duplicated buyers of X and Y bought each brand. Table 9.9 shows there is little variation in the number of times consumers bought a brand, regardless of which other brand they *also* bought in the analysis-period. Thus consumers of Corn Flakes bought them on average 6 times in the 24 weeks, irrespective of whether they also bought Weetabix or Shredded Wheat, and consumers of Puffed Wheat bought that on average about 2 or 3 times.

TABLE 9.9 The Average Number of Purchases by Duplicated Buyers

24 Weeks	The Average Numbers of Purchases of				
	Corn Flakes	Weet-abix	Shred. Wheat	Sugar Puffs	Puffed Wheat
by buyers of					
Corn Flakes	(6)	6	5	4	3
Weetabix	6	(6)	4	3	2
Shredded Wheat	6	6	(4)	4	2
Sugar Puffs	6	6	4	(3)	3
Puffed Wheat	5	5	3	3	(2)
Average	6	6	4	3	2

This kind of result also generalises widely. It therefore illustrates how the duplication law is not the only aspect of the system that follows a simple pattern. It also explains why the average number of purchases could be ignored in the analysis in the last section. The figures in each column of Table 9.9 are more or less constant and therefore cannot affect the other variables.

Similarly, Table 9.10 explains why we could also ignore the extent to which duplicated buyers buy *other* brands. Buyers of each brand made a virtually constant number of cereal purchases in the 24-week analysis-period. Corn Flakes buyers made about 14 purchases, so did Weetabix buyers, and so on.

TABLE 9.10 The Average Number of Purchases of ANY of the Brands, by Buyers of Each Brand of Breakfast Cereals

24 Weeks	By buyers of Corn Flakes	Weet-abix	Shred. Wheat	Sugar Puffs	Puffed Wheat	Ave-rage
The average number of purchases of ANY of the brands	13	14	15	13	15	14

These results illustrate what appears to be the crucial step in analysing multivariate data: to try to split the data into components that can be handled separately, each giving a simple and generalisable result.

9.5 A Breakdown in Generalisation

In many cases however, an initially promising result does not generalise. Returning to the children's height/weight example, if we take *age* into account as a third variable and look at data for Birmingham boys in 1947, we have three paired equations:

$$\log w = .02h + .76, \quad \text{between height and weight,}$$

$$\log w = .04A + 1.43, \quad \text{between weight and age,}$$

$$h = 2.0A + 33.7, \quad \text{between height and age.}$$

But if we analyse data for British boys in 1880 (see Exercises 8F and 8G), we get different results for weight with age and height with age:

$$\log w = .04A + 1.37,$$

$$h = 2.0A + 30.3.$$

Therefore the boys in 1880 were consistently lighter and shorter than those in 1947, by about .06 log lbs and 3 inches. Since the height/weight relationship remains the same for both groups, the boys still had the same "shape": those in 1880 were simply smaller by the equivalent of about $1\frac{1}{2}$ years' growth.

In the system of three different "paired" relationships only the height/weight equation therefore generalises. The power of this equation is, however, increased by the fact that it now also holds for boys who were markedly smaller at any given age, i.e. by the *failure* of the other equations to generalise.

9.6 Relationships in More Than Two Variables

Often similar relationships differ only in one of their coefficients. Then a third variable can be introduced into the relationship to try to account for this variation.

This might be the case with the height/age and weight/age equations, but no work on this problem has yet been done. Another example earlier in Chapter 6 was that the intercept-coefficient of the *height/weight* equation varied somewhat with the apparent nutritional level of the children and therefore a three-variable relationship of the form $\log w = .02h + n$ might result; but here also no analysis on a suitable quantitative measure n of nutrition has yet been carried out.

In contrast, the behaviour of gases provides a well-developed example of a third variable entering into a two-variable relationship. Boyle's Law $PV = C$ relates the pressure and volume of a given body of gas at a constant temperature.

But when the temperature T varies, it has been found that the coefficient C varies proportionally. This gives a more general relationship in *three* variables known as the *Gas Equation*:

$$PV = RT,$$

where R is constant for any given amount of gas.

The Gas Equation contains Boyle's Law as a special case. Thus when temperature T is constant we have again

$$PV = \text{Constant}.$$

Similarly, the Gas Equation contains *Charles' Law*

$$V = KT.$$

This says that when pressure is constant, the volume of any given body of gas expands at a constant proportion of the degree-rise in temperature.

The Gas Equation also contains a third two-variable law

$$P = LT.$$

This says that when the volume of a given body of gas is constant, the *pressure* changes at a constant proportion of the changing temperatures, where L is another constant.

We therefore have a system of three paired equations which differ from those discussed earlier in this chapter. Here the equation for any one pair of variables holds only when the third variable is constant. In the other cases the additional variables did not enter into the paired relationships at all. For example, with height, weight, and age the relationship $\log w = .02h + .76$ held *despite* variations in age. Similarly, in Section 9.2 the relationship between brand usage and any particular attitudinal response (say "Good Value") held regardless of the responses given to other attitudinal variables.

In general, equations with large numbers of variables arise either when the numerical coefficients of a simple equation vary and this variation can be dealt with by relating it in turn to another variable or when the *deviations* from the law can be systematically related to some other variable.

Ultimately, the most general laws of science are freed from arbitrary-looking numerical coefficients by introducing such explanatory variables and by choosing appropriate units of measurement. Thus the Gas Equation $PV = RT$ is only that simple if T is measured as the *absolute* temperature with its zero-point at $-273°C$ (one of the "absolute constants" of physical science). And the equation can become $PV = 2T$, replacing the variable coefficient R by the "absolute" number 2 (or 1.987 calories, to be more exact), when the law is applied to an amount of gas equal to its molecular weight (like 2 grams of hydrogen, 16 grams of oxygen, and so on). Here the choice of units involves introducing a further variable, the molecular weight of the gas.

9.7 Correction Factors

More variables can also be introduced as correction factors, because empirically based relationships are generally oversimplifications that never fit exactly. This is true of all the examples discussed in this book, including the Gas Laws. These hold only for perfect gases, defined, as already mentioned, as substances for which the Gas Laws hold. *Actual* gases follow these laws more and more closely as the pressure of the gas is reduced. But, in general, correction factors have to be used.

Various adjustments have been developed to give a better approximation of actual gases, even under relatively high pressure. The best-known, although still only an approximation, is Van der Waal's equation

$$\left(P + \frac{a}{V^2}\right)(V - b) = RT,$$

where a/V^2 provides a correction for the mutual attraction of the gas molecules, and b is a correction for the volume occupied by these molecules.

One often makes do with *ad hoc* corrections because too little is known about many phenomena to model them explicitly. For example, when using the basic result that children's heights and weights follow $\log w = .02h + .76$, the coefficient .76 is adjusted to values like .74 or .72, depending on the type of child being studied. A similar allowance is made for the fact that teenage girls tend to be relatively heavy by about .04 log lbs. Again, the duplication of purchase law, $b_{XY} = Db_X b_Y$, is widely used even though it systematically overstates the duplication level for very popular brands. One simply corrects the estimated value by subtracting a few percentage points. Until a great deal more is known about such deviations it is usually pointless to develop more explicit mathematical models.

9.8 Summary

The analysis of multivariate data can usually start by examining the relationships between *pairs* of variables. The greater number of variables does not necessarily lead to particularly complex results. The extra information generally provides firmer conclusions than when dealing with only two variables.

If some of the fitted equations do not generalise across different sets of data, additional variables or correction factors have to be introduced to try to account for the differences. This is one way relationships involving more than two variables can be developed.

CHAPTER 9 EXERCISES

Exercise 9A. Following up on the Apple Trees

In the Apple-Tree example of Section 9.1, what should one study next?

Discussion.

The eight trees measured for each rootstock came from one parent. Thus the apparent differences might be due to the specific parents and not to the different rootstocks.

Studies of these rootstocks using trees of different parentage are therefore required. This would also show any generalization of the larger deviations of certain rootstocks from the overall pattern.

Exercise 9B. Following up on Consumer Attitudes

What should one study next in the market-research example in Section 9.2?

Discussion.

The references cited show that the relationship between the attitudinal responses to a brand and its user level is already well-established for many different attitudes and across many product-fields. Therefore three points to be studied next are:

(i) the large occasional exceptions illustrated in Table 9.4
(ii) why the relationship occurs, and
(iii) what determined the numerical value of the coefficient in the relationship.

Exercise 9C. The Nature of the Duplication-of-Purchase Law

Retrace the basic steps and outstanding questions in the development of the duplication of purchase law, $b_{XY} = Db_X b_Y$.

Discussion.

There are three main aspects in the development of any such result: conceptual, empirical and theoretical.

It first had to be decided to analyse consumers' purchasing behaviour in specific time-periods, and to do so by examining separately the number of *people* who buy an item at all and the number of *times* they buy it. (It still has not been established whether the equation or some equivalent holds when the data are viewed in different ways, e.g. for pairs of successive purchases by each consumer, or for the brand-shares of a consumer's total purchases.)

Having noted that the equation $b_{XY} = Db_X b_Y$ gave an adequate fit to one set of data, the next step was to establish whether it generalised across different length time-periods, different product-fields, different countries, and so on. The nature and generality of the exceptions to the relationship also had to be established (e.g. the "large-brand" effect).

Finally, theoretical questions remain, such as

(i) How does the relationship relate to *other* aspects of buyer behaviour?
(ii) How can it be reformulated to account for a consistent exception, such as the "large-brand" effect?
(iii) What determines the numerical value of D, the one coefficient in the model?
(iv) How can a better understanding of the result be reached?

Exercise 9D. Reformulating the Duplication-of-Purchase Law

What does the equation $b_{XY} = Db_X b_Y$ *mean*?

Discussion.

The proportion of consumers of Brand X who also buy Brand Y can be expressed as

$$\frac{b_{XY}}{b_X}.$$

Thus if $b_X = .2$ and $b_{XY} = .05$ (i.e. 20% of all consumers buy X and 5% of these buy X and Y), then $b_{XY}/b_X = .05/.20 = .25$ or 25% of buyers of X also buy Y.

Since $b_{XY} = Db_X b_Y$, we have that

$$\frac{b_{XY}}{b_X} = Db_Y.$$

This says the proportion of consumers of X who also buy Y depends only on b_Y, the penetration of Y. The duplication-of-purchase law therefore says that

> "Consumers of Brand X are *D* times as likely to buy Brand Y as the whole population, where *D* is approximately the same for all pairs of brands in a product-group".

Table 9.11 shows this reformulation of the duplication-law in arithmetical form for the data in Section 9.3.

TABLE 9.11 The Percentage of Buyers of One Brand who Also Buy Another Brand

24 Weeks		Who also Bought				
		Corn Flakes	Weet-abix	Shred. Wheat	Sugar Puffs	Puffed Wheat
% of Buyers of						
Corn Flakes	100%	-	57	41	31	18
Weetabix	100%	71	-	45	35	20
Shredded Wheat	100%	76	67	-	33	27
Sugar Puffs	100%	73	65	42	-	31
Puffed Wheat	100%	69	61	56	50	-
Average	100%	72	63	46	37	24

If the law held *exactly*, the percentages in each column would be identical. The table in fact shows that Corn Flakes were bought by *about* 72% of the consumers of any of the other brands, Weetabix by *about* 63% of the consumers of any other brands, and so on. On the whole, these tendencies to buy one brand do not depend on which other brand is also considered, the deviations are only a few percentage points. With such a simple pattern any exceptions also stand out clearly, like the relatively high duplication between the two Quaker Oats brands, Puffed Wheat and Sugar Puffs. (The message of the duplication-law is that while such groupings or "clusterings" of particular brands might be expected to occur generally, they are in fact the exception.)

TABLE 9.11a The Average 24-week Duplications and Penetration Levels of Each Brand

(D = 48/37 = 1.3)

24 Weeks	Brand					
	Co. Fl.	Weet.	Sh. Wh.	Su. Pu.	Pu. Wh.	Av.
Av. Duplication	72	63	46	37	24	48
1.3 x Penetration	79	64	43	31	21	48
Penetration %	61	49	33	24	16	37

Table 9.11a shows the second part of the relationship, that the duplication level is proportional to the penetration of each brand. Thus with $D = 1.3$, if 50% of the population buy a brand, then about $1.3 \times 50 = 65\%$ of the buyers of any other brand should also buy it. The tendency for the theoretical law to overstate somewhat the observed result for very large brands (like Corn Flakes here) stands out clearly against the general pattern. (The theoretical understatement for the two smaller brands reflects the "Quaker Oats" cluster already mentioned above.)

Exercise 9E. Empirical Variations in *D*

How can one further investigate the nature of the duplication-coefficient *D*?

Discussion.

At this early stage some purely empirical "looking" should be rewarding. Thus when establishing the general validity of the duplication law, values of *D* must have been calculated under many different conditions. Can any patterns be seen in these results?

The references already cited report a general pattern when analysing duplicated purchases for a given set of brands in different length time-periods. It is generally found that, for a particular product-field, the *D*-values increase from almost zero to more than 1, as the first line of Table 9.12 illustrates (Ehrenberg, 1972). In a short time-period such as a week buyers of X are *less* likely to buy Y than the whole population. Buying of one brand inhibits buying the other. But in a year, buyers of X are *more* likely to buy Y than the whole population.

TABLE 9.12 The Duplication-Coefficients for Brands and Varieties in Different Length Time-Periods - An Illustrative Example

	Analysis-Period, in Weeks				
	1	4	12	24	48
D for Brands	.3	.5	1.0	1.2	1.4
D for Varieties	2.4	1.9	1.5	1.4	1.4

The opposite pattern is reported for duplicated purchases between different *varieties* of a product (e.g. different flavours), as shown in the second line in Table 9.12. In a week, buyers of one flavour are far more likely to buy another flavour too. But in the longer time-periods, this tendency decreases.

These results reflect that in a week, i.e. usually on a single shopping-trip, people seldom buy two more or less identical brands (low *D*). But they may well do so on different purchase occasions spread over a longer time-period (high *D*). In contrast, total purchases in one week contain a relatively high proportion of frequent consumers of the product, who may well buy a number of different flavours on the same shopping-trip (high *D*). In longer time-periods, lighter buyers also show up, and they tend to buy *fewer* flavours (a relatively lower *D*).

The variations in the value of D for different length time-periods therefore distinguish between items which are substitutable (i.e. brands of similar or identical product-formulation) and items which are complementary (i.e. different varieties of a product, such as different flavours). This is an example of how additional variables (here length of time-period and type of product) can relate to a coefficient in a given relationship, as was discussed more generally in Section 9.6.

Exercise 9F. The Effects of Advertising

Why is there no need to allow for *advertising* in either the attitudinal or the duplication-of-purchase models?

Discussion.

The results described essentially apply when there are no very marked trends or fluctuations in the sales levels of the different brands. (This is what occurs in most markets most of the time.) It follows that advertising and other marketing variables are then having no *positive* effect on sales. (Their roles are mainly defensive, to maintain the status quo.)

Exercise 9G. The Roles of Other Variables

Boyle's Law, $PV = C$, develops into the more general Gas Equation, $PV = RT$, if temperature T varies, but Age A does not enter into the height/weight relationship, $\log w = .02h + .76$. Why not?

Discussion.

Any empirical law holds only when certain other variables remain the same. Yet other factors may still vary. For example, $\log w = .02h + .76$ holds for children under "normal" conditions: standing upright, effectively weighed without clothing, etc. However, it is found that age can vary without affecting the relationship, and so can the weather! Neither age nor the weather can therefore enter into the relationship.

Boyle's Law holds when there is a fixed amount of gas, no chemical reaction, and effectively constant temperature. If these factors vary, the law breaks down. The effect of varying temperatures is, however, rather simple, namely that the coefficient C varies in proportion. Hence we get $PV = RT$.

Exercise 9H. The Value of More Data

In a recent study of certain leading manufacturers in a number of countries, each company was assessed on six variables A to F (e.g. its Research, its Product-quality, etc.). Relating the percentage scores of the six variables to a measure of overall standing S for each company gave the following working-solutions for two of the countries:

U.K.: $A = .76S - 3$, $B = .58S$, $C = .25S + 5$, $D = .31S + 1$,
Germany: $A = .56S - 1$, $B = .51S - 1$, $C = .44S$, $D = .07S + 3$,
etc.

The fit was generally within a few percentage points, but there were also some very large exceptions (10 to 20 points).

The working-solutions vary mostly from one country to the other. How can this be further analysed?

Discussion.

We need to check that the occasional very large deviations have not unduly affected the working-solutions.

In the present instance it was found that differences between the equations for the U.K. and Germany always appeared to be caused by an exceptional value having affected the fitting process. Further analysis showed that when the *same* equation was fitted to different countries, the fit was almost as good as that of the original working-solutions.

This might not have been seen with just two countries. It became more apparent when a larger number of countries and all six variables were analysed, with the following results:

U.K.:	$A = .6S$,	$B = .5S$,	$C = .4S$,	$D = .3S$,	etc.
Germany:	$A = .6S$,	$B = .5S$,	$C = .4S$,	$D = .3S$,	etc.
France:	$A = .6S$,	$B = .5S$,	$C = .4S$,	$D = .3S$,	etc.
etc.:	$A = .6S$,	$B = .5S$,	$C = .4S$,	$D = .3S$.	etc.

Against these general norms, deviations for particular companies, countries, or variables stand out clearly and can be further investigated.

CHAPTER 10

The Emergence of Theory

The main function of theory is to integrate a wide range of results into a single conceptual framework or model. Thus the discussion in Chapter 9 had an increasingly theoretical orientation because we were interrelating the results for many variables.

Theoretical considerations tend to dominate most analyses, except at the very earliest stages of reducing a new kind of data to summary figures. While a detailed discussion of theory is beyond the intended scope of this book, in this chapter we illustrate how a higher level theoretical result can emerge.

10.1 Different Levels of Theory

Theory operates at many different levels. For example, when we fitted the height/weight relationship, $\log w = .02h + .76$, we were dealing with a low-level theoretical abstraction.

We moved up a step in our "theoretical" approach to data-handling when we used the equation to analyse additional height and weight data. But there was still nothing very advanced about this.

We can move to a still higher level of theory by deriving the height/weight relationship theoretically instead of empirically. For example, from the 1947 data on the Birmingham boys in Chapter 9 we had two equations:

$$h = 2.0A + 33.7, \quad \text{between height and age,}$$

$$\log w = .04A + 1.43, \quad \text{between weight and age.}$$

We derived these two equations empirically, from the data. If we now eliminate the common variable A from both (by multiplying the first equation by .02 and then subtracting the result from the second equation), we again arrive at the height/weight relationship $\log w = .02h + .76$, as we have already been doing in the last chapter.

This derivation is theoretical because we did not here *directly* relate empirical data on heights to data on weights. We did it indirectly or "theoretically" instead. Theoretical work like this is usually simpler than direct numerical analysis of data, as long as one is equipped with the required mathematical expertise.

The Role of Hypotheses

Hypotheses are theoretical formulations or assertions which are not known to be true, i.e. unproven suppositions. Their role is to suggest either new facts to collect or new analyses to carry out. Thus the hypotheses can be tested.

It is mainly through such speculation and hypothesising that one discovers new truths. For example, when we were fitting the height/weight relationship in Chapter 8 we used speculative theory to suggest the hypothesis of a cube-root transformation of weight. Without the "dimensional" theorising that weight tends to be proportional to volume, and that volume is proportional to the cube of a linear dimension like height, we would not have thought of trying a cube-root transformation. It simply is not the kind of idea that springs to mind naturally. In fact, most scientific relationships have forms that are far too complex to be based on mere common sense.

However, hypothesising can easily suggest things which are not true. From the height and age equation $h = 2.0A + 33.7$ for the Birmingham boys in 1947, and the *weight* and age equation $\log w = .04A + 1.37$ for the British boys in 1880, the same kind of theoretical elimination of the common age-variable as above would lead to the hypothesis that height and weight might be related as $\log w = .02h + .70$.

The hypothesis now has to be tested against facts, i.e. data on both height and weight for the same boys. It is then found that it does not hold either for the 1947 or the 1880 boys. Instead of being .70, the intercept-coefficient for these data is .76. The explanation is that the height/age and weight/age relationships in 1947 and 1880 differed (see Section 9.5), and therefore we cannot mix one equation from one set of data and one from the other.

Formulating and testing such a speculative hypothesis would, however, have been a proper and useful thing to do. Often we have only limited data from any particular study, e.g. height and age for 1947, and weight and age for 1880. Scientific progress largely consists of deducing new hypotheses from such incomplete information and then eliminating those which are not true by checking against new facts.

But theoretical arguments and assumptions based on insight or hunch must be sharply distinguished from *validated* theory. The one reflects what we think, the other what we know. It is the failure to make the distinction that has often given "theory" a bad name.

The Main Role of Theory

The function of speculative theory and hypotheses in discovering a particular scientific result is not, however, of lasting importance, except to the history of science. The principal role of successful theory is to link different kinds of *known* results. Thus, the cube-root relationship links the height and weight data with our more general experience of the shape and density of bodies. The link is approximate and suggestive rather than exact and compelling, since children are not rectangular nor perhaps of absolutely constant density as they grow. But this does not detract from the broad conceptual simplification achieved.

Theory is most powerful when it works at a detailed quantitative level and is empirically well-based. In the rest of this chapter we shall illustrate this integrative and explanatory power of more advanced theory with an example.

10.2 A Trend in Purchasing Frequencies

The problem to be tackled is that of finding a model for a series of empirical results which are individually simple to describe but which as a body are complex. Mere inspection of the data or direct "curve-fitting" as practised in the preceding chapters is unlikely to produce useful results.

Instead, the required result can be deduced from *other* findings. To illustrate, we recall the discussion of the five different brands of breakfast cereals in Chapter 9. Each brand was bought at more or less the same average rate per buyer in a 4-week period, as is shown again in Table 10.1. These purchase rates, which are usually called w, vary only by about ± 0.2 from the average 1.7. This variation is small compared with the differences in market-shares of the five brands, which range from 38% down to 5%. (These shares arise from multiplying the penetrations from Table 9.7 and the purchase frequencies from Table 9.9 and then percentaging.)

TABLE 10.1 4-Weekly Purchase Frequencies

4 WEEKS	Corn Flakes	Weet-abix	Shred. Wheat	Sugar Puffs	Puffed Wheat	Ave-rage
% Market-Share	38	31	16	10	5	(100)*
Purchases of the stated brand per buyer of the brand in 4 weeks	1.8	2.0	1.7	1.5	1.6	1.7

* Total sales of the 5 brands

But, in a longer period of 24 weeks, buyers of the different brands bought them at markedly different rates, the values of w ranging from 6 down to 2,

TABLE 10.2 24-Weekly Purchase Frequencies of Different Brands of Breakfast Cereals

24 WEEKS	Corn Flakes	Weet-abix	Shred. Wheat	Sugar Puffs	Puffed Wheat	Ave-rage
% Market-Share	38	31	16	10	5	(100)*
Purchases of the stated brand per buyer of the brand in 24 weeks	6	6	4	3	2	4

* Total sales of the 5 brands

as shown in Table 10.2. These variations seem to be related to the brands' market-shares.

We therefore have a highly discrepant situation: a trend in w in one case and approximately constant values of w in the other. This complex pattern generalises for a wide range of other product-fields, food and non-food, in the U.K. and U.S.A., etc. The data show that the longer the analysis-period, the stronger the trend between average purchase frequency w and market-share (with hindsight we can even discern a very weak trend with market-share in Table 10.1, and this too generalises for other products in short time-periods).

It seems that we need a complex model involving three variables:

(i) the average purchase frequency w of each brand,
(ii) its market-share,
(iii) the length of the analysis-period.

10.3 A Theoretical Model

A simpler answer is reached by introducing another variable altogether, namely the *penetration* of each brand, which is called b. This we have already come across in the last chapter. It is the proportion of the population who buy the brand at least once in the time-period. It varies both with the brand's market-share (the higher the brand's share of total sales, the more people buy it) and with the length of the analysis-period (more people buy a brand in two weeks than in one week, but the value of b does not simply double, since some people buy in *both* weeks). The relationship between penetration b and market-share therefore *varies with the length of the analysis-period.* This is the kind of pattern needed to model the varying trends in w we saw in the purchase frequencies.

The specific form of the theoretical model follows mathematically from three results which have already been described towards the end of Chapter 9. There we had that for any two brands X and Y:

(A) Consumers of Brand X buy the total product-class at about the same average rate as do consumers of Brand Y (Table 9.10).

(B) The proportion of the population who buy both X and Y in the same analysis-period is given by the equation $b_{XY} \doteqdot Db_Xb_Y$ (Table 9.8).

(C) A duplicated consumer buys a brand at about the same rate of purchase as all its other consumers (Table 9.9).

We now explore what follows mathematically from these three results. For simplicity we do so in terms of only three brands, X, Y and Z (the mathematical argument readily extends to more than three brands).

We start with the average rate of purchase of the total product-class by the buyers of Brand X. The number of buyers is Nb_X, i.e. N, the total number of potential consumers in the population, times b_X, the penetration of X. Their purchases of the total product-class are what they buy of Brand X, of Y, and of Z. This can be expressed as

Nb_Xw_X = the number of buyers of X times how often on average they buy X (i.e. w_X),

$Nb_{XY}w_{Y\cdot X}$ = the number of buyers of X who also buy Y(Nb_{XY}), times how often on average they buy Y (where $w_{Y\cdot X}$ denotes the average rate of purchase of Y by those buyers of X who also buy Y),

$Nb_{XZ}w_{Z\cdot X}$ = the number of buyers of X who also buy Z, times how often they buy Z.

Therefore we have that the average purchases of the total product-class by buyers of Brand X is

$$\frac{Nb_Xw_X + Nb_{XY}w_{Y\cdot X} + Nb_{XZ}w_{Z\cdot X}}{Nb_X}.$$

Now from Item (A) above we know empirically that the average rate of purchase of the total product group is the same for buyers of Brand X as for buyers of Brand Y. Therefore

$$\frac{Nb_Xw_X + Nb_{XY}w_{Y\cdot X} + Nb_{XZ}w_{Z\cdot X}}{Nb_X} = \frac{Nb_Yw_Y + Nb_{YX}w_{X\cdot Y} + Nb_{YZ}w_{Z\cdot Y}}{Nb_Y}.$$

But from Item (B) above we also know that

$$b_{XY} = b_Xb_Y \quad \text{and} \quad b_{XZ} = b_Xb_Z, \quad \text{etc.}$$

if we take the simple case where the coefficient $D = 1$. (The more general case where D is not equal to 1 is discussed in Exercise 10D.) From Item (C) above we know empirically that

$$w_{Y\cdot X} = w_Y, \quad \text{and} \quad w_{Z\cdot X} = w_Z, \quad \text{etc.}$$

If we substitute these values in the complex equation above and cancel through by N, we get

$$\frac{b_Xw_X + b_Xb_Yw_Y + b_Xb_Zw_Z}{b_X} = \frac{b_Yw_Y + b_Yb_Xw_X + b_Yb_Zw_Z}{b_Y}.$$

Now if we cancel through by b_X on the left and by b_Y on the right, and eliminate the common term $b_Z w_Z$ from both sides, we get

$$w_X + b_Y w_Y = w_Y + b_X w_X,$$

or, collecting terms in w_X and w_Y,

$$w_X(1 - b_X) = w_Y(1 - b_Y).$$

This is the result we require—an equation which relates how the average rates of purchase w_X and w_Y of Brands X and Y vary to how *another* variable, the penetrations b_X and b_Y, varies from brand to brand.

We can write the equation as $w_X(1 - b_X) = c$, a constant, or dropping the suffix,

$$w(1 - b) = c$$

for the different brands in the given length of analysis-period. In other words, w varies with b, namely as $w = c/(1 - b)$, a very simple result.

How it Works

It is tempting to try to read some direct "meaning" into this theoretical result. But like most laws of science, it merely describes and interrelates observed phenomena and has no obvious "commonsense" meaning. The crucial feature of such a theoretical result is how it works and how it links up with other findings.

Table 10.3 gives the values of $w(1 - b)$ for the five brands in our example in the 24-week and the 4-week periods that have been analysed. The values of $w(1 - b)$ are not *precisely* constant in each time-period, but the variation is relatively small, at about $\pm 10\%$. More important is the fact that the 24-week figures no longer show a trend with market-share. The "correction factor" $(1 - b)$ has therefore accounted for the trend in w.

TABLE 10.3 The Values of w(1-b) for the Five Brands

	Corn Flakes	Weet-abix	Shred. Wheat	Sugar Puffs	Puffed Wheat	Ave-rage
% Market-Share	38	31	16	10	5	(100)
24 weeks	2.2	2.9	2.9	2.5	2.2	2.5
4 weeks	1.2	1.5	1.5	1.4	1.5	1.4

This general no-trend pattern has also been found in many other product-fields. (The rather low values for Corn Flakes here do *not* represent a general feature for market leaders.)

The formula $w(1 - b)$ works in a relatively complex way. It eliminates the strong trend in w in the 24-week period, but does not introduce a contrary pattern for the data in the 4-week period. The formula achieves this because of the mathematical nature of the quantity $(1 - b)$.

In long time-periods, the penetration of leading brands is relatively high, say .4 or .6: i.e. 40% or 60% of the population buy the brands at least once. As the values of b increase the values of $(1 - b)$ decrease at a far greater rate, as Table 10.4 shows. When $b = .6$, $(1 - b)$ is twice as large as when $b = .8$. And when $b = .8$, $(1 - b)$ is 20 times as large as when $b = .99$. Multiplying by $(1 - b)$ therefore has a great effect when b is large.

TABLE 10.4 Values of (1-b) for Large b

b	.8	.9	.99
1-b	.2	.1	.01

In contrast, for short time-periods, the penetrations of most brands are relatively low, say .2 or .1: i.e. only 10% or 20%. Then the values of $(1 - b)$ are all close to unity, as shown in Table 10.4a. Multiplying by $(1 - b)$ in these cases has relatively little differential effect. The difference in $(1 - b)$ for $b = .01$ or $b = .2$ is only about 25%, which is small compared either with the difference in b itself (a factor of 20, i.e. 2000%) or with the differences in sales levels or market-shares.

TABLE 10.4a Values of (1-b) for Small b

b	.01	.1	.2
1-b	.99	.9	.8

The same factor $(1 - b)$ can therefore do two things. It can reflect differences in w which occur in relatively long time-periods because b is relatively high then. And it can reflect the *absence* of large differences in w which occurs in relatively short time-periods, when b is relatively low. The length of the analysis period, one of the variables clearly at work as noted earlier, is therefore taken into account indirectly.

The theoretical importance of the model $w(1 - b) =$ constant is that it interrelates a number of separate empirical results into a single theory. Instead of having an isolated formula for the value of the average purchase frequency of a brand, we have a model that relates this to

(i) consumers' known duplicated brand of purchases and
(ii) their total rates of product usage.

Given that these phenomena exist as they do, the formula $w(1 - b)$ must follow. It is through this integrative function that more advanced theory is built.

Three stages are normally involved in such developments. First a simple regularity is noted, e.g. that w in a short time-period is more or less *constant* (largely ignoring such small variations and even the possibility of a trend as might seem to exist in tables such as 10.1). Secondly, rather different results arise, e.g. that in longer time-periods w *varies* with the market-share. Finally, both types of results can be accounted for by the same model and this follows as a consequence of using *other* findings (e.g. the brand-switching law $b_{XY} = Db_X b_Y$, etc.).

It would be difficult to improve on the words used by Sir Cyril Hinshelwood (1967), uniquely President of the Royal Society and the Classical Association in the same year, to describe this sequence of stages. The first stage he described as

> "Gross over-simplification, reflecting partly the need for practical views and even more a too-enthusiastic aspiration for the elegance of form." (*Constants* indeed!)
>
> In the second stage, "the symmetry of the hypothetical system is distorted and the neatness marred as recalcitrant facts increasingly rebel against uniformity".
>
> In the third stage, "if and when this is obtained, a new order emerges, more intricately contrived, less obvious and with its parts more subtly interwoven, since it is of nature's and not of man's conception".

10.4 Summary

Theory operates at different levels. *Any* abstraction from the observed facts is theoretical, at least at a low level. But as different results interrelate, more advanced theory and understanding begin to grow.

Theoretical arguments can be purely hypothetical and these must be sharply distinguished from *validated* theory. The ultimate role of valid theory is to integrate and interpret results which are well-founded in fact.

CHAPTER 10 EXERCISES

Exercise 10A. Is it Practical?

Of what practical use is a theoretical result like $w(1 - b) =$ constant?

Discussion.

Two practical applications are in assessing sales targets for new and established brands. For simplicity's sake we shall consider relatively short time-periods. The penetrations b of different brands are then all low so that the factor $(1 - b)$ hardly varies, being close to 1 for all brands. The relationship therefore simplifies to $w \doteqdot$ constant. This is illustrated in Table 10.5 for breakfast cereals in a 4-week period.

TABLE 10.5 Purchase Frequencies over 4 Weeks
(Repeated from Table 10. 1)

4 WEEKS	Corn Flakes	Weet-abix	Shred. Wheat	Sugar Puffs	Puffed Wheat	Ave-rage
% Market-Share	38	31	16	10	5	(100)*
Purchases of the stated brand per buyer of the brand in 4 weeks	1.8	2.0	1.7	1.5	1.6	1.7

* Total sales of the 5 brands

A New Brand. Suppose that a new brand of breakfast cereal is to be launched for which a rate of sales of about 2 purchases per housewife per year has been set (for when the sales "settle down" six months or so after the launch). How can this target be achieved?

Two purchases per housewife per year is about .15 purchases per housewife per 4 weeks, or 15 purchases per *100* housewives, expressed on a per hundred basis to keep the arithmetic simple. A 4-weekly sales target of 15 purchases per 100 housewives could be achieved in various ways, e.g. 1% buying 15 times each, 5% buying three times each, 10% buying on average 1.5 times each, or 15% buying just once each in the 4 weeks.

But Table 10.5 shows that over 4 weeks the average purchase frequency w of established brands of breakfast cereal is about 1.7. It follows that about 9% of the population will have to buy the new brand at the predictable average rate of 1.7 purchases each to reach the target of 15 per 100 housewives. (The variation in the w's in Table 10.5 implies some variability in the required penetration, from 8% to 10%.)

We are not here predicting *sales*, the $64,000 question. The prediction is only that, whatever happens, the new brand will tend to be bought 1.7 times in 4 weeks by its buyers. What cannot be predicted is how many buyers there will be, the figure of 9% was the penetration which is required to achieve the given sales *target*.

Instead of having two unknowns b and w, we have a firm prediction for w (worth $64!) and a *target* for b. We now know what to plan for, a penetration of about 9%, and have a yard-stick for assessing test-market and launch results.

The prediction for w is firm unless there is a specific reason to expect the purchase frequency for this particular new brand to be radically different from that of the others in the product-class. The new brand would have to differ more from the existing cereal brands than they do from each other.

An Established Brand. To increase the sales of an *established* brand, like Shredded Wheat, one could in principle aim to increase either b, the number of people buying it, or w, the rate at which existing buyers buy it, or both b and w.

But we now know that the 4-weekly rate of sales of Shredded Wheat cannot be doubled by simply doubling its average purchase rate from the value 1.7 in Table 10.5 to 3.4, since nothing like that 4-weekly rate of purchasing has ever been observed for any cereal brand. It follows that if sales *are* to increase, It has to be through an increase in the number of buyers in 4 weeks. The alternatives have been ruled out by the empirical constraints of consumer behaviour.

Exercise 10B. The Units of Measurement

In the last exercise, sales were equated to the number of buyers multiplied by their rate of purchase, ignoring *how much* they bought per purchase occasion (e.g. whether large or small packs, and how many at a time). How can this be?

When discussing consumers' purchasing behaviour earlier, we sometimes spoke of numbers of purchases and sometimes of numbers of packs bought. Which is it?

Discussion.

The choice of units is often crucial in developing simple and generalisable theory. Thus the various laws of consumer behaviour mentioned generally hold only if the data are expressed as *purchase occasions.* (This allows purchases of different pack-sizes, for example, to be dealt with by *ignoring* the pack-size.) The primary justification of this approach is that it works, in giving simple and generalisable results.

In more detail, the total level of purchasing or of sales in a given period can be decomposed as follows:

Sales = (Number in population) × (Proportion buying at all)
× (Average number of purchase occasions per buyer)
× (Number of packs bought per purchase occasion)
× (Size or price of pack).

Here the number in the population is fixed. The size or price of a pack is usually more or less fixed or known. The average number of packs bought per purchase occasion is also approximately constant (e.g. for different brands, for light or heavy buyers, etc.). This is an *empirical* finding (Ehrenberg, 1972). So is the fact that w is approximately constant in a 4-week period. It follows that the only *variable* component of sales is b, the penetration.

The data in the breakfast cereal example came in the form of number of *packs* bought, so some of the results were referred to like that. But the theory is formulated on the basis of purchase occasions (two packs bought at the same time are treated as *one* occasion). The conflict is resolved because on average just over 1 pack of cereal is bought per purchase, so the distinction between units bought and purchase occasion is in this case numerically trivial, the results hold whichever unit is chosen. In contrast, for a product like petrol, the average purchase is usually just over 3 gallons in Great Britain (and more in the United States say); for this product, theoretical results like $w(1 - b)$ operate *only* when the purchase occasions are used as the analysis unit.

Exercise 10C. The Quantities *b* and *w* in Different Length Time-Periods

Tables 10.1 and 10.2 show that the average consumer purchased Corn Flakes 1.8 times in 4 weeks and 6 times in 24 weeks (or 5.7 to two digits). Why not 10.8 purchases (6 × 1.8) in a period six times as long?

Discussion.

If sales are steady, the total number of purchases made increases in proportion to the length of time-period. Thus the 491 households on which our data here are based (Charlton *et al.*, 1972) made 281 purchases of Corn Flakes in the 4-week period and 1684 purchases in 24 weeks, which is virtually six times 281. (These figures are equivalent to 57 and 340 purchases per *100* households in 4 and 24 weeks.)

But the average purchase frequency w refers to the frequency *per buyer* in the period. Both w and the number of buyers change with the length of the analysis-period. In 4 weeks, 157 of the 491 households bought Corn Flakes, a penetration of 157/491 or $b = .32$ or 32%. These households made an average of 281/157 purchases in 4 weeks or $w = 1.8$.

In 24 weeks, 298 households bought Corn Flakes, a penetration of $298/491 = .61$ or 61%. They made 1684 purchases, so $w = 1684/298 = 5.7$.

As the length of the analysis period increases, the proportional increase in sales is made up of *less than* proportional increases in both b and w, so that

$$\text{4 weeks:} \quad 32 \times 1.7 = 54,$$

$$\text{24 weeks:} \quad 61 \times 5.7 = 348.$$

(The differences from the observed sales levels of 57 and 340 are due to rounding off.)

Exercise 10D. Oversimplification in the Theory

In the theoretical deduction of the model $w_X(1 - b_X) = w_Y(1 - b_Y)$ it was assumed that the coefficient D in the equation $b_{XY} = Db_Xb_Y$ was equal to 1. Does it matter that this is usually not so?

Discussion.

In addition to D generally not being equal to 1, Item (C) used in the deduction is also generally not quite true. The average purchase frequency $w_{X \cdot Y}$ of Brand X per duplicated buyer of X and Y is usually lower than w_X: i.e. $w_{X \cdot Y} = Cw_X$, where C is a constant. The value of C is often about 0.8 (but in the example here it happened to be 1.0).

The more accurate equations $b_{XY} = Db_Xb_Y$ and $w_{X \cdot Y} = Cw_X$ can be used in a theoretical argument exactly like that of Section 10.3. This gives the result

$$w_X(1 - CDb_X) = w_Y(1 - CDb_Y).$$

This is similar in form to $w_X(1 - b_X) = w_Y(1 - b_Y)$. The results are almost identical when D is about 1.3 and C is about 0.8, which is often the case. More generally, the two forms tend to give similar numerical results.

Even though the more exact argument is mathematically soluble, at this stage the formulation $w_X(1 - b_X) = w_Y(1 - b_Y)$ is preferred because it is much simpler. For example, it can be used without first establishing the specific values of C and D for any particular situation.

Using *exact* results in theoretical work commonly leads to more complex mathematics than used here, which is often literally impossible to solve. Simplifying assumptions are therefore commonly introduced, e.g. the notion that the buying of X and Y is simply *independent* rather than linked by $b_{XY} = Db_X b_Y$. It follows that theoretical deductions are often not altogether true, because the oversimplification was too drastic. This is why theoretical deductions always have to be checked against the facts.

Exercise 10E. By Fits and Starts

Medewar (1952) has noted that many scientific papers do not describe how the results were obtained (some facts and then a "theory" to explain them) but are written with hindsight in an unnaturally "logical" form, as if the theory had come first. How was the result $w(1 - b) =$ constant *really* obtained?

Discussion.

First came the empirical result that $w_X \doteqdot w_Y$, in relatively short time-periods. Next, Dr. C. Chatfield noted that if the product-rates of buying were constant (item A in Section 10.3) and if one assumed that buying of one brand was independent of buying another (essentially items (B) and (C)), then it could not be true that w_X equals w_Y. Instead, the w's should show a trend.

However, by this time it had been established empirically that, especially in longer time-periods, w did in fact show some kind of trend with market-share. This led Mr. G. J. Goodhardt to invert Chatfield's earlier argument giving the result $w(1 - b) =$ constant along the lines of Section 10.3. The formula accounted for all the empirical facts, both the trend of w in longer time-periods and its approximate constancy in shorter time-periods.

Exercise 10F. Obvious After the Event?

What does the result $w(1 - b)$ mean?

Discussion.

Table 10.2 shows that consumers of Corn Flakes bought them 6 times in 24 weeks and consumers of Puffed Wheat bought it only twice.

One explanation of such a difference in purchase rates might be that the Puffed Wheat packs are bigger. But this is in effect not so. The same $w(1 - b)$ type of result holds for each separate pack-size and also in product-fields such as petrol where different brands are virtually identical in performance and packaging.

Another possible explanation might be that the sort of households that buy Corn Flakes need more breakfast cereal than Puffed Wheat consumers, and that the households satisfy their total needs with the brand they have chosen. But this is not so either. The *total* purchases of breakfast cereals are much higher, about 14 purchases for the average consumer of each brand (Table 9.10). And in any case they are constant across the different brands.

Basically, the meaning of $w(1 - b) =$ constant lies in how it describes the data and in the form of its theoretical deduction. If consumers buy different brands more or less independently of each other (i.e. according to items A, B, and C in Section 10.2) then the rates at which they buy individual brands *must* follow something like the relationship $w(1 - b) =$ constant.

The temptation to read more into such a law is particularly strong in the present case because not only w_X but also $(1 - b_X)$, the other item in the expression, has a direct meaning of its own: it is the proportion of the population which does *not* buy Brand X in the analysis-period. Why then should the rate of buying per buyer times the proportion of non-buyers be constant for different brands?

The explanation is that the apparent "meaningfulness" of this particular formula is accidental. For example, using the more precise results of Exercise 10D, the expression reads $w_X(1 - CDb_X) = w_Y(1 - CDb_Y)$, and this is sufficiently complex no longer to cry out for any "simple interpretation".

Exercise 10G. Explaining Consumers' Attitudes

Explain the relationship between consumer attitudes and brand usage in Section 9.2, and the occasional large deviations from the relationship.

Discussion.

Since the attitudinal responses are related to the incidence of users, cross-analysis by the "users" and "non-users" of each brand seems a promising next step to explore.

Table 10.6 illustrates typical results. For the attitude "Right Taste", 67% of the users of Brand A say A has it, 62% of the users of Brand B say it about B, etc. In contrast, only 6% of the *non-users* of Brand A say A has the right taste, only 4% of the non-users of Brand B say it about B, and so on.

TABLE 10.6 Percentage of Users and Non-Users of a Brand Holding an Attitude Towards It

	% of the Population Using	"Right Taste"		"Convenient"	
		Users of Stated Brand	Non-Users of the Brand	Users of Stated Brand	Non-Users of the Brand
Brand A	50	67	6	19	3
Brand C	30	62	4	55	48
Brand B	10	69	5	17	2
Brand D	5	60	3	17	2

These results generalise A certain percentage of users of a brand say it has the given attitudinal property, and a much smaller percentage of non-users say so. This holds for many different attitudinal variables and many different brands and product-classes.

Since brands differ in their number of users, a higher proportion of the total population gives a positive attitudinal response for market-leaders than for small brands, the finding in Chapter 9. Thus for Brand A, $(50 \times 67 + 50 \times 6)/100 = 36\%$ of the population should say "Right Taste" about it, and for Brand D only $(5 \times 60 + 95 \times 3)/100 = 6\%$.

The results in Table 10.6 also explain the major exceptions to this general pattern: the results for Brand C and "Convenient" are quite different. The percentage of non-users of Brand C who regard it as "Convenient" is only fractionally lower than that amongst users (48% and 55%), and both percentages are far higher than for the other brands. This accounts for the exceptionally large proportion of all consumers who regarded Brand C as "Convenient" in Table 9.4.

Such exceptional responses generally occur when a brand differs physically in the relevant respect from the other brands. Thus an indigestion of headache remedy which can be taken as a tablet is more "Convenient" than one which requires water and a glass. Consumers, both users and non-users, notice this and say so. In contrast, when different brands are similar in product-formulation, consumers' attitudinal responses are essentially different ways of saying that they know of or use or like that type of product, as exemplified by the brand.

Although the original undigested data in Table 9.3 in the last chapter looked fairly complex, the results after analysis are beginning to "hang together". For example, it now seems that behaviour here influences attitudes rather than that attitudes cause behaviour. The popular notion of different brands generally differing in their "images" cannot be rooted in any discernible facts. Consumers seem to give the product "image" to the brand they use, i.e. they see it as typical of the product-class. Only when there is a real difference is this reflected in the image, but among both users and non-users. Such conclusions affect our view of the nature of competition among similar products and of the way advertising works in a consumer society (e.g. Ehrenberg, 1974a).

PART III: STATISTICAL VARIATION

This part of the book concentrates on ways of summarising and interpreting irregular variation within a single set of readings.

Chapter 11 discusses how the individual readings in a set of data can be arranged as a frequency distribution and be further summarised by measures of their average size and scatter. Observed frequency distributions can take different shapes and these can often be described by mathematical functions. Chapter 12 discusses the main examples: the "Normal", the "Poisson", and various "Binomial" distributions.

The concept of probability is often helpful in analysing irregular variation. Chapter 13 outlines this approach, in which the observed variability of the data is regarded as if it were random. This leads to models of a probabilistic or "stochastic" kind.

Chapters 14 and 15 describe and evaluate techniques of correlation, regression and factor analysis which are widely used in statistical analyses. These methods aim to describe the way different variables vary together within a single set of readings. (In Part II we considered how different variables vary together for *more* than one set of readings.)

The topics covered in this part of the book are of more restricted practical relevance than those in Parts I and II. For example, some subject-areas almost never use frequency distributions or probability methods, but others do so a great deal. Experienced readers will know which topics are most relevant to their areas of interest. But many readers will need to gain some broad familiarity with all the topics covered here, and this they should be able to extract.

CHAPTER 11

Summary Measures

So far we have been dealing with variations between different sets of data. This is "controlled" variation. For example, when we were analysing heights and weights, we first took boys of different age-groups and looked at the systematic variation between them. We controlled the sex and age differences.

But if we simply took a group of children and knew nothing else about them except their heights, the differences in the readings would be "uncontrolled" variation. We now concentrate on this kind of uncontrolled variation within a single set of readings.

Reporting uncontrolled data in full is clumsy, especially when there are many readings. This chapter discusses how to look at such data as frequency distributions and to reduce these to meaningful summaries of average size and scatter.

11.1 Empirical Frequency Distributions

As a small numerical example, consider the following set of eight readings:

3, 4, 8, 3, 6, 1, 3, 4.

To see these better, we can arrange them in order of size:

1, 3, 3, 3, 4, 4, 6, 8.

To summarise further, we can express the readings as a *frequency distribution*, one 1, no 2's, three 3's, etc., as in Table 1.1.

TABLE 11.1 The Frequency Distribution of the 8 Readings

	Value								The Total Number of Readings
	1	2	3	4	5	6	7	8	
Frequency:	1	0	3	2	0	1	0	1	8

This kind of frequency distribution might be a sufficient description of an isolated set of readings, but it is still too clumsy if we want to compare different sets of data.

One factor which usually differs is the number of readings in each set of data. For example, Table 11.2 compares our original 8 readings with a second set of 40 readings. The eye cannot immediately take in the detailed similarities and differences in the two sets of data.

TABLE 11.2 Two Sets of Readings to Compare

	Value								The Total Number of Readings
	1	2	3	4	5	6	7	8	
1st set	1	0	3	2	0	1	0	1	8
2nd set	5	1	14	10	1	4	1	4	40

We can eliminate the visual confusion by showing the observed frequencies as *proportions* of the total number of readings in each set, as in Table 11.2a.

TABLE 11.2a The Two Frequency Distributions Expressed as Proportions

	Value								Total
	1	2	3	4	5	6	7	8	
1st set	.12	.00	.38	.25	.00	.12	.00	.12	0.99*
2nd set	.12	.03	.35	.25	.03	.10	.03	.10	1.01*

* Not 1.00 due to rounding errors

Now it is clear that the two sets of readings have similar properties. Table 11.2b further simplifies the visual comparisons by expressing the frequencies as percentages and avoiding the clumsy decimal point. (But in mathematical work it is simpler to work in *proportions*.)

TABLE 11.2b The Frequency Distributions Expressed as Percentages

	Value								Total
	1	2	3	4	5	6	7	8	
1st set %	12	0	38	25	0	12	0	12	99
2nd set %	12	3	35	25	3	10	3	10	101

Table 11.2c simplifies the comparison even more by arranging the frequencies in broad groups (although here, with only four effective categories, the grouping has possibly been overdone, especially for the large 3–4 category). But the table is still not succinct enough to provide a memorable summary of the data. To reduce the data still further we need a summary like an *average*.

TABLE 11.2c The Two Frequency Distributions in Grouped Categories

		Values				
		1-2	3-4	5-6	7-8	Total
1st set	%	12	63	12	12	99
2nd set	%	15	60	13	13	101

11.2 Summaries of Average Size

Averages are the main tool of statistical analysis. But some averages are good and some can be misleading. To judge whether simple averages describe the data adequately, one first needs to calculate them and then check the figures against the data. This also helps one to see what the data themselves are like. The golden rule in looking at data is "Average *before* you look".

The three main types of average are the *mode*, the *median*, and the *mean* (or arithmetic average). We illustrate these with our original set of eight readings, repeated in Table 11.3.

TABLE 11.3 The Frequency Distribution of the 8 Readings

(From Table 11.1)

	Values								
	1	2	3	4	5	6	7	8	Total
Frequency	1	0	3	2	0	1	0	1	8

The *mode* is defined as the most frequent reading. For the data in Table 11.3 the mode is 3, there are more 3's than any other reading. Sometimes the mode is difficult to determine precisely because there is more than one "most frequent" reading, or the grouping obscures the mode, as in Table 11.2c.

The *median* is the value that is exceeded by half the readings. This is not always clear. For example, in Table 11.3, 4 out of 8 readings are greater than 3, and 4 out of 8 are less than 4, so that each of these could be the median, as defined. In such cases the median is conventionally taken at 3.5, half-way between the largest and smallest possible values.

Finally, the *mean* or arithmetic "average" of the readings is their sum divided by the total number of readings. In our example this is $32/8 = 4$.

Thus for the 8 readings in Table 11.3 we have

the Mode at 3,
the Median at 3.5,
the Mean at 4.

Although the three measures are conceptually quite different, they take similar values in this example. Such cases occur mainly with distributions that are approximately symmetrical and have their mode or "hump" in the middle. Figure 11.1 gives an illustration for the heights of a group of 10-year-old boys.

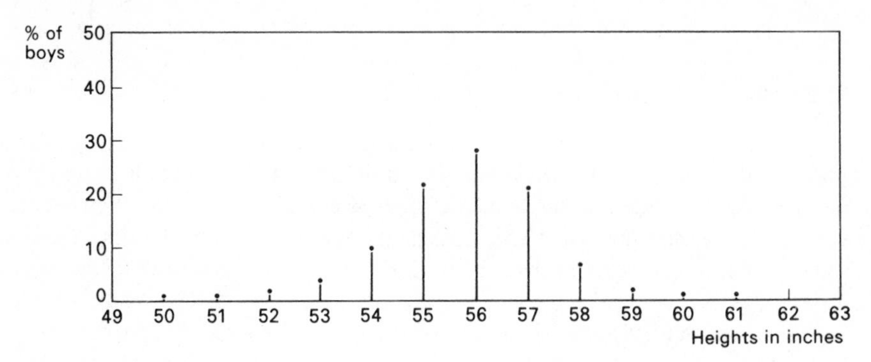

Figure 11.1 The Heights of 10-Year-Old Boys

These kinds of symmetrical distribution are exceptionally important. They are virtually the only kind of statistical data that can be described in a simple standardised way. The mean, median and mode will always be approximately equal. Therefore a single number has all the different meanings of these measures and most of the readings will fall close to this value.

This situation should be implied whenever an unqualified average is reported. For example, if the average age of a group of people is described as 30 years, the implication should be that about half are younger than 30, and that most of them are close to 30. If the data are more complex than this, one needs to say so.

The Average of Non-symmetrical Distributions

As a contrast, Table 11.4 sets out the reading frequencies of a monthly magazine. Most people either read all or almost all of the issues, or else they read none or almost none. Very few people read about half of the issues.

TABLE 11.4 The Number of Issues of the Monthly Magazine X Read per Year

(Percentages from Figure 11.2)

	Number of Issues Read													All Adults
	0	1	2	3	4	5	6	7	8	9	10	11	12	
% of Adults	40	15	5	2	1	1	1	0	1	1	3	10	20	100

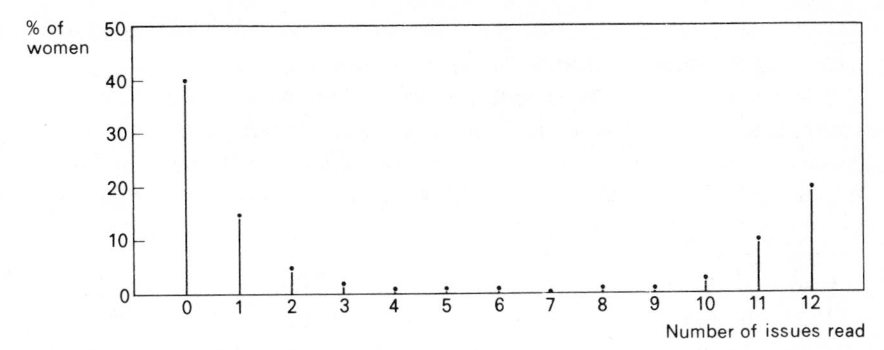

Figure 11.2 The Number of Issues of a Monthly Magazine Read per Year

The data therefore fall into a U-shaped distribution, as Figure 11.2 illustrates.

The median of the distribution is 1, since 40% of the people read no issue and 45% read 2 or more. If we regard the mode as a "locally" most common reading, this distribution has two modes, 0 and 12. Finally, the average number of issues read per person per year is 440/100 = 4.4. We therefore have

two Modes	at 0 and 12,
the Median	at 1,
the Mean	at 4.4.

Much of the confusion about averages arises from situations like this. The mean, median and mode take very different values, but sometimes one type of average is interpreted as if it were another. The mean here is in no sense a typical or modal reading: hardly anybody reads 4 or 5 issues and nobody reads 4.4 issues. Again, half the population does *not* read more than the mean number of issues, 63% read 4 or fewer issues and only 37% read 5 or more.

There is no simple routine way to describe such data. They require a *tailor-made* description: e.g. that about 55% of the people read no or only 1 issue and about 30% read 11 or 12 issues. This would summarise the main characteristics of these particular data quite well.

The Pre-eminence of the Mean

The mean is the most commonly used measure of average size. This is because it provides a summary which can generally be used in further analyses of the data where the median and mode cannot. For example, when combining two sub-groups of readings into a single group, the mode and the median of the sub-groups cannot be related to the corresponding values of the total group. Instead, one has to go back to the individual readings to determine the mode or median of the total group.

There is no such difficulty with the mean. Suppose we have one set of 8 readings with a mean of 4 and another set of 12 readings with a mean of 6. Since a mean value is the sum of the individual values divided by the number of readings, the mean of the combined sets will be $(8 \times 4) + (12 \times 6)$ divided by $(8 + 12)$ or $104/20 = 5.2$. No such calculations are possible for the median or the mode.

However, the latter two measures are not meaningless. It can be helpful to know the most frequent or "modal" reading, and also to see where the 50:50 or median division comes in a set of data. But the median and the mode do not provide summaries which can be widely used. The distinction is between interesting concepts and usable summaries.

Another property of the arithmetic mean is that the *total* value of the readings can easily be calculated by multiplying the mean by the number of readings. Such totals can sometimes be of practical relevance. For example, an average rainfall of 2 inches per month adds up to 24 inches in the whole year. (There is no corresponding benefit in knowing that the average height of a group of boys is 57 inches, "laid end to end . . ."; nor does it follow from a mean readership of 4 magazine issues per person that a total of 2 million copies of magazine X were sold in a city of 500,000, since some copies will have been read by more than one person.)

11.3 Measures of Scatter

When summarising uncontrolled variation one must also describe how far the readings are from the mean. There are several different statistical measures for this. One is the *mean deviation*, already widely used in Parts I and II.

The Mean Deviation

In our example of eight readings

$$1, \quad 3, \quad 3, \quad 3, 4, 4, 6, 8,$$

the individual readings deviate from the mean, 4, by

$$-3, -1, -1, -1, 0, 0, 2, 4$$

units respectively. The total or average of these deviations is necessarily zero because the negative deviations balance the positive ones. This provides a useful check on one's arithmetic.

To summarise the scatter of the deviations we need to measure their average size. One simple procedure is to ignore the negative signs and take the average of the numbers, which is 12/8 or 1.5. This is the *mean deviation*. (Sometimes this is also called the "mean absolute deviation", where "absolute" signifies that one is considering the sheer *size* of the deviations, irrespective of plus or minus signs. The *modulus* symbol, two vertical lines as in $|x|$, described verbally as "mod x", is also used to signify absolute values.)

The mean deviation is a very simple and useful measure in numerical analyses of data. For example, in all the analyses in Parts I and II of this book it seemed natural to look at the avergae (absolute) size of the deviations from the various "models" that were fitted.

The Standard Deviation and the Variance

The standard deviation and its square, the variance, are the conventional measures of scatter used in statistical theory. They are much easier to manipulate mathematically and thus more useful in theoretical work, but are conceptually less obvious than the mean deviation and more complicated to compute with pencil and paper.

In our example of eight readings, we are typically faced with positive and negative deviations from the mean:

$$-3, -1, -1, -1, 0, 0, 2, 4.$$

The standard deviation is based on the idea of squaring all the deviations to eliminate any negative signs. Three steps are involved: (1) squaring the deviations, (2) taking the average of the squares, and (3) taking the square root of this average in order to return to the original scale of measurement. Expressed as a formula, we have

$$\text{Standard deviation} = \sqrt{\left\{\frac{\text{Sum of (Deviations)}^2}{\text{No. of readings}}\right\}}.$$

With the eight deviations above, the squares are

$$9, \quad 1, \quad 1, \quad 1, 0, 4, 16.$$

The average of these squared deviations is $32/8 = 4$. The standard deviation is the square root of this, or $\sqrt{4} = 2$. (It is sometimes also called the "root mean square".)

The formula for the standard deviation is generally slightly adjusted for technical reasons in the theory of statistical sampling. Thus the sum of the squared deviations is divided by one less than the number of readings or $n - 1$, i.e.

$$\text{Standard deviation} = \sqrt{\left\{\frac{\text{Sum of (Deviations)}^2}{n - 1}\right\}}.$$

This makes no difference to the numerical result if the number of readings, n, is at all large. Even for our small numerical example with $n = 8$, the standard deviation using the divisor $(n - 1)$ is only about 5% different, 2.1 instead of 2. To all intents and purposes we can therefore think of the standard deviation as an average, but *use* the formula with $(n - 1)$.

The square of the standard deviation is defined as the *variance*, and is a useful mathematical quantity. It is the intermediate stage in calculating the standard deviation, i.e. the average of the squared deviations:

$$\text{variance} = \frac{\text{Sum of (Deviations)}^2}{n - 1}.$$

The Coefficient of Variation

The coefficient of variation is a related measure that is sometimes used to express the standard deviation as a percentage of the mean instead of in the original units of measurement:

$$\text{Coefficient of Variation} = \frac{100 \times \text{Standard Deviation}}{\text{Mean}}.$$

For example, with a mean of 4 and a standard deviation of 2, as in the case of our 8 readings, the standard deviation is half the size of the mean or 50%:

$$\text{Coefficient of Variation} = \frac{2 \times 100}{4} = 50\%.$$

This approach is most useful when the scatter in different sets of readings increases proportionally with the mean values of the readings. For example, in Table 11.5, instead of having to report and remember quite different numerical values of the standard deviation (e.g. .5 for set A, 1.2 for set B, etc.), we can adequately summarise the size of the scatter with a single "constant" figure of about 25%.

TABLE 11.5 Approximately Constant Coefficients of Variation in Five Different Sets of Data

	Sets of Data				
	A	B	C	D	E
Mean Value	2	5	10	40	120
Standard Deviation	.5	1.2	3	9	31
Coefficient of Variation	25%	24%	30%	22%	26%

In contrast, Table 11.5a describes five other sets of data where the size of the scatter increases only minimally with the mean. Here it is simpler to report the approximately constant standard deviations of 5 units than the different coefficients of variation.

The method of reporting does not affect the *interpretation* of the data. It is still easy to see that Set L with a mean of 4, has a relatively large standard deviation of 4 (the coefficient of variation being about 100 %) whereas Set Q, with a mean of 80, has relatively small scatter (the coefficient of variation being only about 8 %).

TABLE 11.5a Approximately Constant Standard Deviations in Five Other Sets of Data

	Sets of Data				
	L	M	N	P	Q
Mean Value	4	8	12	25	80
Standard Deviation	4	3	5	5	6
Coefficient of Variation	100%	37%	42%	20%	8%

The Range

The final measure of scatter to consider is the *range*. This is defined as the difference between the highest and lowest readings in the data, an obvious, common-sense type of measure. For the eight readings

$$1, 3, 3, 3, 4, 4, 6, 8,$$

the range is $8 - 1 = 7$. Reported together with the mean of 4, the range of 7 helps to give a good feel of these data; one likes to know how different the largest and smallest readings in a group are (although giving the two extreme values of 1 and 8 seems even more informative).

But like medians and modes, the range has serious disadvantages in detailed analytic work.

(i) It depends on the two extreme values and is therefore very sensitive to odd outlying readings. (But it provides a good check on the *occurrence* of any unusually high or low values.)

(ii) With large numbers of readings not already ordered by size, searching for the highest and lowest values is laborious.

(iii) The numerical value of the range depends on the number of readings. (Adding another reading may *increase* the range but can never decrease it.) This makes it difficult to use when comparing the scatter of sets of data with different numbers of readings. (In contrast, the mean and standard deviations are *averages*, and thus independent of the number of readings.)

(iv) The range of some combined set of readings cannot be calculated from the ranges of each of the sub-sets being grouped (or vice versa). Instead, one has to go back to the raw data and look for the highest and lowest values in each set.

A variation on the range is to exclude one or more of the extreme readings. This can give the measure more stability. The best-known example is the "inter-quartile range", the difference between the two quartiles. (These are the two values below and above which 25% of the readings lie, and thus are akin to the median). But such measures are seldom used in practice.

The Descriptive Meaning of Measures of Scatter

In our example of 8 readings the mean is 4 and the range is 7. If we reported just these two figures, the description should imply that the readings are more or less systematically distributed about the mean of 4, from 0 or 1 to 7 or 8. But the readings *could* extend from 3 to 10 and still have a mean of 4 and a range of 7, as follows

$$3, 3, 3, 3, 3, 3, 4, 10.$$

If this were the case, it would be misleading to report the mean and the range alone. One would need to describe the data in more detail, i.e. mostly 3's, with one high value at 10.

Similar considerations apply to the use of other measures of scatter. For example, the mean deviation and standard deviation tell us the "average" size of the deviations from the mean, but do not indicate how the individual deviations are distributed. Are most of the deviations of about this average size, or are half much greater and half less, or are most of them small with one or two very large deviations, or what?

The only simple description arises with symmetrical humpbacked distributions, like the one illustrated in Figure 11.1. The mean value here is 56 and the mean deviation is 1.2. Most of the readings fall quite close to the mean. Well over half the individual deviations from 56 are smaller than the *mean* deviation, and only about 10% are bigger than *two* mean deviations.

This is a common pattern for symmetrical humpbacked distributions and it is discussed in more detail in Chapter 12. For such distributions either the mean deviation or the standard deviation adequately summarizes the proportion of readings that lie in any particular range of values. But for distributions that are *not* symmetrical and humpbacked (like the U-shaped one in Figure 11.2), it is not possible to describe in any general way how many readings lie within one mean deviation of the mean, or more than two mean deviations away, and so on. More elaborate methods are needed to describe such variation. These also are discussed in the next chapter.

11.4 Summary

The "shape" of a set of readings can be seen by arranging them in order of size and grouping them in convenient intervals. It is easier to compare sets of data with different numbers of readings if the frequencies of each value are expressed as proportions or percentages of the total number of readings in each set.

The data can be summarised further by calculating the average size of the readings. The mean or arithmetic average is the preferred measure because it can readily be used in further analyses, e.g. when combining or separating different sets of data. Two other measures of average size are the mode (the most frequent reading) and the median (the value exceeded by half the readings). These are helpful concepts but do not provide summaries which in practice are usable in further analyses.

The mean, the mode and the median tend to be approximately equal in humpbacked, symmetrical distributions because most of the readings lie relatively close to the mean. Such symmetrical distributions are therefore easy to summarise. But quoting the mean value on its own can be misleading with other kinds of distributions.

The mean deviation, the standard deviation and its square, the variance, are commonly used measures to describe the *scatter* of readings. These are different ways of summarising how far the individual readings differ from the mean. The mean deviation is the average of all the deviations, ignoring any minus signs. It is arithmetically and conceptually easy to use. The variance eliminates the minus signs by squaring the deviations and *then* averaging. The standard deviation is the square root of the variance. These two measures are easier to use in mathematical analysis.

CHAPTER 11 EXERCISES

Exercise 11A. A Logical Sequence

What is a logical sequence in describing statistical data?

Discussion.

The successive steps in describing data are best kept independent of each other. A typical sequence is

- n the number of readings;
- m the mean, which is independent of n;
- s the standard deviation of the readings from the mean, which is independent of n and m;
- f the "shape" of the distribution (to be summarised by some mathematical function), which is independent of the values of n, m, and s.

Exercise 11B. The Greek Sigma Notations

Statistical tests often use the Greek symbols σ (small sigma) and Σ (capital sigma). What do these mean?

Discussion.

A common notation is to use the Greek letters for population values and Roman letters for sample values. For example, σ means the standard deviation of a set or "population" of readings, and s means the standard deviation of a sample of n readings from that population.

If sampling is not involved, then regardless of the number of readings, the data effectively represent the whole population. Strictly speaking, Greek letters are then the appropriate ones to use. However, Roman letters are simpler to use in practice. The context should make the situation clear.

Capital sigma, Σ, is used for a completely different purpose. It denotes the *sum* of the relevant readings. Thus for n readings of x, say x_1, x_2, x_3, etc. up to x_n,

$$\Sigma x = x_1 + x_2 + x_3 + \cdots + x_n.$$

Sometimes this is written as

$$\sum^{n} x$$

when there is possible doubt about how many terms are being summed, or as

$$\Sigma x_i,$$

where x_i stands for the ith reading, i taking all possible values from 1 to n. This may be written still more explicitly as

$$\sum_{i=1}^{n} x_i,$$

i.e. the summation of x_i for all values of i from 1 to n.

Exercise 11C. Deviations from the Mean

Prove that in a set of readings the average deviation from the mean is zero.

Discussion.

If you have n readings of x, the sum of the deviations from the mean $\bar{x}$, can be written as $\Sigma(x - \bar{x})$. Here $\bar{x} = \Sigma x/n$, or the total of the readings, Σx, divided by their number, n.

$$\begin{aligned} \text{Now } \Sigma(x - \bar{x}) &= (x_1 - \bar{x}) + (x_2 - \bar{x}) + (x_3 - \bar{x}) + \cdots + (x_n - \bar{x}) \\ &= (x_1 + x_2 + x_3 + \cdots x_n) - n\bar{x} \\ &= (\Sigma x) - n\bar{x} \\ &= \Sigma x - n \Sigma x/n \\ &= 0. \end{aligned}$$

Exercise 11D. Combining Two Averages

In Section 11.2 we noted that the mean of two combined sets of readings could be calculated from the two separate means. Express this in algebraic terms.

Discussion.

If we have two sets of n_1 and n_2 readings with means m_1 and m_2, the numerical *totals* of each set are n_1m_1 and n_2m_2. The total of the combined set of readings is therefore $(n_1m_1 + n_2m_2)$. Thus the mean is

$$\frac{n_1m_1 + n_2m_2}{n_1 + n_2}.$$

(There are three commonly used symbols for the mean of a set of n readings:

(i) $\bar{x}$, described as "x bar", for any set of readings,
(ii) m, usually for a sample mean,
(iii) μ, the Greek "mu", for the mean of a population.

Because the suffices in our example referred to different sets of data rather than individual readings, we used m instead of $\bar{x}$.)

Exercise 11E. A Computing Short-cut

The expression $\{\Sigma(x^2) - n\bar{x}^2\}/(n-1)$ is a short-cut formula for calculating the variance. Prove and discuss it.

Discussion.

The variance of a set of n readings is the average of the squared deviations from their mean. Consider the sum of the squared deviations:

$$\Sigma(x - \bar{x})^2 = (x_1 - \bar{x})^2 + (x_2 - \bar{x}) + \cdots + (x_n - \bar{x})^2.$$

Now

$$\begin{aligned}(x_1 - \bar{x})^2 &= (x_1 - \bar{x})(x_1 - \bar{x}) \\ &= x_1^2 - x_1\bar{x} - \bar{x}x_1 + x^2 \\ &= x_1^2 - 2x_1\bar{x} + \bar{x}^2.\end{aligned}$$

Similarly

$$\Sigma(x - \bar{x})^2 = \Sigma(x^2 - 2x\bar{x} + \bar{x}^2)$$

Therefore

$$\begin{aligned}\Sigma(x - \bar{x})^2 &= (x^2 - 2x\bar{x} + \bar{x}^2) \\ &= \Sigma(x^2) - \Sigma(2x\bar{x}) + \Sigma(\bar{x}^2) \\ &= \Sigma(x^2) - 2\bar{x}\,\Sigma x + \Sigma(\bar{x}^2) \\ &= \Sigma(x^2) - 2\bar{x}n\bar{x} + n\bar{x}^2 \\ &= \Sigma(x^2) - 2n\bar{x}^2 + n\bar{x}^2 \\ &= \Sigma(x^2) - n\bar{x}^2.\end{aligned}$$

The sum of squared deviations can therefore be written as

$$\text{Sum}\,(x^2) - n\bar{x}^2$$

or, because $nx^2 = n\,\Sigma(x)^2/n^2$, as

$$\text{Sum}\,(x^2) - (\text{Sum}\,x)^2/n.$$

Both forms are usually easier to calculate than the basic expression Sum $(x - \bar{x})^2$, especially when using a desk or pocket calculating machine.

The reason is that the standard formula requires calculation of all the deviations, $(x - \bar{x})$, before squaring them. With the short-cut formulae one can square the original readings x, sum these squares, and sum the x-readings (all in one single operation on a machine with a "revolution counter").

For example, consider the 5 readings

$$4, \quad 25, \quad 6, \quad 9, 12.$$

The mean is $x = 11.2$, and the deviations are

$$-7.2, 13.8, -5.2, -2.2, .8.$$

If we use the standard formula for the variance, $\Sigma (x - \bar{x})^2/(n - 1)$, we first have to write down the deviations, then square them, and then sum these squares, giving $51.84 + 190.44 + 27.04 + 4.84 + 0.64 = 274.80$. (The last two steps can be done in one operation on a calculating machine.) Hence we get a variance of $274.80/4 = 68.7$, and a standard deviation of 8.3.

But if we use the short-cut formula we avoid computing the deviations and introducing negative numbers with their extra error possibilities. Squaring the original readings and summing them gives $16 + 625 + 36 + 81 + 144 = 902$. (The same computation will also give the total of the readings, 56, if there is a "revolution counter".) The sum of squared deviations is therefore $902 - (56)^2/5 = 274.8$, as before.

An additional short-cut which is convenient on many calculating machines is to calculate $\Sigma (x^2) = 902$ and $\Sigma x = 56$ in one operation, then multiply $\Sigma (x^2)$ by n $(902 \times 5 = 4{,}510)$, and then subtract $(\Sigma x)^2 = 9{,}510 - (56)^2 = 1{,}374$. This gives $n \Sigma (x - \bar{x})^2$ or 5×274.8 in this case. To obtain the variance we therefore divide by $n(n - 1) = 20$, giving 68.7 as before. The only number that needs to be written down is the final answer.

A Warning. While the short-cut formula avoids calculation of the individual deviations, it also by-passes any chance of *looking* at these deviations to get the "feel" of the data, to spot unusually large deviations, etc.: all the things which need to be done when first handling new data. The short-cut formula should therefore only be used when one is familiar with the kind of data in question and when the resulting standard deviation agrees with prior expectations (implying no aberrant outliers in the new data).

Exercise 11F. The Standard versus the Mean Deviation

What are the relative advantages and disadvantages of these two measures?

Discussion.

The advantages of the standard deviation (and of the variance) are that they are much easier to use in mathematical theory than the mean deviation.

When calculating the mean deviation it is easy for the human mind to eliminate the minus signs before averaging, but it is very difficult for formal mathematics to cope with this. It involves examining each deviation separately and then doing something different according to each deviation's sign. This operation is difficult to incorporate in any mathematical operation.

In contrast, squaring each deviation when calculating the variance is complex to the human mind unless the numbers are very simple. Yet it is *mathematically* easy because the same process (squaring) occurs regardless of the sign of the initial number. The outcome is always a positive number.

For example, it is possible to have a short-cut formula for the variance but no corresponding simplification is feasible for the mean deviation. The same applies to many other mathematical applications of these measures (e.g. the "analysis of variance" discussed in Chapters 18 and 19).

But conceptually and numerically the mean deviation is the simpler measure. In the numerical example in the last exercise, the mean deviation is the average of the numbers 7.2, 13.8, 5.2, 2.2, 0.8, which is 29.2/5 = 5.8. It is even simpler to compute if the deviations are rounded to the nearest whole number. But the corresponding pencil and paper calculations for the standard deviation involve squaring and taking square-roots. The numerical analyses in Parts I and II would have been much more laborious if carried out in terms of the standard deviation.

There is, however, no major problem of having to choose between the two measures. In the most important case, for *Normal* Distributions, the two measures are directly equivalent to each other:

$$1 \text{ standard deviation} = 1.25 \text{ mean deviation},$$

as will be seen in the next chapter. One can therefore calculate the mean deviation but translate it into standard deviations when necessary (like changing from inches to centimetres). In other cases, *neither* measure has a direct descriptive meaning.

Exercise 11G. Outliers

With an exceptionally high or low value in a set of readings, how sensitive are

(i) the mean, median, and mode?
(ii) the mean deviation, standard deviation, variance, and range?

Discussion.

Consider two simple sets of 1,000 readings A and B:

A: 100 3's, 800 4's, 100 5's,
B: 100 3's, 799 4's, 100 5's, 1 1,000.

The "outlier" at 1,000 in B might be a measurement or recording error. These are often very dramatic and are important to spot and to eliminate from the main analysis.

The mean, median and mode of the two sets of data are

	A	B
Mean	4	5
Median	4	4
Mode	4	4

Only the mean is at all sensitive to the outlier. The single reading of 1,000 in Set B increased it by 25%, but even here the effect is not dramatic. Unless the means of other such sets of data are generally very close to 4, one would probably not react to the value of 5 as implying an aberrant value.

The mean deviation, standard deviation, variance and range of the two sets are approximately

	A	B
Mean Deviation:	0.2	2
Standard Deviation:	0.4	3
Variance:	0.2	10
Range:	2	997

In terms of absolute increase, the standard deviation is a little more sensitive than the mean deviation. The variance is even more sensitive because the odd outlying value is *squared* before averaging, and thus becomes more dominant. The range is clearly too sensitive to act as a measure of "average" or typical scatter at all, but it is highly efficient for actually spotting outliers.

Exercise 11H. The Divisor ($n - 1$) for the Variance

Discuss the practical and theoretical implications of using the divisor ($n - 1$) instead of n in the variance formula, i.e.

$$\frac{\Sigma(x - \bar{x})^2}{n - 1} \text{ instead of } \frac{\Sigma(x - \bar{x})^2}{n}.$$

Discussion.

From a "practical" point of view, n is the better divisor because it is easier to comprehend. One can see the formula represents the average of the squared deviations from the mean. In contrast, the expression Sum $(x - \bar{x})^2/(n - 1)$, has to be accepted as a "formula".

Yet we have seen that using the divisor ($n - 1$) instead of n makes virtually no difference *numerically*, except when the number of readings is much smaller than 10.

The reasons for using the ($n - 1$) formula are entirely theoretical. They arise especially in statistical sampling theory, and also in procedures like the "analysis of variance" (discussed in Chapters 18 and 19) where the mathematics is far simpler when using ($n - 1$).

By using the divisor ($n - 1$) the value of the variance becomes independent of the number of readings. (In statistical sampling theory this is like having unbiased estimators.) As an illustration consider a small set of 3 readings

2, 4, 6.

The mean is 4 and the sum of the squared deviations from the mean is

$$(2 - 4)^2 + (4 - 4)^2 + (6 - 4)^2 = 8.$$

Dividing by ($n - 1$) the variance is 4, Dividing by (n) the variance is $2\frac{2}{3}$,

Now, consider a different number of readings of the same kind of data, for example all possible sub-groups of *two* readings from the data:

2 and 4, 4 and 6 and 2 and 6.

These three sets have the following variances:

	Divisor	
	$(n - 1) = 1$	$n = 2$
2 and 4	2	1
4 and 6	2	1
2 and 6	8	4
	4	2

Using the divisor $(n - 1)$, the average value of the variances is 4, the same as the variance of the original set of three readings. But using the divisor n, the average variance of the pairs of readings is somewhat smaller than that for the original set when using n there. This result generalises to any set of data.

The theoretical advantage of $(n - 1)$ is a very strong one in much advanced work, but in practice one could still approximate and use the conceptually simpler divisor, n. However, even in practice, writing the formula with $(n - 1)$ is now very widespread.

Exercise 11I. Degrees of Freedom

Can the theoretical considerations for the divisor $(n - 1)$ be explained in commonsense terms?

Discussion.

In general, n variable quantities $x_1, x_2, x_3, \ldots, x_n$ can vary in n different ways, i.e. each variable can take any value, independently of the other variables. One can thus say the data has n ways in which to vary, or n "degrees of freedom", a notion due to Sir Ronald Fisher.

If we take the n deviations from the mean $\bar{x}$:

$$(x_1 - \bar{x}), (x_2 - \bar{x}), (x_3 - \bar{x}), \ldots, (x_n - \bar{x}),$$

only $(n - 1)$ of these quantities can vary independently. The last one is determined by the others because the deviations from the mean have to add to zero (see Exercise 11C),

As an illustration consider two readings, x_1 and x_2. Each can vary any way one likes. Now consider two particular values, $x_1 = 3$ and $x_2 = 7$. The mean is 5 and the deviations are

$$-2 \text{ and } +2.$$

Given the first deviation of -2, the other *must* be $+2$. Only one of the two deviations can vary independently.

Similarly, with three readings, only two of the deviations from the mean can vary independently. They determine the third deviation. This generalises for any number of readings. Having calculated the *mean*, we have effectively "used up" the independence of one of the readings.

Since there is no reason to identify one particular reading as being "used up", the same idea is expressed as having used "one degree of freedom" in the whole set of data.

More generally, one degree of freedom has to be subtracted for each constant or coefficient in a model that is fitted to the data. This idea is very helpful in various parts of statistical theory, as we shall see in Parts IV and V. It also provides a kind of reason why the average of the squared deviations from the mean is formed as if there were only $(n - 1)$ readings.

CHAPTER 12

Frequency Distributions

Many phenomena are highly irregular when observed at the individual level, but when analysed in groups the patterns often become systematic and generalisable. For example, we cannot predict an individual's income unless we know other facts about him, such as his age, occupation, etc. But for any *group* of people a regular pattern tends to appear. In general *most* will have relatively low incomes, some will have higher incomes, and a small number will have very high incomes, as illustrated in Figure 12.1. This kind of "skew" pattern occurs generally with income data for groups of fairly similar people, and in that sense it is predictable.

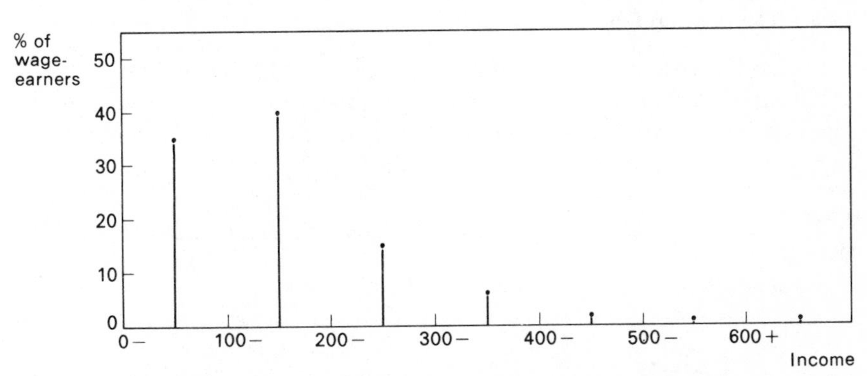

Figure 12.1 Wage-Earners' Incomes (Grouped in Units of 100)

We saw another case of this in the last chapter, with the number of monthly magazine issues read in a year. It may be impossible to predict the number read by any one person, but for any group of people the U-shaped distribution in Figure 12.2 occurs widely. It is the occurrence of such *statistical regularities* which makes the study of individually irregular data important.

When the same type of observed frequency distribution arises often, it is worth modelling by a mathematical formula. Then the theoretical formula can be used to compare and summarise different sets of data.

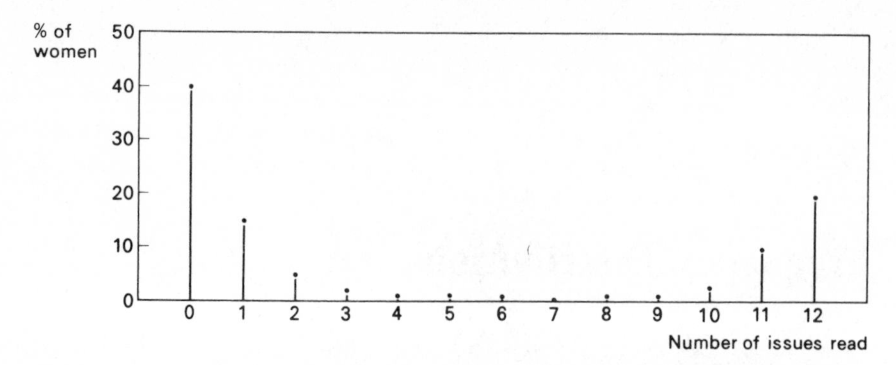

Figure 12.2 The U-Shaped Distribution for Magazine Readership (Figure 11.2)

The number of theoretical frequency distributions in common use is fairly small. This chapter deals with three main cases: the Normal, the Poisson, and various versions of the Binomial. Sometimes the mathematics is fairly complicated, but this need not matter much in practice. One mostly uses numerical tables or computer programmes, and simple verbal shorthand like "it's a Normal Distribution" often effectively summarizes the mathematics.

12.1 The Normal Distribution

Figure 12.3 shows a set of readings, the heights of 10-year-old boys, which are grouped symmetrically around a single modal value, where most of the readings lie close to this central point. Such data can often be approximated by a particular mathematical formula called the Normal Distribution.

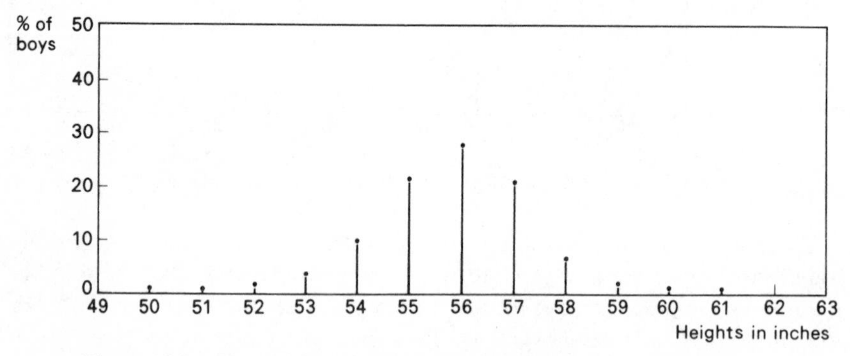

Figure 12.3 The Heights of 10-Year-Old Boys (Figure 11.1 repeated)

If the variability of the observed phenomenon is due to a large variety of independent factors, then an approximately Normal Distribution must result. This is a very common situation. For example, errors of measurement often follow this distribution. In the 18th Century it was called "the normal curve of errors", and the name Normal stuck. (The distribution was discovered in 1711 by de Moivre in England. Sometimes it is called the Gaussian Distribution, after the 19th Century German mathematician, C. F. Gauss.)

The Normal is an exceptionally simple distribution because it always takes the same shape. Describing this shape in terms of the standard deviation as a measure of scatter, a Normal Distribution has

68 % of its readings between ±1 s.d. from the mean,
95 % of its readings between ±2 s.d. from the mean,
99.7 % of its readings between ±3 s.d. from the mean.

Figure 12.4 shows how a Normal Distribution is composed. Thus mean values and the size of scatter may differ, but just about two-thirds of the readings will *always* lie between ±1 standard deviation from the mean of the data. This makes it unnecessary in practice to refer to the mathematical formula for the distribution (see Exercise 12K).

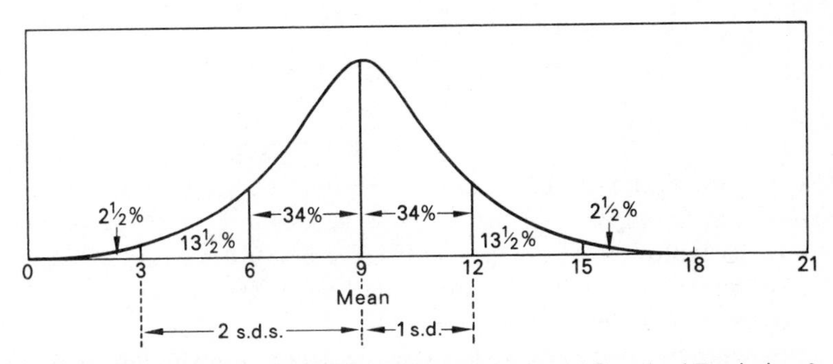

Figure 12.4 A Normal Distribution with Mean 9 and Standard Deviation 3

Theoretically the Normal Distribution ranges from minus infinity to plus infinity. But because only 0.3 % of the readings lie more than ±3 standard deviations from the mean, the distribution can approximate data that lie within some fairly restricted range.

In all Normal Distributions there is a simple relationship between the standard deviation and the mean deviation:

mean deviation = 0.8 standard deviation.

Thus the Normal Distribution can also be described as

58 % of the readings between ±1 m.d. from the mean,
90 % of the readings between ±2 m.d. from the mean,
98 % of the readings between ±3 m.d. from the mean.

The remarkable thing about the Normal curve is that it can be described for most purposes by three values like this, or the equivalent ones for the standard deviation. (In statistical sampling theory, Chapter 18, the 1% and 0.1% values of ±2.6 s.d. and 3.3 s.d. are conventionally referred to as well.) One can therefore fully summarize appropriate data by reporting that it follows a Normal Distribution with a certain mean and standard deviation (or mean deviation).

Numerical Examples

The Normal Distribution approximates so much observed data that it is important to be familiar with its role in summarising and comparing different sets of data. The following numerical examples illustrate this.

Table 12.1 reproduces the 8 readings discussed in the last chapter. The mean was 32/8 = 4 and the mean deviation 12/8 = 1.5. Even with such a small and "jumpy" set of data, the Normal Distribution works adequately.

TABLE 12.1 The Frequency Distribution of 8 Readings from Chapter 11

	Value							
	1	2	3	4	5	6	7	8
Frequency	1	0	3	2	0	1	0	1

It is easy to see that 5 out of 8 readings, of about 62%, lie within ±1 mean deviation and about 87% of the readings lie within 2 mean deviations of the mean. (A typical grouping problem is whether or not to count the reading of "1" within the 2 mean deviation limit. Such problems can usually be resolved on commonsense grounds.)

In this case the two observed percentages are close to the theoretical values of 58% and 90% for the Normal Distribution, and hence there is useful correspondence. Even with such a small number of readings, the data can therefore be described as being approximately Normal, with a mean of 4 and a mean deviation of 1.5. From this summary the full data could be reconstructed to within a close degree of approximation. About 60% of the readings must take the values 3, 4 or 5, and so on.

Table 12.2 gives another set of readings. The mean is 180/30 = 6 and the mean deviation is 48/30 = 1.6.

These data are also approximately Normal because 16/30 (53%) of the readings lie within ±1 mean deviation and 28/30 (93%) lie within ±2 mean deviations of the mean. These are again within a few percentage points of the theoretical figures for a Normal Distribution. The degree of approximation involved is usually insignificant considering the typical practical use of the theoretical distribution in comparing two such sets of readings.

TABLE 12.2 A Set of 30 Readings

	Value									
	1	2	3	4	5	6	7	8	9	10
Frequency	0	0	4	3	6	6	4	3	2	2

Table 12.3 expresses the frequencies in the two sets as percentages, to eliminate the differences in the total numbers of readings. Even so, the direct comparison is still not easy to do. Just how do the two sets of readings differ and in what ways are they similar?

TABLE 12.3 Comparison of the Two Sets of Data from Tables 12.1 and 12.2
(Frequencies expressed in percentage form)

	Value									
	1	2	3	4	5	6	7	8	9	10
8 readings from T. 12.1 %	13	0	37	25	0	13	0	13	0	0
30 readings from T. 12.2 %	0	0	13	10	20	20	13	10	7	7

The fact that both sets of readings are approximately Normal makes the comparison much easier, as shown in Table 12.4. The degree of approximation in the fit of the Normal Distribution does not affect the conclusion that the readings in the second set are generally about 2 units higher than those in the first.

TABLE 12.4 Summary of the Two Sets of Data in Table 12.3

	Distribution (Approximate)	Mean	Mean Deviation
8 readings from T. 12.1	Normal	4	1.5
30 readings from T. 12.2	Normal	6	1.6

TABLE 12.5 Summary of the Two Sets of Data and Comparison with the Normal Distribution

	% of readings within	
	± One m.d.	± Two m.d.
The 8 readings	62%	87%
The 30 readings	53%	93%
The Normal Distribution	58%	90%

Table 12.5 illustrates the similarity of both observed distributions to the theoretical Normal Distribution, after allowing for the difference in means. Saying that both sets of readings are Normal with mean deviations of about $1\frac{1}{2}$, but that their means differ at 4 and 6, clearly communicates the essence of the data.

12.2 The Poisson Distribution

We now turn to data that count the number of times some event occurs in a given interval of time or space. The simplest theoretical distribution for this type of counting data is called the Poisson, named after an 18th Century French mathematician.

Poisson Distributions can arise when the events occur more or less independently of each other with a constant average frequency of occurrence. A classical illustration concerns the number of Prussian soldiers kicked to death by a horse (von Bortkewitsch, 1898). The chance of such a death happening is rare, and on the whole one case will not be connected with another. In records for ten army corps over twenty years there was a total of 122 such deaths, on average .61 per corps per year.

Out of the total of 200 possible annual readings there were 109 cases where *no* deaths occurred in a corps, 65 cases of one death, 22 cases of 2 deaths, etc. The detailed distribution of deaths can be closely fitted by a Poisson Distribution with a mean of .61, as Table 12.6 shows. We therefore have a succinct summary of the data.

TABLE 12.6 **The Number of Soldiers Killed Annually by Horse-Kicks in a Prussian Cavalry Corps, and the Theoretical Poisson Distribution**

(10 Corps over 20 years)

	Number of Deaths per Year						No. of "Corps-Years"
	0	1	2	3	4	5+	
Observed	109	65	22	3	1	-	200
Poisson	109	66	20	4	1	.1	200

Another example of a Poisson Distribution, reported more recently, is the incidence of "major strikes" per week in the United Kingdom from 1948 to 1959 (Kendall, 1961). In that 626-week period there were 563 strikes, or about .90 per week. Table 12.7 shows there were 252 weeks with no strike, 229 weeks with one strike, 109 weeks with two, 28 weeks with three, 8 weeks with four, and none with more. These numbers approximate a theoretical Poisson Distribution with a mean of .90.

Poisson Distributions differ from each other only in one parameter, their mean values, usually denoted by *m*. A particularly simple property of the

TABLE 12.7 The Fit of the Poisson Distribution for the Occurrence of 'Major Strikes' in the U.K. per Week in 1948-1959

	Number of strikes per week						Total number of weeks
	0	1	2	3	4	5+	
Observed	252	229	109	28	8	0	626
Poisson	254	229	103	31	8	1	626

Poisson is that its variance equals its mean, i.e.

$$\text{mean} = m = \text{variance}.$$

A simple way to estimate the theoretical Poisson frequencies is to equate the number of readings taking the value r to (m/r) times the number taking the value $(r - 1)$. For example, if the number of weeks with zero strikes is 252 out of a total of 626, the theoretical proportion of weeks with one strike is $252(.90/1) = 227$. The number of weeks with 2 strikes is $227(.90/2) = 102$, and so on. These estimates are almost identical to the theoretical frequencies in Table 12.7. (The figures in the table were calculated by a fractionally more accurate method outlined in Exercise 12F. This shorter formula works well if the number of zeros is relatively large, as here.)

The Poisson Distribution can take different shapes, depending on the value of the mean. It can be reverse-J-shaped, like our two examples, or humpbacked if the mean is high. This makes it impossible to give any routine description in terms of the standard deviation, as one can do with the Normal Distribution. But if the observed mean and variance of a set of data are approximately equal, then it is worth calculating the theoretical Poisson frequencies to see if they fit. For example, with the data on strikes, the observed mean is .90 and the observed variance is .86. These are close enough to indicate that a Poisson Distribution with a mean $m = .90$ should give a good fit.

The Normal Approximation to Poisson. Poisson Distributions with relatively high means take a humpbacked and increasingly symmetrical shape, tending to resemble a *Normal* Distribution. This makes the data even easier to handle.

TABLE 12.8 The Distribution of the Number of Yeast Cells in Each of 400 Squares, and the Poisson Distribution

	Number of Cells														No. of Squares
	0	1	2	3	4	5	6	7	8	9	10	11	12	13+	
Observed	-	20	43	53	86	70	54	37	18	10	5	2	2	-	400
Poisson	4	17	41	63	74	70	54	36	21	11	5	2	1	1	400

Table 12.8 gives an example with a mean of about 5. It sets out the number of yeast cells counted in each of 400 squares into which a square millimetre was divided (Fisher, 1950). The observed mean is 4.7 and the variance is 4.5, so the observed distribution should closely fit a Poisson, which it does.

The observed distribution has a mode at 4 which falls close to the mean at 4.7 as does the median (with 202 readings below 4.7 and 198 above it). These characteristics indicate that a Normal Distribution should also fit the data. To check this quickly, we can calculate the theoretical standard deviation by taking the square root of the mean, which equals the variance in a Poisson. This is 2.2. (The standard deviation of the data could have been calculated directly, giving the value 2.1, but theory is faster!) According to the main characteristics of a Normal Distribution we should have

		Normal	Observed
±1 s.d., i.e. between	2.5 to 6.9:	68%	263/400 = 66%
±2 s.d., i.e. between	0.3 to 9.1:	95%	391/400 = 98%
±3 s.d., i.e. between	−1.9 to 11.3:	99.7%	398/400 = 99.5%

These observed data can therefore be adequately modelled by either the Poisson or the Normal Distribution, even though the Poisson is discrete (taking whole numbers only) and the Normal is continuous (taking fractional numbers also). This is typical of the way different kinds of approximate solutions can be used.

A variety of natural phenomena follow the Poisson Distribution to a fair degree of approximation. Apart from our examples, the Poisson describes certain cases in cosmic radiation in physics, breakdowns in telephone equipment, and certain other forms of "accident". However, in many situations the simple conditions for the Poisson Distribution, that successive events occur independently of each other but at a constant average rate, do not quite hold. In fact, the very simplicity of the Poisson Distribution with its single parameter makes it rather inflexible. The main importance of the Poisson is as a constituent of more complex models.

12.3 The Negative Binomial Distribution

Not all cases of accidents follow a Poisson Distribution because every person does not have the same chance of an accident. For example, in the case of the Prussian soldiers, if some belonged to cavalry corps and others to infantry corps, they would have different exposure to horses and *systematically* different degrees of risk. The incidence of deaths then might *not* follow a Poisson Distribution because the corps might have different mean number of deaths over the years, and the Poisson requires a constant mean rate.

Again, some people might be accident-prone, *inherently* pre-conditioned to have more accidents than others, regardless of how many accidents they had previously. Alternatively, having one accident might make a person more likely to have another, which is called a "contagious" or "learning" type of phenomenon, because something like it also occurs in the spread of certain diseases and in psychological learning situations. In certain circumstances both these situations could lead to a frequency distribution called the Negative Binomial Distribution, discovered by de Montmort in about 1700. (The name Negative Binomial stems from a mathematical technicality which usually has no direct practical implications.)

One example occurs with consumer purchases of frequently bought branded goods. Table 12.9 shows the number of times a sample of housewives bought Corn Flakes in a given six-months period (Charlton *et al.*, 1972). Thirty-nine per cent of the sample did not buy Corn Flakes at all, 14% bought once, 10% bought twice, and so on.

TABLE 12.9 The Numbers of Purchases of Corn Flakes Made in 24 Weeks
(Number of Households out of a Sample of 491)

Number of Purchases	0	1	2	3	4	5	6	7	8	9	10	11	12	13
Households buying	193	71	49	28	20	22	14	16	11	9	12	7	7	4
Number of Purchases	14	15	16	17	18	19	20	21	22	23	24	29	36	37
Households buying	1	3	5	4	2	2	2	2	2	1	2	1	1	-

The basic statistics (to 3 digits for working purposes) are that the means = 3.43 and the variance = 26.9. Since the variance is so much bigger than the mean, the distribution certainly is not Poisson. However, the Negative Binomial Distribution (NBD) gives a good fit, as shown in Table 12.9a. This occurs very generally for such data (e.g. Ehrenberg, 1972).

TABLE 12.9a The Fit of the Negative Binomial Distribution

24 Weeks		Number of Purchases										
		0	1	2	3	4	5	6	7	8	9	10+
Cornflakes buyers	Observed %	39	14	10	6	4	4	3	3	2	2	11
	NBD %	35	16	10	7	6	5	3	3	2	2	11

The Negative Binomial Distribution is either reverse-J-shaped, as here, with most of the readings at 0, some at 1, less at 2, etc., or, if the mean is large, humpbacked with a long positive tail to the right. If the mean of the NBD is denoted by m, the variance is defined as

$$\text{variance} = m(1 + m/k),$$

where the quantity k is always positive but generally not integral. Thus the scatter of the readings in an NBD is always larger than the mean.

The value of the parameter k can be calculated by equating the variance of the observed distribution to the expression $m(1 + m/k)$. For the data in Table 12.9 this gives $k = .50$. Given the value of k, one can estimate the theoretical frequencies of the NBD with an easy-to-use formula. It relates the proportion p_r of r occurrences to the proportion p_{r-1} of $(r - 1)$ occurrences:

$$p_r = \left(\frac{m}{m + k}\right)\left(1 - \frac{1 - k}{r}\right)p_{r-1}.$$

Using the data in Table 12.9 as an example, we can estimate p_1, the frequency of *single* purchases of Corn Flakes, from p_0, the proportion of non-buyers, by computing

$$p_1 = \left(\frac{3.43}{3.43 + .50}\right)\left(1 - \frac{1 - .50}{1}\right).39 = .17 \text{ or } 17\%,$$

and so on for p_2, p_3, etc. (This works well only if the observed p_0 is reasonably large.) The results are fairly similar to those in Table 12.9a, which were calculated by a somewhat more complex method (discussed in Exercise 12I).

The Negative Binomial Distribution has had a fairly wide range of practical applications, e.g. in analysing bus drivers' accidents and in the ecological distribution of the number of species of a particular type of plant. In the case of consumer purchasing, the underlying explanation seems to be proneness rather than a learning type of phenomenon. That is, some households are consistently more likely than others to buy Corn Flakes, instead of each purchase of Corn Flakes increasing the tendency to buy it again. Underlying models will be discussed further in Chapter 13. Here we are concerned with using such mathematical distributions simply to describe and summarise observed frequency distributions.

12.4 The Binomial Distribution

Theoretically, the number of occurrences or events being counted in the Poisson and Negative Binomial Distributions can extend indefinitely to plus infinity, but in most cases the theoretical frequency of high numbers is very small. Thus observations extending only to some finite upper limit can also be accommodated by the distributions.

However, there are situations where the maximum number of occurrences has a clear-cut and attainable upper limit. For example, the number of patients who improve in a clinical trial of a drug cannot exceed the total number of patients in the trial.

Such data can often be represented by the Positive Binomial Distribution, usually referred to simply as the Binomial Distribution. (Sometimes this is also called the Bernouilli Distribution, after the 17th Century Swiss mathematician Jacob Bernouilli.) The phrase "bi-nomen" means "two names". In a binomial classification the observed items are sorted into two categories, such as Success or Failure, Boy or Girl, Yes or No, etc.

A more general case is the Multinomial Distribution, where the classification consists of more than two classes: e.g. single, married, widowed or divorced; or Yes, No, or Don't Know responses to a question; etc. The statistical theory of Multinomial Distributions is similar to the Binomial. The basic mathematics used is permutations and combinations. (The resemblance between the Positive and the *Negative* Binomial Distributions lies only in the form of certain basic mathematical formulae.)

An example of a Binomial Distribution is the incidence of boys and girls in families of a given size. If the sex of successive children in a family is determined independently for each child and there is no difference in the tendency towards boys or girls among different families, then the proportion of families with n children who have r boys should be given by the Binomial formula

$$\frac{n!}{(n-r)!r!}p^r q^{n-r}.$$

Here p is the overall proportion of boys and $q = 1(-p)$ is the overall proportion of girls. (The "factorial" symbol ! in $n!$ stands for the product $n(n-1)(n-2)\ldots 3 \times 2 \times 1$. Similarly $(n-r)!$ stands for $(n-r)(n-r-1)(n-r-2)\ldots 3 \times 2 \times 1$. The expression 0!, e.g. the value of $(n-r)!$ when $r = n$ boys, equals 1.)

Table 12.10 gives data on the observed percentages of some 3-children families who had either 3, 2, 1, or no boys (Geissler, 1889). The overall proportion of boys is $p = .51$, and n equals 3. The theoretical values calculated by Geissler from the above formula fit well (based on his value of $p = .5147676$, and multiplying by 100 to give percentages).

Table 12.11 illustrates the distributions for $n = 1$, 3, and 5. With one-child families, 51% are boys, but with 5-child families, only 3.7% are all boys.

TABLE 12.10 The Binomial Distribution of the Number of Boys in 3-children Families

	Number of Boys in Family				Average Proportion of Boys
	3	2	1	0	
Observed %	14	38	36	12	.51
Binomial %	14	39	36	11	.51

While the three observed distributions are very different, they can all be closely fitted by theoretical Binomial Distributions with a proportion $p = .51$ (or Geissler's value .5147676). Only the value of n varies. The data can therefore be succinctly summarised as being Binomial with an average of 51% boys.

TABLE 12.11 The Distribution of Boys in 1-, 3- and 5-child Families and the Theoretical Binomial Frequencies
(Expressed as Percentages)

Number of Children in Family	Number of Boys in Family						Average Percentage of Boys
One			One	None			
Observed %			51.5	48.5			51
Binomial %			51.5	48.5			51
Three		Three	Two	One	None		
Observed %		13.9	38.3	36.2	11.6		51
Binomial %		13.6	38.6	36.4	11.4		51
Five	Five	Four	Three	Two	One	None	
Observed %	3.7	16.7	32.2	30.5	14.0	2.9	51
Binomial %	3.6	17.0	32.1	30.3	14.3	2.7	51

The Binomial Distribution depends on the two parameters p and n. These differ in their status. The size of the individual groups being examined, n, is generally determined before the observations are collected, whereas p is determined by the data itself. If the data's two categories (boy:girl, success:failure, etc.) are scored 1 and 0, the mean and variance of the distribution are

$$\text{mean} = np,$$

$$\text{variance} = npq.$$

Since q is a proportion less than 1, the variance of a Binomial is always less than its mean.

When the parameter p is at or near .50, the distribution will be approximately symmetrical. But when p markedly differs from .50, the Binomial Distribution is quite skew. For example, with an industrial batch production process, each batch might consist of 20 items, $n = 20$, and an average of 10% of the items might be faulty, so $p = .1$. If one fault occurs independently of another and with a consistent probability of .1, the theoretical Binomial Distribution shown in Table 12.12 will result.

TABLE 12.12 The Binomial Distribution for Faults in Batches of n = 20 Items, with on average p = .1 Faults

	Number of Faults per Batch								
	0	1	2	3	4	5	6	7	8-20
% of all batches	12	27	29	19	9	3	.9	.2	.05

Twelve batches per 100 will be free of faults, 27 will contain 1 fault, and so on, with virtually no batches (5 in 10,000) having 8 or more faults. This kind of result is used in industrial quality-control inspection schemes.

The Poisson and Normal Approximations to the Binomial

In certain cases the Binomial Distribution approximates the Normal and Poisson Distributions. When the Binomial parameter p is very small, for rare events, then $q = (1 - p)$ will be nearly 1. Thus the variance of the Binomial Distribution, npq, will be virtually equal to its mean, np. This is characteristic of the *Poisson* Distribution and in such cases the Binomial closely approximates the Poisson.

When the Binomial parameter n is large, or even for small n when the proportion p is near the .50 mark, the distribution is approximately symmetrical. Then it can usually be well represented by a *Normal* Distribution. (It was in this context that de Moivre first discovered the Normal Distribution, one of the fundamental steps in theoretical statistics.) For example, for families with $n = 5$ children in Table 12.11, the mean is about 2.6 and the theoretical standard deviation is $\sqrt{(5 \times .51 \times .49)}$ = about 1.1. About 63% of the readings lie within one standard deviation of the mean and 93% lie within two standard deviations. This compares with the Normal values of 68% and 95%. In such cases, one can use the simpler calculations of the Normal Distribution to describe the data.

12.5 The Beta Binomial

Close examination of Table 12.11 shows that in the 3- and 5-child families, the observed numbers of *all* boys or *all* girls are slightly higher than the theoretical values. In his classic book *Statistical Methods* (1950), Fisher quoted the data for 8-child families (shown here in Table 12.13) and noted a similar excess. The original data (Geissler, 1889) show that this excess in fact generalises for all family sizes, i.e. the discrepancies are systematic. Thus the fit of the Binomial Distribution to these data is close, but not perfect.

TABLE 12.13 The Distribution of Boys in 8-child Families

	Number of Boys in Family								
	8	7	6	5	4	3	2	1	0
Observed %	.6	3.9	12.4	22.2	27.9	19.8	9.9	2.7	.4
Binomial %	.5	3.7	12.2	23.1	27.2	20.5	9.7	2.6	.3

The Binomial Distribution should fit if the sex of each baby is determined independently and if the average incidence of boys shows no systematic trends (e.g. between different types of families, first-born and later children, winter and summer babies, etc.). But it is not self-evident that the sex of babies behaves like this. For example, the sex of a first child might have affected the chemical balance of the mother's hormones, in turn affecting the conception or survival of a subsequent baby of the opposite sex. Or there might have been social or economic pressures to have at least one boy, leading to a higher proportion of boys in *smaller* families. The evidence does not suggest that either of these possibilities occurred in this data, but the excess of all-boy or all-girl families shows that *some* special factor was at work.

One possible factor is the incidence of identical twins, but Fisher noted these did not occur frequently enough to account for all the observed discrepancies. Another possibility is that the underlying incidence of boys might vary among families. It could be .60 for one, .52 for another, .43 for a third, and so on, instead of .51 for all.

If this were true, the sex of babies could still be determined independently and at a constant proportion *within each family*, so that the incidence of boys and girls would still follow a Binomial Distribution for each family. But the required model would then consist of a "mixture" of Binomial Distributions, each with a different proportion p. These different values of p would then follow a frequency distribution across different families. If the distribution of p-values were of the so-called "Beta" type, the resulting distribution of the number of boys would be of the *Beta-Binomial* form (also known as the Negative Hypergeometric Distribution).

Table 12.13a shows the fit of the Beta-Binomial to the data for 8-child families.

TABLE 12.13a The Fit of the Beta-Binomial for the 8-child Families

	Number of Boys in Family								
	8	7	6	5	4	3	2	1	0
Observed %	.6	3.9	12.4	22.2	27.9	19.8	9.9	2.7	.4
Beta-Binomial %	.6	4.0	12.5	22.9	26.7	20.4	9.9	2.8	.4

To one decimal place there is no excess of all-boy or all-girl families, but working to two decimal places there is still a small systematic excess of about .04, that occurs consistently for all family sizes from 2 to 12. This must be due to an additional factor (perhaps *now* identical twins). The Beta-Binomial therefore provides an improved, but still not perfect model for the data.

The Beta-Binomial Distribution has three parameters (see Exercise 12J for technical details) which makes it more flexible than the simple Binomial. Depending on the values of the parameters, the Beta-Binomial can take a variety of different shapes. For example, it can be reverse-J-shaped with a mode at 0, it can be humpbacked, or it can be U-shaped with two modes at 0 and *n*. Such a variety of shapes can occur in the same empirical context, e.g. for the number of different episodes of a television programme which viewers see (Goodhardt *et al.*, 1975).

Table 12.14 illustrates a U-shaped distribution. The readings show how many different issues of the weekly magazine *Woman* were bought in 12 weeks.

TABLE 12.14 A U-Shaped Distribution and the Fit of the Beta-Binomial

	Number of Issues Bought												
	0	1	2	3	4	5	6	7	8	9	10	11	12
Observed %	85	4.2	1.7	.8	.4	.3	.2	.4	.3	.5	1.1	1.2	3.8
Beta-Binomial %	87	2.0	1.1	.8	.7	.6	.5	.6	.6	.6	.7	1.1	3.8

The distribution is rather a *skew* U: 85% of the sample bought none, 4% bought 1 issue, 1.7% bought 2, and less than 1% bought from 3 to 9 issues. Then the frequencies increase again, with 1.1%, 1.2%, and 3.8% buying 10, 11 or all 12 issues. Such a pattern is very common in readership data. People either buy all or almost all of the issues, or else they buy none or almost none; very few readers buy about half. None of the other theoretical distributions discussed in this chapter can describe such U-shaped data. While the Beta-Binomial gives a close fit there are some significant discrepancies. For example, more people bought 1 or 2 copies than the model predicts. Thus, the data are more complex than even a relatively sophisticated distribution like the Beta-Binomial can fully describe. There may be another, yet more complicated, mathematical function which could fit such data better.

But there is also another approach to such data. Perhaps some of the complicating factors could be handled by more *direct* analysis, instead of by fitting more complicated mathematical models. For example, there are two kinds of purchasers, those buying regularly by subscription and those buying occasional copies at newsagents or newstands; such groups could be analysed separately. Again, whilst the Beta-Binomial could be expected to hold if all twelve issues sold the same number of copies, there may have been one or two "bumper issues" in the 12-week period, explaining the "excess" of purchasers of one or two copies.

12.6 Other Distributions

There are many other mathematical distributions that can be fitted to empirical data, but few have been applied as widely as the examples already described.

One distribution of particular theoretical importance is the Gamma-Distribution. This is related to the Beta-Distribution (which is the ratio of two Gamma-Distributions), and has various applications in the theory of statistical sampling and for "stochastic" models of data. This will be discussed in later chapters.

Sometimes the scale of measurement of an observed variable can be transformed so that apparently complex data can be modelled by one of the simpler distributions. The main example is the Lognormal Distribution (e.g. Aitcheson and Brown, 1957). This arises with certain kinds of skew data with only a few high values (like income distributions), where logarithms of the readings may follow a Normal Distribution.

12.7 Summary

Statistical frequency distributions are theoretical formulae that describe observed distributions of readings. They are particularly useful when the same form of distribution occurs for different sets of data.

The most widely used distribution is the Normal, which describes symmetrical, humpbacked distributions. It is exceptionally simple because it always takes the same shape, e.g. 68% of the readings lie within ± 1 standard deviation of the mean.

The Poisson, Binomial, and Negative Binomial Distributions are useful in cases where the occurrence of an event is counted. They refer to data with different degrees of scatter. The variance is equal to the mean for the Poisson Distribution, but it is always smaller than the mean for the Binomial and larger than the mean for the Negative Binomial.

CHAPTER 12 EXERCISES

(*Exercises 12F onwards deal with relatively technical matters in fitting frequency distributions.*)

Exercise 12A. The Use of a Distribution

What is the point of fitting complicated formulae like the Negative Binomial Distribution?

Discussion.

The pay-off comes when the same formula describes different sets of data. To illustrate, Table 12.15 gives the distributions of purchases of Corn Flakes and Puffed Wheat over 12-week and 24-week periods (Charlton *et al.*, 1972).

TABLE 12.15 Frequency Distributions of Purchase of Corn Flakes and Puffed Wheat in Different Length Time-Periods

(% of households buying)

	Number of Purchases											
	0	1	2	3	4	5	6	7	8	9	10+	Total
Corn Flakes												
% buying in 24 weeks	39	14	10	6	4	4	2	3	2	2	14	100
" " " 12 "	51	15	8	6	5	5	2	3	1	2	2	100
Puffed Wheat												
% buying in 24 weeks	84	9.6	2.4	1.0	.6	.6	.4	.2	0	.2	1.0	100
" " " 12 "	90	6.3	1.4	.6	0	.4	.2	.2	0	.4	.2	100

The four distributions differ markedly, yet they can all be fitted by an NBD (as was illustrated in Table 12.9a for the 24-week Corn Flakes data). Only the means and variances differ, as Table 12.16 shows. Since the occurrence of the NBD for purchasing data is a very general finding (e.g. Ehrenberg, 1959, 1972), these two parameters are all one needs to summarise and distinguish the different sets of data.

TABLE 12.16 The Means and Variances of the Four Distributions

	Distribution (approx.)	Mean	Variance
Corn Flakes			
24 weeks	NBD	3.4	27
12 weeks	NBD	1.8	8
Puffed Wheat			
24 weeks	NBD	.4	2.7
12 weeks	NBD	.2	.9

Exercise 12B. The Parameters of a Distribution

Although the means of the NBD distributions in Table 12.16 are not "typical" values, they do have a physical meaning. They represent the relative sales levels or market-shares of the brands. But the *variances* of these skew distributions have no descriptive meaning. Are there other characteristic values which would be more useful in describing the data?

Discussion.

The parameters of a distribution are values which serve to identify a particular distribution. Different distributions of the same type can thus be distinguished by the numerical values of these parameters, e.g. two Normal Distributions can have different means and/or variances. (Different *types* of distributions, e.g. a J-shaped Poisson and a symmetrical Normal, are more difficult to compare. It is like trying to compare a straight line with a curve.)

The NBD has two parameters, in the sense that two numbers will differentiate one NBD from another. But one does not have to use the mean and variance: various other aspects of the data can also be used. Table 12.17 sets out three alternatives for the breakfast cereal data.

The first is the percentage of the sample buying at all (i.e. the penetration "b" from Chapters 9 and 10, or 100 minus the percentage of zeros) and the second is the average number of purchases per buyer, or w. These parameters give one a good descriptive "feel" of the data—how many people buy the item at all in the time-period, and how often on average they do so.

TABLE 12.17 Other Parameters of the Four Distributions

	% of sample buying at all	Av. number of purchases per buyer	k
Corn Flakes			
24 weeks	61	5.6	.50
12 weeks	49	3.7	.55
Puffed Wheat			
24 weeks	16	2.6	.08
12 weeks	10	2.2	.06

Furthermore, these parameters follow useful relationships. Multiplying b by w for a brand reproduces its overall mean level of purchases, as given in Table 12.16 (on a per 100 household basis). Again, the values of b and w for different brands are linked, because $w(1 - b)$, or $w(100 - b)$ for percentages, is approximately constant as we saw in Chapter 10. Finally, the results in time periods of different lengths are linked, as will be shown in Chapter 13.

The third parameter, k, is a more abstract quantity that arises from the mathematical formula for the frequency p_r of the NBD. But it too has a useful descriptive property. It appears that for a given brand the value of k hardly varies in different length time-periods. It is about .5 for Corn Flakes and about .07 for Puffed Wheat (the variation in the values is small compared with the other differences in the data). This is a general property of k which makes it a very simple parameter to use, generally only *one* k-value has to be specified for each brand, irrespective of the length of time-period analysed.

Thus different parameters can have different descriptive advantages. Each type of distribution has a minimum number of parameters that need to be determined in order to "fix" it. But one can use more than this minimum number for different purposes and these numerical values will then be interrelated.

Most of the common distributions have a minimum of two parameters. The Poisson is unusual because it is fully specified by a single characteristic value. This could be the mean m, or the variance (which is equal to m) or the proportion of zeros (e^{-m}). Clearly, all these different values are mathematically equivalent to each other.

The fact that the Poisson Distribution has only one parameter makes it particularly simple, but also rather inflexible. It can only fit if the observed variance is (approximately) equal to the mean. In the NBD, the variance is always *greater* than the mean, and the parameter k determines this difference since the variance is equal to $m(1 + m/k)$.

In the ordinary (or positive) Binomial Distribution, the variance, npq, is always *less* than its mean, np, by a factor q, which depends directly on the mean since $q = 1 - p$. Thus the Binomial Distribution is also relatively inflexible. There are not many practical situations, outside artificial games of chance, where it gives a good fit. The Binomial has its widest applicability as an ingredient of more general distributions, such as the Beta-Binomial in Section 12.5.

Exercise 12C. Deviations from a Model

Discuss the use of frequency distributions in summarising the deviations of observed data from a theoretical model.

Discussion.

Summarising the irregular deviations from a model is one of the most common uses of statistical methods. Examples occurred with the quarterly readings in the four areas in Chapters 1 and 2, and with the deviations of the age-group means from the relationships like $\log w = .02h + .76$ in Part II.

Such deviations can commonly be summarised by the Normal Distribution. The deviations are often due to a large variety of independent factors or "errors", which is the situation in which the Normal Distribution tends to arise.

As an example, the 30 quarterly deviations for 1969 and 1970 in Tables 2.2 and 2.2a (excluding the two exceptional QIII values) were

$-7, -6, -5, -4$, 5 – 3's, 2 – 2's, 4 – 1's, 2 0's, 3 1's, 2 2's, 3 3's, 1 4, 2 5's, 2 6's.

They have a mean of 0 and a mean deviation of 3. If they follow a Normal Distribution, we would expect about 95% of the readings to lie within ± 6 and 58% to lie within ± 3. Because of rounding, the deviations are grouped at integer values, and there is also a particular clustering at -3 and at $+3$. This has to be allowed for by "smoothing" the data. Thus the percentage of observed values lying between ± 6 are 97% and 88% (depending on whether the 6's are included or excluded), an average of 92% which is close to 95%. Similarly, the observed percentages between ± 3 are 70% and 43%, averaging at 56%, which is close to the theoretical 58%.

Given that direct empirical analysis has shown the deviation to be apparently *irregular*, a description of the deviations as approximately following a Normal Distribution with mean 0 and mean deviation 3 would therefore allow one to reconstruct the data rather closely.

Exercise 12D. Exceptional Deviations

How can exceptionally large deviations be dealt with?

Discussion.

If the deviations from a model follow a generalisable pattern such as a Normal Distribution, this can be used to judge values which appear exceptional.

Thus in Chapter 1, two readings (for QIII in the East and West) gave deviations of 25 and 27 which were about 8 and 9 times the mean deviation of 3 of the remaining 30 readings. Since with a Normal Distribution only 1 in 1,000 readings lie more than even just *four* mean deviations from the mean, the data can no longer be described as being approximately Normal.

This does not prove that the two readings are necessarily wrong. But they *are* exceptional. It is easier to describe the data by saying that there are 30 readings which are approximately Normal (as usually happens) with a mean zero and mean deviation 3, plus the two large exceptions.

Only *exceptionally* large deviations need to be reported separately. A "border-line" deviation, say 3 or 4 times the mean deviation, would not markedly affect the Normal approximation and therefore need not be excluded. If there are more than "a few" exceptions they have themselves to be summarised statistically, unless they form generalisable patterns (e.g. that QIII in the East and West is *always* about 25 units high, every year).

Exercise 12E. Fitting the Positive Binomial Distribution

Illustrate the numerical calculation for the Binomial Distribution.

Discussion.

For a Binomial Distribution with parameters n and p, the proportion of readings taking the value r is

$$\frac{n!}{(n-r)!r!}p^r q^{n-r}.$$

There is no short-cut to working out these values, except that one can usually use the Normal approximation for large n and the Poisson approximation for very small p.

In the past, extensive tables were published giving the Binomial proportions for different values of n and p, but now the proportions are usually generated as needed by simple computer programmes. However, working out small examples by hand helps to provide better understanding.

As such an example, we shall calculate the theoretical proportions for 3, 2, 1, and 0 boys in three-children families for the observed data in Table 12.4. Because the Binomial formula involves many multiplications of p by q, rounding-off errors tend to build up. It is therefore better to work with p-values to 3 or 4 digits in the detailed calculations.

Taking Geissler's value of $p = .5148$ (rounded from .5147676 to 4 digits), and remembering that both 0! and any number raised to the power of 0 are equal to 1, we have for the theoretical proportion of families with

3 boys

$$p_3 = \frac{3!}{0!3!}(.5148)^3(.4852)^0$$

$$= \frac{3 \times 2 \times 1}{1(3 \times 2 \times 1)} \times .1364 \times 1$$

$$= .136, \quad \text{or } 13.6\%.$$

The proportion of families having 2 boys is

$$p_2 = \frac{3!}{1!2!}(.5148)^2(.4852)^1$$

$$= \frac{3 \times 2 \times 1}{1(2 \times 1)} \times .2650 \times .4852$$

$$= 386, \quad \text{or } 38.6\%.$$

Similarly, $p_1 = 36.4\%$ and $p_0 = 11.4\%$. A simple check of the calculations is that the sum of the proportions, $p_3 + p_2 + p_1 + p_0$, should equal 1.

Exercises 12F onwards deal with relatively technical matters

Exercise 12F. The Mathematics of the Poisson Distribution
What is the mathematical formulation of the Poisson Distribution?

Discussion.
In Section 12.2 we noted that in a Poisson Distribution with mean m, the proportion of readings taking the value r is m/r times the proportion taking the value $(r - 1)$. We can write this as the "recurrence formula"

$$p_r = \frac{m}{r} p_{r-1}.$$

It follows that if the proportion of zeros $p_0 = \mathrm{k}$, some empirical constant, then the various proportions p_0, p_1, p_2, etc. must take the form

$$p_0 = \mathrm{k},$$
$$p_1 = m\mathrm{k},$$
$$p_2 = m^2\mathrm{k}/2,$$
$$p_3 = m^3\mathrm{k}/3 \cdot 2 \cdot 1 = m^3\mathrm{k}/3!$$

$$p_r = m^r\mathrm{k}/r\,!, \quad \text{etc.}$$

The sum of all the proportions $p_0, p_1, \ldots$ must be unity, so that $\Sigma\,(\mathrm{k} + m\mathrm{k} + m^2\mathrm{k}/2 + \cdots + m^r\mathrm{k}/r! + \cdots) = \mathrm{k}\,\Sigma\,(1 + m + m^2/2 + m^3/3! + \cdots + m^r/r! + \cdots) = 1$, The series in brackets is well-known in elementary algebra as the exponential series. The sum equals the expression e^m, where e is an absolute constant, approximately 2.718, which arises in certain kinds of mathematics (e.g. in connection with logarithms).

It follows that $ke^m = 1$, so that k must equal e^{-m}. The Poisson frequencies are therefore

$$p_0 = e^{-m},$$
$$p_1 = m\,e^{-m},$$
$$p_2 = m^2\,e^{-m}/2,$$
$$p_3 = m^3\,e^{-m}/3!, \quad \text{etc.,}$$

with the rth term being

$$p_r = m^r\,e^{-m}/r!.$$

The mean value of the Poisson Distribution, i.e. the readings 0, 1, 2, etc. multiplied by the proportion of times they occur, is therefore given by

$$(0 \times e^{-m} + 1 \times m\,e^{-m} + 2 \times m^2\,e^{-m}/2 + 3 \times m^3\,e^{-m}/3! + \cdots$$
$$+ r \times m^r\,e^{-m}(r! + \cdots)$$
$$= (0 + m\,e^{-m} + m^2\,e^{-m} + m^3\,e^{-m}/2! + \cdots + m^r\,e^{-m}/(r-1)! + \cdots)$$
$$= m\,e^{-m}(1 + m + m^2/2! + \cdots + m^{r-1}/(r-1)! + \cdots),$$

taking $m\,e^{-m}$ outside the brackets. The new terms inside the brackets are again an exponential series and hence add to 1. Therefore the mean of a Poisson is $m \times 1$, i.e.

$$m.$$

To determine the variance of the Poisson Distribution, we need to calculate $\Sigma\,(r - m)^2 p_r$ for all values of r from 0 upwards. By arguing along the lines of Exercise 11E, we can show this equals $\Sigma\,(r^2 p_r - m^2)$. The average of $r^2 p_r$ can be seen to be $m^2 + m$, if we write it as $\{r(r-1) + r\}p_r$ and work along the same lines as we did for the mean in the previous paragraph. Thus the variance of the Poisson is $m^2 + m - m^2$, i.e.

$$m.$$

Exercise 12G. Calculating the Poisson Frequencies

What is the best way to calculate the numerical values of the Poisson frequencies, $p_r = m^r\,e^{-m}/r!$?

Discussion.

The only complex part of the Poisson formula is the exponential expression e^{-m}, where $e = 2.718$. This can be worked out using logarithms. For example, if $m = .61$, as in the case of the Prussian soldiers (Table 12.6), we look up the logarithm to base 10 of 2.718, which is .434 and multiply it by $-m = .61$, giving $-.265$. The antilogarithm of this number (written in the usual logarithm form as $\bar{1}.735$) is given in logarithmic tables as .543. Once we have the value of e^{-m}, the rest follows simply. We can note that $p_0 = e^{-m}$ and multiply by m/r for successive values of r. Thus $p_1 = .61 \times .543 = .331$, $p_2 = .61 \times .331/2 = .101$, etc.

One can shorten the calculations by using a table like Table 12.18, giving values of e^{-m} for selected values of m. Interpolation for intermediate values of m is easy because of the additive property of exponents.

TABLE 12.18 Values of e^{-m}

m	e^{-m}	m	e^{-m}	m	e^{-m}
.01	.990	.1	.905	1	.368
.02	.980	.2	.818	2	.135
.03	.970	.3	.741	3	.050
.04	.961	.4	.670	4	.018
.05	.951	.5	.607	5	.007
.06	.942	.6	.549	6	.002
.07	.932	.7	.497	7	.001
.08	.923	.8	.449	8	.000
.09	.914	.9	.407	9	.000

Thus for $m = .61$, we can write

$$\begin{aligned} e^{-0.61} &= e^{-0.6} \times {}^{-0.01} \\ &= .549 \times .990, \\ &= .544, \end{aligned}$$

which is the same as before (within rounding errors).

We used a slightly different method to fit the Poisson Distribution to the strike data in Section 12.2. There we simply started with the observed number of zeros and multiplied this by m to obtain an estimate of p_1, by $m/2$ to obtain an estimate of p_2, and so on. This method is easier because it avoids calculating e^{-m}, but it gives slightly different results.

A theoretical distribution rarely fits perfectly (especially with sample data). Thus the two methods of fitting will not give identical results because the observed and theoretical numbers of zeros will not be exactly equal. The quicker recurrence formula gives estimated frequencies p that do not quite add to 1.0. This method is simpler but not as accurate, and only works well if the proportion of zeros is high.

The shorter method can usually not be used for Poisson data with a mean above 1 (like that in Table 12.8) because the theoretical number of zeros is small and the *observed* number might even be nought. In such cases it is better first to calculate e^{-m} from the mean and $e^{-m}m^r/r$ for some value of r near the mean, and then to use the recurrence formula for both higher and lower values of r. This reduces the effects of rounding-off errors. When the mean is much larger than 1, the Normal approximation can be used, as with the yeast-cells data in Table 12.8.

Exercise 12H. The Exponential Distribution

What is the time interval between successive events in a Poisson Distribution?

Discussion.

The Poisson Distribution gives the frequencies with which different numbers of events will occur in time intervals of a given length. Since this number varies (sometimes 2 events a week, sometimes none, sometimes 1 or 3, etc.) the amount of time between *successive* events must also vary. This is an important concept in many practical applications, e.g. for queues (the "waiting-time" for patients at a hospital or for aircraft landing at airports), breakdowns in equipment (the "life" of electronic equipment), learning processes (the time taken to learn particular repetitive tasks), and reaction times in chemistry.

It can be shown mathematically that if the occurrences follow a Poisson Distribution with mean m, the time interval t from any given instant till the next event follows a distribution of the form

$$1 - e^{-t/m}.$$

This is the Exponential Distribution, which is so-called because the variable occurs as an *exponent* (here of the number e). The distribution of time intervals between successive events can then be deduced, the *average* of the distribution is $1/m$, and its variance $1/m^2$.

The particular characteristic of exponential functions is that they transform *additive* properties into multiplicative ones. Exponential distributions occur for example in biological growth situations ("exponential growth"), where in successive equal time intervals things often grow proportionately to their size.

Exercise 12I. The Negative Binomial Distribution

Why are there different ways of estimating the theoretical frequencies of the Negative Binomial Distribution?

Discussion.

In Section 12.3 we calculated the NBD parameter k by equating the variance of the observed distribution to the theoretical variance formula $m(1 + m/k)$, where m was the observed value of the mean. But the theoretical frequencies in Table 12.9 were calculated by using a different and somewhat more complex method of estimating k.

Different fitting methods will give the same results if the observed data follow the theoretical distribution *exactly*. But this rarely occurs. Most theoretical models are at best close approximations to the data, and with *sample* data additional fluctuations occur.

Fitting distributions by the mean and the variance is a well-established procedure in statistics, particularly for the Normal Distribution. It is known as the *method of moments* (the mean and variance being technically known as the first two "moments" of a frequency distribution). But with a highly skew distribution an occasional exceptional value can markedly influence the variance (see Exercise 11G). Consequently, use of the variance can lead to relatively unreliable estimates of other parameters, such as k in the NBD.

An alternative way of estimating k is by equating the observed proportion of zeros p_0 to the theoretical NBD value for the number of zeros, which is $(1 + m/k)^{-k}$. The resulting equation cannot be solved for k by direct algebra, but simple methods are described in the literature (e.g. Chatfield, 1969; Ehrenberg, 1972). This method is statistically more efficient for data with many zeros than the method of moments.

The general formula for the proportion p_r of values r in an NBD with mean m and parameter k is

$$p_r = \frac{(k + r - 1)!}{(r + 1)!(k - 1)!}\left(\frac{m + k}{k}\right)^{-k}\left(\frac{m}{m + k}\right)^2.$$

(Strictly speaking, factorial expressions like $(k + r + 1)!$ should be expressed as Gamma-functions, since k is usually non-integral and factorials are not defined for such values.) The parameter k is called the "exponent" because the above expression arises as the expansion of the second term in the binomial expression

$$\left(\frac{m + k}{k}\right)^{-k}\left(1 - \frac{m}{m + k}\right)^{-k},$$

where the exponent k has a negative sign. Hence the name of the distribution. (The expansion of a binomial expression will be discussed more fully for the positive Binomial in Exercise 13L.)

Exercise 12J. The Beta-Binomial Distribution

By reference to the statistical literature, set out the basic formulae of the Beta-Binomial Distribution discussed in Section 12.5.

Discussion.

The Beta-Binomial Distribution has three parameters. They are n, the fixed size of the phenomenon being examined (e.g. the number of children in a family in Table 12.13a, or the number of weeks in Table 12.14), and two quantities α and β, which depend on the observed data.

The formula for the Beta-Binomial proportion of observations where there are r occurrences out of n (e.g. r boys in a family of n children) is:

$$\frac{n!}{(n - r)!r!}\,\frac{(\alpha + r - 1)!(n + \beta - r - 1)!}{(n + \alpha + \beta - 1)!}\,\frac{(\alpha + \beta - 1)!}{(\alpha - 1)!(\beta - 1)!},$$

(where the factorial should strictly be written as Gamma-functions for non-integral α and β). The mean and variance are given by

$$\text{mean} = n\alpha/(\alpha + \beta)$$

$$\text{variance} = n\alpha\beta(n + \alpha + \beta)/(\alpha + \beta)^2(1 + \alpha + \beta).$$

The theoretical frequencies in Table 12.13a were found by equating these values to the observed mean and variance (using the method of moments), and solving for α and β with $n = 8$.

An alternative way of fitting the distribution is to equate the theoretical expressions for the mean $n\alpha/(\alpha + \beta)$ and the number of zeros, $\{(n + \beta - 1)! \cdot (\alpha + \beta - 1)!\}/\{(n + \alpha + \beta - 1)!(\beta - 1)!\}$, to the observed values and solve for α and β. If the incidence of zeros is large, as in Table 12.14, this method is statistically more efficient but cumbersome to deal with (Chatfield and Goodhardt, 1970).

Exercise 12K. The Normal Distribution

What is the mathematical formula for the Normal Distribution? Comment on its use.

Discussion.

For the Normal Distribution with mean μ and standard deviation σ, the proportion of readings in the interval $\mu - k\sigma$ to $\mu + k\sigma$ (for any positive value k) is given by the integral

$$\frac{1}{\sqrt{(2\pi\sigma^2)}}\int e^{-(x-\mu)^2/2\sigma^2}\,dx.$$

from $(\mu - k\sigma)$ to $(\mu + k\sigma)$.

Table 12.19 gives some selected values of this function (more detailed tables are given in most textbooks of statistics).

TABLE 12.19 The Descriptive Characteristics of the Normal Distribution

(Selected values)

± Distance from the mean	The proportion of readings lying within the stated limits
± .5 s.d.	40%
± .8 s.d.	58%
± 1.0 s.d.	68%
± 1.6 s.d.	90%
± 2.0 s.d.	95%*
± 2.6 s.d.	99%*
± 3.0 s.d.	99.7%
± 3.3 s.d.	99.9%*

*Conventional 5%, 1% and .1% significance levels

Thus about 40% of the readings lie within .5 σ on either side of the mean, and only 1 in a thousand lie more than 3.3 σ away.

Practical statisticians or data analysts virtually never deal directly with the theoretical formula or even with the detailed numerical results in the table. The reason is that all Normal Distributions take the same shape, therefore given their mean, their standard deviation, and something like the values in the table, one can describe the data for almost all practical purposes.

Exercise 12L. The Combination of Frequency Distributions

What is the frequency distribution of $(x + y)$ when variables x and y each follow a specified frequency distribution?

Discussion.

In general there is no simple answer. For example, adding two Normal variables x and y with means $\mu_x = 5$ and $\mu_y = 20$ will generally give a complex frequency distribution with two humps or modes, one at 5 and the other at 20.

Conversely, if one comes across such a complex-looking distribution in practice, it can often be best analysed by separating the data into two sub-groups, each of which follows a simple, unimodal distribution.

However, there are cases where "mixtures" of different distributions lead to a simple result. One case was the Beta-Binomial in Section 12.5, where different Binomial Distributions for individual families were combined. A similar case arose with the Negative Binomial Distribution of purchases in Section 12.3, which was a mixture of different Poisson Distributions. (This is outlined further in Chapter 13.)

Exercise 12M. Adding Different Variables

What is the frequency distribution of $(x + y)$, when x and y are two different variables for the same items?

Discussion.

Adding the two variables here is like adding peoples' salaries in successive weeks, or their salaries plus their *other* forms of income.

The combination of different variables is one of the more important parts of statistical theory and practice. In general, there are no straightforward answers. But cases where simple answers occur are exceptionally important. They arise particularly where the variables are *independent* of each other. Examples are:

(i) Two independent Normal Distributions for x and y with means μ_x and μ_y, and standard deviations σ_x and σ_y. The sum $(x + y)$ will be Normally distributed with mean $(\mu_x + \mu_y)$ and standard deviation $\sqrt{(\sigma_x^2 + \sigma_y^2)}$. This is a fundamental result. In particular, it follows that the distribution of the *mean* of $x + y$, i.e. $(x + y)/2$, will be Normally distributed with mean $(\mu_x + \mu_y)/2$ and standard deviation $\sqrt{\{(\sigma_x^2 + \sigma_y^2)/4\}}$. If $\sigma_x^2 = \sigma_y^2 = \sigma^2$, the standard deviation of $(x + y)/2$ will therefore be $\sigma/\sqrt{2}$. This is the basis of sampling theory, as discussed in Part IV.

(ii) Two independent Poisson Distributions for x and y with means μ_x and μ_y. The sum $(x + y)$ will also follow a Poisson Distribution, with mean $(\mu_x + \mu_y)$.

(iii) Two independent Negative Binomial Distributions with means μ_x and μ_y and parameters k_x and k_y. If $\mu_x/k_x = \mu_y/k_y = \mu/k$, then $(x + y)$ will be distributed as an NBD with mean $(\mu_x + \mu_y)$ and a second parameter equal to μ/k.

CHAPTER 13

Probability Models

The concept of probability is linked to the notion of randomness. Both are theoretical abstractions and cannot be directly observed. The two concepts are difficult to pin down precisely, but we all have some idea of what they mean and they have many useful practical applications.

Probabilities apply to individual items or events that occur on a more or less chance or random basis. This means that individually the items or events occur in no discernible pattern, but in the aggregate, or over a long term, they tend to occur in certain proportions. For example, the sex of any one baby seems to be determined in a random way; in any sequence of births we can see no girl-boy-girl order or the like. But in the aggregate we generally find that 51% of babies are boys.

In everyday terms probabilities are mainly used in predicting events that are unknown and uncertain, such as the sex of an unborn child or whether it will rain tomorrow. But there are several other distinct uses of probability mathematics. These mostly occur in relatively advanced forms of analysis. However, some familiarity with probability and the notions of randomness and chance is useful at many levels of statistical work. In this chapter we shall be primarily concerned with the use of probabilities in describing and interrelating facts which are known but irregular.

13.1 The Probability Concept

The probability of a particular event occurring can be any number from 0 to 1. If the event is part of a steady series, then the probability of the given outcome should equal the proportion of times that even occurs in the series. That is how the numerical value of the probability is often arrived at in the first place.

Compared with the use of theoretical frequency distributions and proportions discussed in Chapter 12, probability models are simply an alternative descriptive language. For example, we used the theoretical Binomial Distribution to say that 14% of three-child families have three boys, 39%

have two boys, 36% have one boy, and 11% have no boys. Alternatively, we could express this in the language of probabilities and say that the probability of any such family having three boys is .14, two boys .39, etc. The crucial difference is that proportions are characteristics of a group of readings, while probabilities are corresponding statements about each individual reading.

Probabilities have two practical advantages over proportions. Firstly, they are easier to interrelate mathematically, particularly in complex situations. (Since proportions are always expressed in terms of a particular set of data, they need to be redefined if that set is subdivided, combined, etc.) Secondly, while theoretical frequency distributions merely *describe* the data, probabilities can also imply an underlying process since they are tied to the notion of randomness. Thus looking at data in terms of the individual readings often allows one to develop some deeper theoretical understanding of why the observed phenomena occur the way they do.

To use the probability concept one must first satisfy the essential condition of knowing (or assuming) how the probabilities of different events interrelate. The simplest case is when the probabilities are taken to be independent of each other. For instance, if the probabilities of successive children being boys are independent and always .51 (or, more precisely, .5148), then it follows that the probability of a family having 3 boys should be about $.51 \times .51 \times .51 = 51^3$. Thus 51 out of the 100 first-born will on average be boys; for 51% of these, i.e. about 26, the *second*-born will be boys; and for 51% of these, i.e. 13, the *third*-born will be boys. But the data need not behave like that. Whether probability statements provide useful descriptions, or what *kinds* of probability statements do so, depends on the nature of the particular data.

Suppose 60% of buyers of Corn Flakes buy them again on their next purchase of breakfast cereals. It is then not necessarily useful (or correct) to say that all consumers of Corn Flakes have the same probability of .6 of buying it the next time. If all the 60% buy Corn Flakes again on their *third* purchase, then it begins to look as though there are two kinds of people: some (60%) who *always* buy Corn Flakes and some (40%) who always switch.

But suppose instead that the data show that only 60% of those who bought Corn Flakes on both their first and second purchases buy them again on their third purchase, and that, similarly, of those who bought Corn Flakes on their second purchase only (i.e. not on their first) 60% buy them again on *their* third purchase. Then we might summarise a relatively complex situation by saying that *any* buyer of Corn Flakes has a 60% chance, or a .6 probability, of buying them again the next time. This would have to be true whether the buyer belongs to a group who already bought them 100 times in the past or only the once. Attaching a probability statement to each individual is then easier and simpler than making statements about all possible groups.

A situation where the probability of an event depends only on the directly preceding event, is called a simple first-order Markov process (after the Russian mathematician, A. A. Markov). This is one of the simpler *stochastic* processes, which are defined as more or less random or probabilistic forms of behaviour that involve linked sequences of events, especially ones linked over time. The word "stochastic" comes from the Greek *stochos*, meaning guess, i.e. "what will hapen next?". (In practice, consumer behaviour follows a more complex stochastic pattern where the probability of repeat-buying is related to the frequency of purchase.)

Irregular or "as-if random" phenomena take many forms. In the next section we briefly outline how probability models are used to describe irregular phenomena with independent probabilities.

13.2 Independent Probabilities

When analysing irregular data we deal with three distinct entities. One is the actual observations which combine into an empirical distribution. The second is a theoretical or mathematical frequency distribution which is fitted to describe the data to within some degree of approximation. The third is a probability model which speaks in terms of the individual observations and implies an underlying process to account for the theoretical distribution.

For example, in Chapter 12 we looked at data on the incidence of major strikes per week in the United Kingdom from 1948 to 1959. The observations approximated a theoretical Poisson Distribution with a mean of .90. As already mentioned in Section 12.2, Poisson Distributions can arise when events occur independently of each other and with a constant average frequency (i.e. with no trend in their probability of occurrence). Therefore we can infer that the process underlying the incidence of major strikes might be one of more or less independent events with constant probability, where the occurrence of a major strike is not influenced by when the previous one occurred. This is called a Poisson Process. By using this probability model we may gain more understanding of the system, for example, whether the occurrence of a "run" of several strikes is no more than one might expect to occur occasionally, on the basis of chance in a Poisson Process, or whether it implies some special causative factor.

As another example, suppose that an observation can be affected by a large number of small irregular factors acting independently of each other, e.g. various sources of "error". Then it can be shown mathematically that as the number of such factors increases, different observations of this type tend to follow a Normal Distribution. This theoretical result is known as the

Central Limit Theorem and is of exceptional importance in statistics. If we observe a Normal Distribution, the theorem suggests a possible underlying mechanism, of many independent chance errors. And since many phenomena are known to consist of or to be influenced by diverse and more or less independent factors, it also explains why the Normal Distribution occurs so widely.

Again, we know that the observed incidence of boys and girls in families approximates a Binomial Distribution. This implies that it follows a Bernouilli Process, named after Jacob Bernouilli. Here the requirements are that for a fixed number of possible events (e.g. the number of children in a family), the occurrence of a particular event (a child being a boy) must have a constant probability and that the outcomes of different events must be independent of each other. Thus the process implies that the sex of successive children in a family is determined independently and that the probability of either sex is constant.

We therefore have a choice of descriptions for the incidence of boys and girls. We can model the observed data either with a Binomial Distribution, simply describing the data in terms of groups and proportions, or with a Bernouilli Process, speaking in terms of individuals and probabilities and also implying an underlying mechanism. But we can only apply the probability model to aspects of the data that behave irregularly, with no systematic patterns. Thus one must first establish *empirically* that to the limits of current knowledge the events effectively occur in an apparently chance or random manner.

There is no inherent logical reason why a Bernouilli Process should occur for such data. If instead of looking at *family* grouping, we observed children at different schools, or groups of children playing together, the sexes of successive children would not appear to be independent. One school might be all boys, another all girls, and a model of independent probabilities for successive children in each school would not describe such data at all. Similarly, children mostly play together in groups of the same sex, at least until the age of about 15. After that, groups of 2 tend to be mixed more often than a 50:50 model would predict, whilst *larger* groups observed talking, or eating together, or playing games tend still to be predominantly of one sex, The sex distributions in such cases therefore do not follow anything like a Bernouilli Process.

Even for the sex distribution in *families* the Binomial Distribution does not fit exactly. We saw in Chapter 12 that the theoretical values slightly, but consistently, underestimated the number of all-girl or all-boy families, Thus it is purely an *empirical* finding that a model of independent probabilities approximately describes the distribution of boys and girls in families. We can only say *empirically* that the sex of successive children in families acts almost as if it operates independently with constant probabilities.

When a probability model fits well this still does not prove that the underlying physical process is *really* probabilistic or *really* random. All we have is a situation where the data appear sufficiently irregular for it to be useful to describe them *as if* they were random. We have no base in theory for any assumptions of independence or randomness. Both independence and randomness are abstract concepts which can never be fully established in observational data but which can supply a concise "as-if" model.

So-called games of chance are a common example of confusion here. There is no inherent reason why tossing a coin should be a chance phenomenon and follow a Bernouilli Process. Heads do not necessarily come up randomly and independently in successive throws with a constant probability of $p = \frac{1}{2}$ (or some other fixed value if the coin is biased).

If a coin is placed tails-up on one's hand, tossed gently so it turns over just once, and then caught horizontally, it will show heads every time. But if the coin is tossed so that it turns a large number of times, differing on each occasion, heads or tails cannot be predicted. It is an *empirical* finding that no one has yet developed the skill to make the outcome regular under such conditions, or to predict the variations from one toss to the next (otherwise they would be demonstrating their skill on television). Therefore tossing a coin is only an "as-if" random process, and this description depends on empirical observations that under the stated conditions there are no regular patterns in the results.

13.3 Stochastic Models

Irregular events which do not appear to follow simple *independent* probability processes have to be described by more complex stochastic models. These use the idea of conditional probabilities, where the probability of an event may be influenced by a previous event or depend on other factors. For example, one labour strike could trigger off others (although this does not usually seem to happen), or there could be some common factor (a large increase in the cost of living, some new governmental legislation, or a politically-inspired "plot") which causes more strikes to occur at certain times than would occur "by chance" under a Poisson Process with independent probabilities. The Poisson *Distribution* would then not give a good fit to the data.

The Beta-Binomial Distribution mentioned in Section 12.5 was an example of a more complex stochastic model and arose because the simple Binomial Distribution did not entirely give a good fit. The possible underlying process suggested was that the probability of boys varied among families instead of being constant at .51 for all families.

Other stochastic models arise with the Negative Binomial Distribution which has already been mentioned. Models involving the NBD have been used in a variety of applications, including the occurrence of accidents, the spread of animals or plants in ecology, and buyer behaviour (e.g. Greenwood and Yule, 1920; Fisher *et al.*, 1943; Ehrenberg, 1972). Once such a stochastic model has been successfully fitted, it is easier to interrelate many different aspects of the observed data because one is dealing with probabilities and individual readings rather than with proportions tied to specific sets of data.

For example, we have observed that the number of purchases of any particular brand of frequently bought goods in a time period follows an NBD, as noted in Table 12.9a. However, there are at least two different underlying processes that might cause an NBD to occur. One is the "contagious" or "learning" type of situation. Here everybody starts with the same probability, but once the event occurs for someone the probability of recurrence (of buying again) increases for that individual.

The second possible underlying process is the "heterogeneous" or "proneness" model. Here the event occurs with different probabilities for different individuals, but the probabilities do not change over time. To differentiate the two possibilities, one must look at other characteristics of the observed data.

To illustrate this procedure, we consider the "heterogeneous" process which can lead to an NBD. The model involves a mixture of different Poisson Distributions (as do various stochastic processes, e.g. Haight, 1967). It assumes that an individual's purchases of a brand over successive periods of time (e.g. weeks) follow a Poisson Processs, and that different consumers' long-run average rates of purchasing μ (Greek "mu" = the means of the different Poissons) follow a Gamma-Distribution, as shown in Table 13.1. Thus the distribution in each row is assumed Poisson (with mean μ_A, μ_B, etc.) and the distribution of the μ's in the last column is assumed Gamma.

TABLE 13.1 **Schema of the Poisson-Gamma Model Leading to Negative Binomial Distributions**

	Successive Weeks									Long-run	
Consumer	1	2	3	4	5	6	.	.	.	Averages	Distribution
A	x	x	x	x	x	x	x	.	.	μ_A	Poisson
B	x	x	x	x	x	x	x	.	.	μ_B	Poisson
C	x	x	x	x	x	x	x	.	.	μ_C	Poisson
D	x	x	x	x	x	x	x	.	.	μ_D	Poisson
.	x	x	x	x	x	x	x	.	.	.	.
.	.	.	.	.	.	.	.	.	.	.	.
.	.	.	.	.	.	.	.	.	.	.	.
Distribution	NBD	NBD	NBD	NBD	.	.	.	.	.	Gamma	.

This model cannot be validated *directly*, since in practice one cannot observe the individual consumers' long-run averages, μ, and check the form of their distribution. (Nor can one fully check out the Poisson assumption, as applying in the long run). However, from this mixed Poisson–Gamma model a variety of theoretical deductions can be made mathematically. If these approximate the relevant aspects of the observed data, they support the validity of this particular theoretical model and at the same time make it a *useful* one, in that it describes and integrates all these different aspects of the data in one theoretical formulation.

The first theoretical deduction is that the number of purchases made by the different consumers in any given time-period should follow a Negative Binomial Distribution. This is of course the observation we started off with, that consumer purchases tend to follow an NBD. We have here the typical backwards-and-forwards or chicken-and-egg process in theory-building. We were considering the Poisson–Gamma model only because we already *knew* that an NBD tends to fit.

But there are additional deductions. For example, the model implies that the distribution should be an NBD in *any* time-period, e.g. a week as for any column shown in Table 13.1, or for a month as in aggregating the data for 4 weeks, and so on. And this also is what is found in practice (with some deviations in very short time-periods).

Next, the Poisson–Gamma model says that the NBD parameter k should be constant in different length time-periods. This is what is found in practice (e.g. Table 12.17 in Exercise 12B). In contrast, the "contagious" model mentioned above, whilst also leading to an NBD, says that k should vary in direct proportion with the length of the time-period. This therefore provides a very sharp differentiation between the two types of underlying processes.

A further deduction is that in different length time-periods the average purchase frequency of the brand should increase less than proportionately to the length of the time-period, as we saw in Chapter 10. The quantitative

TABLE 13.2 Observed Values of Average Purchase Frequencies and Theoretical NBD Predictions from the 24-Weekly Values

	Average Purchase Frequency per Buyer in				
	24 weeks	12 weeks		4 weeks	
	Obs.	Obs.	**Theo.**	Obs.	**Theo.**
Corn Flakes	5.7	3.5	**3.7**	1.8	**2.1**
Weetabix	5.7	3.8	**3.7**	2.0	**2.1**
Shredded Wheat	4.4	3.0	**2.9**	1.7	**1.8**
Sugar Puffs	3.4	2.4	**2.4**	1.5	**1.6**
Puffed Wheat	2.6	2.1	**1.9**	1.6	**1.4**
Average	4.4	3.0	**2.9**	1.7	**1.8**

details can be summarised by the approximate formula $(w_T - 1)(w_t - 1) = (T/t)^{.82}$, where w_T and w_t are the average purchase frequencies per buyer in time-periods of length T and t. Table 13.2 compares this theoretical formula with the observed data for breakfast cereals.

Other examples arise if we examine repeat buying in successive equal time-periods. For instance, consumers who buy a given brand in the second period but not the first ("new buyers") generally buy it about 1.4 times on average, as the Poisson–Gamma model predicts (e.g. Ehrenberg and Pyatt, 1971, pp. 25 and 70; Ehrenberg, 1972).

These examples illustrate three major points about stochastic models. Firstly, a simple model can both describe and interrelate a great variety of different results.

Secondly, although the model is probabilistic (e.g. a mixture of Poisson Processes) the general user needs no explicit probabilistic mathematics to apply the results, but just straightforward averages or the like (as in Table 13.2).

Thirdly, although all the results were reached by using a stochastic or probabilistic theory, this does not imply that a consumer actually makes purchase decisions on a random or chance basis. It only says that purchases by a variety of different consumers appear to be sufficiently irregular so that in the aggregate they can be successfully summarised by a probability model, *as if* they were in certain respects probabilistic.

13.4 Probability and Uncertain Events

Until now we have concentrated on the use of probability mathematics to model different kinds of irregular variations in observed data. Another use is to try to quantify "degrees of belief" about some uncertain hypothesis or assertion, e.g. that Homer was blind, that it will rain tomorrow, or that one's next child will be a boy.

Instead of interrelating observed data on the apparently irregular incidence of boys and girls, we may have to say or do something before the fact is observed, i.e. before the child is born. We may then attach a probability of .51 to the child being a boy.

This is an assertion about one's state of uncertainty, not about the child. The statement can be interpreted empirically by referring to the frequency with which similar statements would be correct. Thus in asserting that an uncertain event has a probability of .51, one may imply that such a statement (about *any* event) should prove correct on just over half the occasions.

If a topic has been studied extensively, like the incidence of male and female babies, there is clearly an empirical basis for determining such predictive probabilities. But in other cases it is difficult to establish the correct probability values to use. This difficulty is recognised by the term "subjective probabilities" that is widely used in this context (i.e. "guessed" probability levels rather than objectively established ones). An even greater problem is knowing whether the different uncertain events being studied are independent or, if not, how their probabilities of occurrence are interrelated.

Using probability mathematics to deal with uncertain events requires extensive prior knowledge about the kind of events involved and their interrelations. Without such knowledge probability applications would involve making arbitrary assumptions. (The so-called "Bayesian" approach to decision-theory can involve many such difficulties, but its discussion is outside the intended scope of this book.)

Statistical sampling (discussed in Part IV) applies probability mathematics to uncertainty in a different way. Here the chance element and the independence of successive observations are deliberately introduced into the data by the physical operation of random sampling.

13.5 Summary

The probability concept can be useful when dealing with irregular or uncertain phenomena. In this chapter we have discussed its use for describing irregular observed data, such as statistical frequency distributions. The main steps are to assign a probability value to the individual observation and to specify the independence or interdependence between the probabilities of different observations.

The descriptive use of probabilities does not imply that the phenomena really occur by chance, but only that they appear so irregular that they can be successfully described *as if* they were random. Simple models of independent events can provide an underlying rationale for observed frequency distributions like the Normal, Poisson, and Binomial. Other stochastic processes can be used to model more complex phenomena, where the probability of one event is influenced by the occurrence of another event.

CHAPTER 13 EXERCISES

(*Exercises 13L onwards illustrate some of the more technical uses of probabilities in dealing with frequency distributions and stochastic models.*)

Exercise 13A. Exclusive Events

If the probability of rain on a certain day is .4, what is the probability of dry weather?

Discussion.

If the only possibilities are "rain" and "dry", the probability of dry weather must be .6 since the probability of *something* happening must be 1, i.e. Probability of Rain + Probability of Dry = 1.

If additional categories were possible, e.g. "mild drizzle" with a probability of .05 and "no record available" with a probability of .01, the probability of a dry day would have to be only .54 (so that .40 + .05 + .01 + .54 = 1.00). These probabilities should mean that in a long run of days, about 40% tend to have rain, about 5% have some drizzle, about 1% have no record, and about 54% are known to be dry.

Whether it is useful to think of any specific day as having such individual probabilities depends on the absence of predictable patterns. For example, these probability statements would make no sense in a tropical country that had a rainy season of about 150 consecutive days (40%) and drought the rest of the year. Whether the loss of records on about 1% of days is effectively random is also open to question. (Perhaps records are mainly lost during exceptionally heavy downpours or are not as regularly kept at weekends.)

Exercise 13B. Independent Events

If the probability of having arthritis in a lifetime is .3 and that of having measles is .8, and if the two events are independent, what is the probability of having both?

Discussion.

If a proportion .8 of the population have measles, then on the independence criterion, .8 (80%) of those having arthritis should also have measles. The probability of having both is therefore .8 of .3, or .24. Thus 24% of the population should have both. Independence here means for example that the incidence of the one phenomenon does not affect the incidence of the other.

The criterion of independence is relatively straightforward mathematically, but it can never be fully established empirically since it means independence in all *possible* respects, i.e. in every possible sub-group of the population. However, if all *available* cross-analyses have shown no dependence between the two variables, a probabilistic model of independence could provide a useful description. (One kind of complication that can arise is that the incidence of an illness generally depends on how long one lives and that different illnesses tend to occur at very different ages.)

Exercise 13C. Non-independence

If in the duplication-of-purchase law $b_{XY} = Db_X b_Y$ of Section 9.3 the coefficient D is 1, is purchasing of Brand X independent of purchasing of Y? (Here b_X and b_Y are the proportions of the population who buy X and Y at least once in the analysis-period, and b_{XY} is the proportion who buy both X and Y, each at least once.)

Discussion.

If the coefficient D is greater than 1, b_{XY}/b_Y is greater than b_X, i.e. the proportion of buyers of Y who also buy X is greater than the proportion of the whole population who buy X. There is then a tendency for buying of X to go with buying of Y, and so buying of X is not independent from that of Y. Similarly, if D is less than 1, buying of Y *inhibits* buying of X (b_{XY}/b_Y is less than b_X) and the two are not independent.

But if $D = 1$, then the proportion of buyers of Y who also buy X equals the proportion of the whole population who buy X, so that purchasing of the brands *is* uncorrelated in this respect, but it is not necessarily independent in *other* respects (e.g. how often people buy).

Exercise 13D. Conditional Probabilities

Suppose that the incidence of boys and girls is 50:50 and independent for successive children in a family. What is the probability that a two-child family has two boys?

Is this probability of two boys affected if we know that
one child is a boy?
the child nearest the *door* is a boy? (Posed by G. J. Goodhardt.)

Discussion.

The probabilities are 1/4, 1/3 and 1/2.

With independent probabilities of .5 of being a boy, the probability that both children are boys is $.5 \times .5 = .25$, or 1/4.

Any additional information affects the probabilities because here probabilities essentially reflect a lack of information, Knowing that one child is a boy must increase the probability of an all-boy family. But since we do not know whether the first- or second-born child is the boy, we can no longer say that *each* child has a .5 probability of being a boy. For *one* of the children this is already known and hence the probability is 1, but we do not know which.

A more basic approach is needed. There are four possible outcomes for a two-child family, each equally likely: boy-boy, boy-girl, girl-boy, girl-girl. But if one child is known to be a boy, the girl-girl possibility no longer exists for that family. The probability of the other child being a boy, i.e. two boys in all, is therefore 1 out of 3.

If we have more *specific* information, namely that the child standing nearest the door is a boy, the probability of the family having two boys is even greater: 1/2. This is the case because with independent probabilities of .5, the probabilities of the *other* child being a boy is .5.

This value is not affected by telling the first child to move away from the door. The critical part of the information is that it has identified a *particular* child. (The situation might seem more obvious if the *first-born* child were known to be a boy.)

Exercise 13E. The Frequency Interpretation

The argument in the last exercise is in terms of theoretical probabilities. What do they refer to in empirical terms?

Discussion.

Suppose we have a large number of two-child families (e.g. 100,000), with boys and girls being divided 50:50 and distributed independently as far as one can judge (e.g. 50% of first-borns are boys, 50% of these have *brothers*, and so on).

The first question in Exercise 13D refers to all 100,000 families. Since 25,000 have two boys, we can say the probability of a family having two boys is 25,000/100,000 = 1/4.

The second question concerns those families for whom we know one child is a boy. This excludes the girl-girl families and leaves 75,000. Of these, 25,000 still have two boys. Hence the probability of one of these families having two boys is 1/3.

The third question concerns those families where one *particular* child is known to be a boy, e.g. the one nearest the door. Excluding the possibility that both children are equally close to the door, this refers to 50,000 families. In the other 50,000 a *girl* will be nearest the door, on the supposition that boys and girls are distributed independently in *all* respects. Of the first 50,000 families, 25,000 have two boys. Thus the probability of one of these families having two boys is 1/2.

Comparison of the discussions in this and the last exercise indicates how the language of probabilities can be more concise than that of proportions.

Exercise 13F. An Unusual Occurrence?

A family with six children has all boys. Given that the probability of boys is .51, does this imply some special causal factor?

Discussion.

If 51% of children are boys, and if the occurrence of boys amongst successive children is independent, one would expect about 1.8% of all six-children families to have all boys (i.e. $.51^6$).

If approximately this percentage is found on examining large numbers of families, the occurrence of six boys in one particular family would not necessarily signify any special factor. It is what one expects to see happen "by chance" (in 1.8% of all families) if the incidence of boys is random.

However, the model of the independent random occurrence of boys or girls is only a theoretical abstraction. The sex of the individual child is presumably determined by specific factors and the facts have shown (Chapter 12) that a simple Bernouilli Process is not quite true, We therefore cannot strictly *prove* by an appeal to probabilities that the occurrence of 6 boys in a particular family is not due to some special factor.

While the argument is therefore slightly less powerful, the empirical facts still show that the incidence of boys and girls in families of all sizes closely approximates Binomial Distributions with $p = .51$. Except for the small extent to which the Binomial model does not fit, the incidence of boys is therefore consistent with the hypothesis that it occurs on an *as-if random* basis. Thus no special factors have to be invoked to account for a particular family-pattern such as six boys (compared with the factors which account for *mixed* families in their observed proportions and for all-girl families).

Exercise 13G. Deviations from a Model

Discuss the way observed deviations from a theoretical model can be described in probabilistic terms.

Discussion.

Deviations from a model are often irregular and tend to follow a Normal Distribution, as discussed in Exercise 12C. However, there is no defined set or "population" of such readings; each deviation is distinct. For example, in the data in Chapter 1 the deviation in QI in the West was -3, and this was considered quite separately from the deviation of 6 for QIII of the next year, which was analysed in Chapter 2.

This is a situation where the use of probabilities rather than proportions is particularly appropriate (i.e. statements about individual readings rather than about groups of readings). In attempting to describe the Normality of the data we do not have to say that in some arbitrary group of readings, 68% lie within $\pm$ one standard of the mean. Instead. we can make the following sequence of assumptions.

(i) That any particular deviation can be regarded as coming from a certain type of probability distribution (say Normal) with mean 0 and standard deviation σ.

(ii) That the probability distributions for each of the other deviations takes the same form (e.g. Normal with mean 0 and standard deviation σ).

(iii) That the individual deviations are independent of each other.

We can test these assumptions by checking whether *any* group of deviations broadly fit a Normal Distribution with mean 0 and standard deviation σ. (By assumption (iii) it does not matter which group we select for this, as long as the selection does not depend on the observed values themselves, e.g. excluding all large positive values.)

The use of probabilities in dealing with deviations from a model leads to statements which are logically far more precise (and hence easier to manipulate) than the use of proportions and relative frequencies.

Exercise 13H. Theoretical Assumptions

Reference to statistical analyses in the scientific literature shows that in discussing a theoretical model (e.g. the simple straight line $y = ax + b$) it is often assumed that the deviations from the model are independent and Normally distributed, with mean 0 and a certain standard deviation which is constant all along the line. Is such an assumption justified?

Discussion.

The assumption is justified only if it is consistent with the facts, i.e. if it is based on direct empirical analysis of the specific data, or of a range of previous similar data.

But the assumption is often made before any data have been analysed. Assuming independence of the deviations, or even a zero mean, could then be quite inappropriate since the theoretical model itself might not fit. For example, if a linear model has been specified but the empirical relationship is curved, the deviations from any linear equation fitted will not be independent of each other, or Normal, or have a zero mean (except more or less accidentally), or have constant scatter all along the line.

Exercise 13I. "At Haphazard"

In discussing the strength of heredity, Fisher (1950, p. 191) said he assumed for the sake of the argument that "any environmental effects are distributed at haphazard". What is involved?

Discussion.

When natural phenomena are considered to act haphazardly, the usual implication is that one knows little about them. There is therefore no reason to believe they lack patterns, let alone that they are "random".

But if detailed analysis of the observed phenomena has shown them to be effectively irregular, they can be usefully described by a stochastic (i.e. quasi-random) model. However, the randomness is only a property of the model, not of the empirical phenomena, and can therefore not be assumed *a priori.*

Exercise 13J. What are Random Independent Events?

Ask a friend to call heads or tails when throwing a coin in an effectively random manner for about 10 throws. Comment on the results.

Discussion.

General experience suggests that your friend will not make the same call, e.g. "heads", more than about five times in succession. He will change to "tails" for one or more throws, and then probably back again to heads, and so on. Yet with random throws of an unbiased coin, calling "heads" all the time would tend to be right about half the time, and there is no way of consistently doing better.

People tend to vary their calls because

(i) they do not believe the quasi-randomness of the throwing and/or the lack of bias of the coin; or
(ii) they want to demonstrate free will; or
(iii) they want to make a "game" of it; or
(iv) they do not understand the nature of randomness and, in particular, that of independent random events (the outcome of one event being independent of the outcomes of previous ones).

Exercise 13K. The Central Limit Theorem

Can you prove that the sum of a large number of small, independent, random variables will tend to be Normally distributed?

Discussion.

Suppose for simplicity that all the random variables have zero mean. It is then easy to see at an intuitive level that the sum of these variables must follow a humpbacked and more or less symmetrical distribution. Since a large number of variables are involved, about half the values in any particular instance will be positive and about half negative. Their sum will therefore tend to be near zero. Relatively large positive or negative values will occur only rarely.

However, it is very difficult to prove mathematically that the distribution tends to the *Normal* form. The reason for this difficulty is easy to see.

Our starting-point was very general: the sum of a large number of independent random variables that can follow any form of distribution. In contrast, the final result is very specific: that the probability of taking the value x should be proportional to $e^{(x-\mu)^2/2\sigma^2}/\sqrt{(2\pi\sigma^2)}$, the so-called "probability density function" of a Normal Distribution with mean μ and variance σ^2. The connection between the two must therefore be complex. (Although already surmised by the great French mathematician Laplace around 1800, the first rigorous proof of the Central Limit Theorem was only given in 1901, by the Russian mathematician Liapounov. It has since been extended and refined.)

Exercises 13L onwards deal with relatively technical matters.

Exercise 13L. The Binomial Expansion

The Binomial Frequency Distribution gives the probabilities of r occurrences out of n observations, for an event with probability p, by expanding the Binomial formula $(p + q)^n$. What does this mean?

Discussion.

Consider $n = 2$. Then $(p + q)^2 = (p + q)(p + q) = p^2 + 2pq + q^2$. The three terms in this expression equal the probabilities that in two observations the event with probability p occurs both times (p^2), only once ($2pq$, either in the first or the second observation, hence the factor 2), or *neither* time (q^2).

More generally, the terms of the corresponding expansion of the Binomial expression $(p + q)^n = p^n + np^{n-1}q + [n(n - 1)/2]p^{n-2}q^2 + \cdots + [n!/(n - r)!r!]p^{n-r}q^r + \cdots + q^n$ give the probabilities of $n, (n - 1), (n - 2), \ldots, (n - r)$ occurrences out of n observations, as already noted in Section 12.4. (The expression $n!$ stands for $n(n - 1)(n - 2)\ldots 3 \times 2 \times 1$.)

The expression $(p + q)^n$ is clearly a very neat way of summarising the terms of the Binomial Frequency Distribution. But there are some drawbacks.

In a Binomial situation, the two probabilities p and q are interrelated, since the probability q of the event not occurring is $1 - p$, so that $p + q = 1$. In considering the expression $(p + q)^n$, we could therefore write $(p + q)^n = (1)^n = 1$. For example, with $p = .6$, we have $(.6 + .4)^n = 1^n = 1$. This does not get us anywhere, and the mathematician says "It is not what I am interested in". Instead, he expands $(p + q)^n$ into its constituent terms, as outlined above, *before* taking note that $p + q = 1$. (The sum of these terms then necessarily adds to 1, which merely reflects that the probability of *some* outcome, $n, n - 1, n - 2, \ldots, 3, 2, 1$, or 0 occurrences, is 1.)

What the mathematician wants us to focus on is the individual term in the expansion, which for the rth term happens to give the probability of $(n - r)$ occurrences and r non-occurrences. But "picking-out" this rth term can be a rather clumsy thing to do, especially when we note that in any *real* situation the terms are all simply *numbers*. For exmaple for $n = 3$ and $p = .6$, the four binomial terms in expanding the expression $(.6 + .4)^3$

are

$$.216, .432, .288, .064.$$

These have to be written in a clearly established sequence (or appropriately printed out by a computer) if we are to know which terms refer to 3, 2, 1 and 0 occurrences. The numbers lack any identifying labels.

The situation gets worse in more complex cases, such as when there are *two* binomial characteristics. Consider for example families of n children, where the incidence of boys and girls have probabilities p and q, and the incidence of being born on a weekday and at the weekend have probabilities a and b. Then the expansion of the product of two binomial expressions, i.e. $(p + q)^n(a + b)^n$ will give terms like

$$\frac{n!}{(n - r)!r!} p^{n-r} q^r \frac{n!}{(n - s)!s!} a^{n-s} b^s$$

for the probability of a family having r girls and s children born at the weekend. Here "picking out the term for r and s" from all possible $n \times n$ terms begins to be quite complex, especially when in any practical instance all the terms are merely numerical. The only direct way to differentiate the terms is by tabulating them in some neat manner. Doing *theoretical* mathematics with an expression like $(p + q)^n(a + b)^n$ is equally difficult. Some way of labelling the terms seems to be called for.

Exercise 13M. Probability Generating Functions

Is there a way of overcoming the problem of identifying terms in the expansion of $(p + q)^n$?

Discussion.

What is needed is a form of mathematical *labelling*. For a single-child family, consider the expression

$$pu^1 + qu^0$$

where p is the probability of the child being a boy and $q = (1 - p)$ is the probability of its being a girl. This type of expression is called a "probability generating function" (p.g.f.) because it "generates" the probabilities for the incidence of boys and girls. The quantity u is a mathematical labelling device technically called a "dummy variable". It takes no real values, but makes it easy to pick out the terms wanted. The coefficient of u^1 gives the probability of 1 boy, that of u^0 the probability of 0 boys.

Similarly, for families of 2 children we can write $(pu^1 + qu^0)^2 = p^2u^2 + 2pqu^1 + q^2u^0$, where the coefficient of u^r gives the probability of r boys, with $r = 2, 1$ or 0. More generally we can write the probability generating function of the Binomial with parameters p and q either as

$$(pu^1 + qu^0)^n,$$

or more concisely (if less explicitly)

$$(pu + q)^n,$$

since $u^1 = u$ and $u^0 = 1$.

There is now no problem in picking out terms. Thus for $n = 3$ and $p = .6$, as in the last exercise, we have $(.6u^1 + .4u^0)^3 = .216u^3 + .432u^2 + .288u^1 + .064u^0$. The dummy variable "$u$" with its exponent r acts as a clear label for $r = 3, 2, 1$, and 0.

Although the function $(pu + q)^n$ might seem more complex than the straightforward Binomial expression $(p + q)^n$, the introduction of the dummy variable in fact *simplifies* the ensuing mathematics. Writing $(pu + q)^n$ also eliminates the temptation to add $p + q = 1$ in $(p + q)^n$.

For the situation with *two* Binomial variables we write the p.g.f. with two different dummy variables, say u and v:

$$(pu + q)^n(av + b)^n.$$

Here the general term, for $(n - r)$ boys and r girls, and for $(n - s)$ weekday births and s weekend ones, is

$$\left\{\frac{n!}{(n - r)!r!}p^{n-r}q^r u^{n-r}\right\}\left\{\frac{n!}{(n - s)!s!}a^{n-s}b^s v^{n-s}\right\}.$$

It is now simpler to identify the corresponding probability as the coefficient of $u^{n-r}v^{n-s}$.

Exercise 13N. The Negative Binomial Distribution

What is the probability-generating-function of the Negative Binomial Distribution?

Discussion.

In Exercise 12I we noted that the Negative Binomial Distribution arose from expanding the second term in the expression

$$\left(\frac{m + k}{k}\right)^{-k}\left(1 - \frac{m}{m + k}\right)^{-k}.$$

Although the formula looks different, it is directly equivalent to the positive Binomial except that the sign of the exponent is changed. This can be seen by multiplying the first term into the second, giving

$$\left(\frac{m + k}{k} - \frac{m}{k}\right)^{-k},$$

where $(m + k)/k$ is equivalent to p and m/k to q. As before, the inside of the bracket adds to 1. It is useful to express this as a probability generating function by introducing a dummy variable, say u

$$\left\{1 + \frac{m}{k} - \frac{m}{k}u\right\}^{-k}, \quad \text{or} \quad \left\{1 + \frac{m}{k}(1 - u)\right\}^{-k}.$$

Expanding this expression in powers of u gives the probability of observing r occurrences as the coefficient of u^r.

The usefulness of the p.g.f. approach can be illustrated by a very simple and powerful extension of the above formula to more than one time-period.

Consider consumer purchasing data in *two* time-periods, of length T_1 and T_2. Then the expression

$$\left\{1 + \frac{m}{k}[T_1(1 - u_1) + T_2(1 - u_2)]\right\}^{-k}$$

is the p.g.f. of the *bivariate* NBD, where m refers to the average rate of purchasing in a period of some unit length (e.g. a week). When this expression is multiplied out in powers of u_1 and u_2, the coefficient of the term $u_1^r u_2^s$ will give the probability of a consumer making r purchases in the first period and s purchases in the second period.

More generally still, the p.g.f. of the *multivariate* NBD is given by the succinct expression

$$\{1 + m\Sigma\, T_i(1 - u_i)/k\}^{-k}.$$

Here the summation Σ is over t different time-periods of lengths $T_1, T_2, \ldots, T_t$. The results mentioned in Section 13.3 (for the NBD as such, for k, for repeat-buying, for "new" buyers, and for the average purchase rates w_T in different length time-periods) all stem from this single expression.

Exercise 13P. The Poisson–Gamma Hypothesis

What is the justification for the Poisson–Gamma formulation of the NBD model outlined in Section 13.3?

Discussion.

The important justification of such a model is *indirect*. The model works in the sense that deductions such as those illustrated in Section 13.3 fit the facts. This has been found to a close degree of approximation in many thousands of different cases. Certain systematic deviations also occur (especially relating to very short time-periods) which are also increasingly well-understood. The model therefore provides a workable and highly generalisable summary of complex data.

But none of this is a *direct* justification of the model's two basic assumptions of Poisson and Gamma Distributions. As already noted in Section 13.3, these assumptions cannot be checked directly because one cannot have empirical data extending over an indefinitely long time-period (as the model specifies).

The Poisson assumption can, however, be checked on a *sample* basis, i.e. for limited succession of time-periods like the four quarters of a year. Here the Poisson assumption of independent random events has been shown to hold well. This fits in with commonsense experience; *precisely* how many purchases of Corn Flakes, say, one makes in one quarter will hardly depend on precisely how many one made in the previous quarter (and it is the variation about one's *average* number of purchases per quarter which the Poisson aims to model). Exceptions occur in very short time-periods like a week or less where one is less likely to buy in the middle of the night than during the day, or *just* after one has already bought the item. But in many respects the NBD model is not sensitive to such deviations from the Poisson assumption (Ehrenberg, 1972; Chatfield and Goodhardt, 1973).

The Gamma-Distribution assumption, although also not testable directly, has been given a mathematical justification by Goodhardt and Chatfield (1973). This is derived from other kinds of empirical observations altogether. Thus, given that purchasing of one brand is approximately independent of purchasing of another brand (as illustrated by the duplication of purchase law, $b_{XY} = Db_X b_Y$, with D near to 1) and given that the average rate of product purchase is independent of the brand bought (Table 9.10), it follows from some powerful mathematics that different consumers' average purchasing frequencies for any one brand *must* follow a Gamma-Distribution. This kind of result justifies the initial assumption and also links up the Poisson–Gamma model for a single brand with a range of other results concerning switching between different brands. Nonetheless, the direct application of the whole Poisson–Gamma model rests not on this kind of justification at all, but on the extent to which it actually works in practice.

CHAPTER 14

Correlation and Regression

In this chapter we discuss two statistical techniques which are often used in analysing relationships between variables: correlation and regression.

The data dealt with differ from those in Part II. There we always had two or more sets of data for analysis, e.g. the heights and weights of groups of children of different ages, sexes, races, etc., as illustrated in Figure 14.1A. In contrast, correlation and regression generally deal with a *single* set of readings. The individual readings are differentiated only by their values in the two variables, as shown in Figure 14.1B.

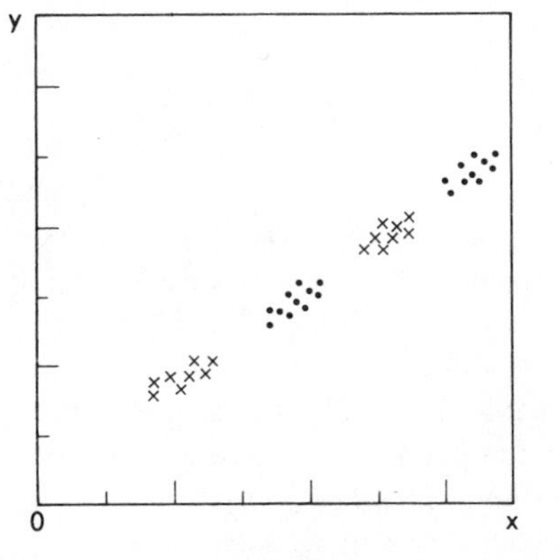

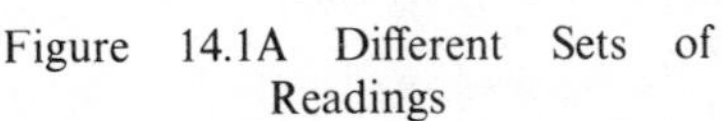
Figure 14.1A Different Sets of Readings

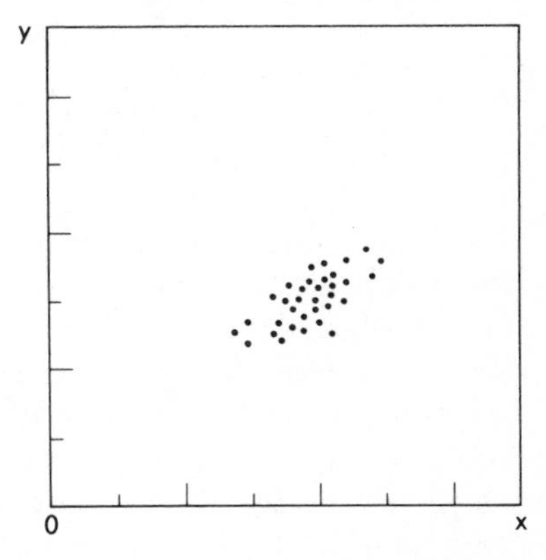

Figure 14.1B A Single Set of Readings

14.1 The Correlation Coefficient

Correlation coefficients are indices that measure the *strength* of a relationship. (The general idea of correlation was largely developed by the biologist Sir Francis Galton in the 1880s, but the particular form now in general use, the "product-moment correlation coefficient", was introduced by Karl

Pearson in 1898.) To provide a simple numerical illustration, we consider five pairs of readings in Table 14.1. Clearly there is some tendency for high values of y to go with high values of x, as Figure 14.2A also shows. The correlation coefficient aims to *measure* this tendency.

TABLE 14.1 Five Pairs of Readings in the Variables x and y

x:	1,	2,	2,	4,	6
y:	17,	11,	23,	19,	30

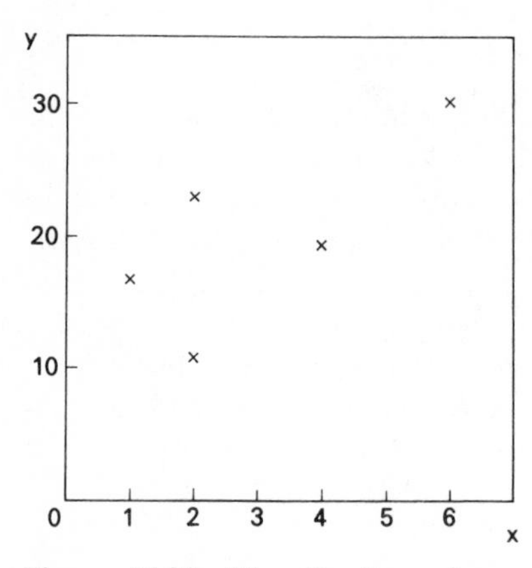

Figure 14.2A The Readings from Table 14.1

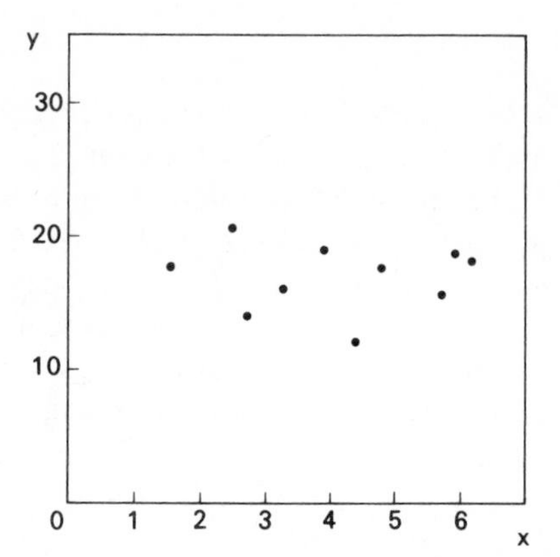

Figure 14.2B Uncorrelated Readings

The possible values of the correlation coefficient range from +1 to −1 for complete positive or negative correlation. Complete correlation describes a situation where all the readings lie exactly on a straight line having either a positive or a negative slope. (A negative slope means that the lower values of y go with the higher values of x.) Correlation coefficients near 0 represent situations where there is no particular tendency for the x and y values to vary together linearly, as illustrated in Figure 14.2B.

The formula for the correlation coefficient, usually denoted by r, is

$$r = \frac{\text{Covariance of } x \text{ and } y}{\sqrt{(\text{variance } x)}\sqrt{(\text{variance } y)}}$$

The covariance is the product $(x - \bar{x})(y - \bar{y})$ of the deviations of each pair of readings x and y, from the means $\bar{x}$ and $\bar{y}$, averaged across all pairs of readings, or Sum $(x - \bar{x})(y - \bar{y})/(n - 1)$. If a pair of x and y readings are both greater or both smaller than the corresponding means, then the product of $(x - \bar{x})(y - \bar{y})$ will be positive. But if x, say, is greater than the mean and

y lower, the product will be negative. The average of the products therefore reflects the extent to which high x goes with high y and low x with low y.

For the five pairs of readings in Table 14.1 the means are $\bar{x} = 3$ and $\bar{y} = 20$. The covariance of x and y is therefore the average value of

$$\begin{aligned}(1-3)(17-20) &+ (2-3)(11-20) + (2-3)(23-20) \\ &+ (4-3)(19-20) + (6-3)(30-20) \\ &= (-2)(-3) + (-1)(-9) + (-1)(3) + (1)(-1) + (3)(10) \\ &= 6 + 9 - 3 - 1 + 30 \\ &= 41.\end{aligned}$$

To find the average for n pairs of readings we divide by $(n - 1)$, just as we do when computing variances, as discussed in Chapter 11. In our example thiş gives

$$\text{Covariance } (xy) = 41/4 = 10.25.$$

The numerical value of the covariance depends on the scales of measurement of x and y (e.g. inches versus feet, as can be seen by multiplying the x-values in Table 14.1 by 12 and recalculating the covariance). This effect is eliminated by dividing the covariance by the standard deviations of x and y. This results in the correlation coefficient as defined in the formula above. For the five pairs of readings in Table 14.1, the variance of x is $\text{Sum}(x - \bar{x})^2/(n - 1) = \text{Average}\{(1-3)^2 + (2-3)^2 + (2-3)^2 + (4-3)^2 + (6-3)^2\} = 4$, and the variance of y is 50. The correlation coefficient for the five pairs of readings is therefore

$$r = \frac{10.25}{\sqrt{4}\sqrt{50}} \doteq \frac{10}{14} \doteq .7,$$

or .72, to be more precise.

The correlation coefficient is relatively tedious to calculate, especially with more extensive data. Exercise 14Q gives a convenient computing formula for use with a desk machine, but with extensive data it is now customary to use a computer.

14.2 Interpreting the Correlation Coefficient

To see how the correlation coefficient reflects the strength of the relationship between x and y, suppose the linear equation $y = 3x + 10$ has been fitted to the data in Table 14.1. For any particular value of x, say $\hat{x}$, the equation gives an estimated value $\hat{y}$, namely $3\hat{x} + 10$. For example, when $\hat{x} = 1$, $\hat{y} = 13$. There will then be deviations between the observed and the estimated values of y, i.e. $(y - 3x - 10)$, which are called the "residuals". The variance of these deviations is called the "residual variance". For $y = 3x + 10$, the deviations for the five y readings are 4, -5, -7, 3, -2 and the residual variance is 25.75, or a residual *standard deviation* of 5.1.

If the relationship between x and y is strong, the residual variance will be small compared to the variance of y (i.e. the average squared deviation of the observed y-values from their *overall* mean $\bar{y}$). This is then reflected by a high value of the correlation coefficient. The connection is that the square of the correlation equals 1 minus the ratio of the residual variance to the variance of y, i.e.

$$r^2 = 1 - \frac{\text{Residual variance}}{\text{Variance of } y}.$$

This can be rewritten as

$$\text{Residual variance} = \text{Variance of } y(1 - r^2).$$

Thus $(1 - r^2)$ measures the extent to which the relationship has reduced the variance of the y-readings. This is often referred to as x having "accounted for" a proportion r^2 of the variance of y, leaving $(1 - r^2)$ "unaccounted for".

To use a less abstract measure of scatter than the variance we can take square roots. This gives

$$\text{standard deviation of residuals} = (\text{standard dev. of } y)\sqrt{(1 - r^2)}.$$

(Variances and standard deviations are more useful in the mathematics of correlation and regression analysis than the *mean deviations* which were used in Part II.)

The variance of the y-data in Table 14.1 is 50, so that the standard deviation of y is 7.1. We have already calculated the correlation coefficient as .72, so the residual standard deviation should be

$$7.1\sqrt{(1 - .72^2)} = 4.9 \doteqdot 5.$$

This is close to the value of 5.1 that we worked out directly from the data. (Strictly speaking, the residual variance or standard deviation can only be derived from the correlation coefficient when the deviations are from a "regression" equation, as described in Section 14.3 below. But, in practice, the derivation holds approximately for almost any reasonable equation that is fitted to the data.)

The "unexplained scatter" of the y readings about the equation $y = 3x + 10$ is therefore smaller than the original scatter of y: a standard deviation of about 5 compared to one of 7. But the reduction is not very large. Correlations need to be very high to reduce the residual variation dramatically. This is illustrated in Table 14.2. A relationship with a correlation coefficient of .5 reduces the y-scatter by only about 13%. Even a correlation of .95 reduces the y-scatter by only about 70%.

TABLE 14.2 The Standard Deviation of the Residuals as a Percentage of the Standard Deviation of y

	Correlation r						
	.1	.3	.5	.7	.9	.95	.99
$100\sqrt{(1 - r^2)}$	99	95	87	71	44	31	14

Comparing Different Correlation Cofficients

A major drawback of using correlation coefficients is that although they are measures of scatter, they do not actually show whether or not two different sets of data have the *same* scatter. This is because correlations measure the residual variance relative to the *total variation* in the data, and the two sets may differ in this respect. If the two correlations are the same, it does not follow that the residual scatter is the same. And if the two correlations are different, it does not follow that the residual scatter is different.

The older literature on correlation analysis gives correction formulae to allow for differences in the total scatter of y (or in x). But it is much easier to compare the residual variances in the different sets of data directly, without using correlations at all. For example, in Figure 14.3A the average size of the residual scatter is clearly the same, even though the two correlations differ radically.

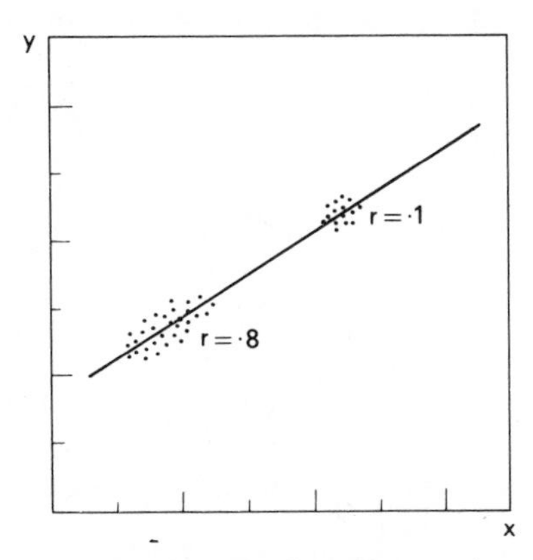

Figure 14.3A Similar Scatter but Different Correlations

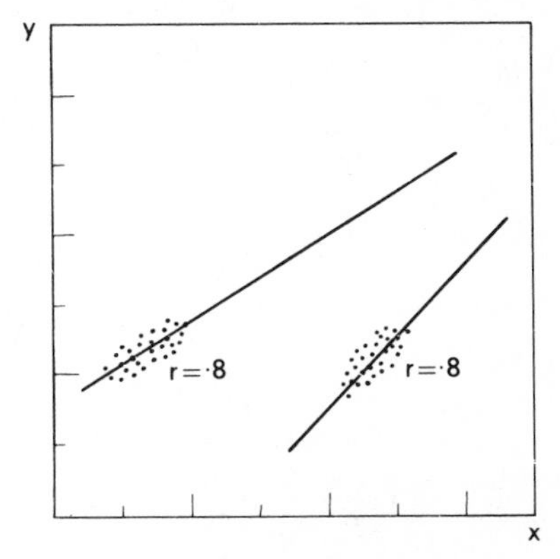

Figure 14.3B Equal Correlations but Different Equations

The correlation coefficient also does not tell us whether the actual *relationships* between x and y are the same in two different sets of data. In the two sets of data in Figure 14.3B for example, the two correlations and the sizes of the residual scatter are the same, but the two relationships are different.

Having calculated the correlation coefficient for a given set of data, it is not at all clear what one can usefully do with it. In particular, it is of no help for prediction or for comparing different sets of data.

Usually it is not difficult to see that in a given set of data there is a positive relationship, with some scatter. What then does it add to say that the numerical value of the correlation coefficient is .6? It does not tell us what the relationship is. It also does not tell us how big the scatter is, except that it is relatively small compared with the observed variation in y, whatever *that* may be. (If the latter is reported as well, it is simpler for predictive or comparative purpose to give the size of the residual scatter directly and let anyone who wants to do so take its ratio to that of the y-variation.)

14.3 Regression Equations

The primary need in analysing a relationship between two variables is to describe how y actually varies with x. This means we have to specify an equation that somehow describes this variation. With scattered readings we also have to describe the scatter about the equation. Furthermore, various equations can then give reasonable fits, as Figure 14.4A illustrates. One criterion for choosing among such alternatives is by the degree to which they fit the data.

With the five pairs of readings in our example, the equation $y = 3x + 10$ had a residual standard deviation of about 5, or 5.1 to be more exact. But

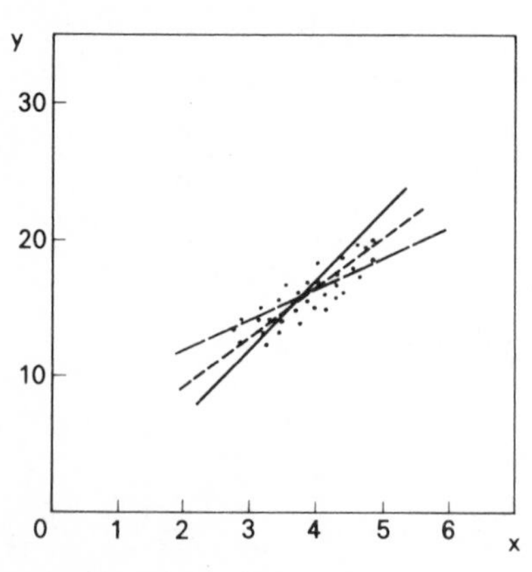

Figure 14.4A Three Possible Equations

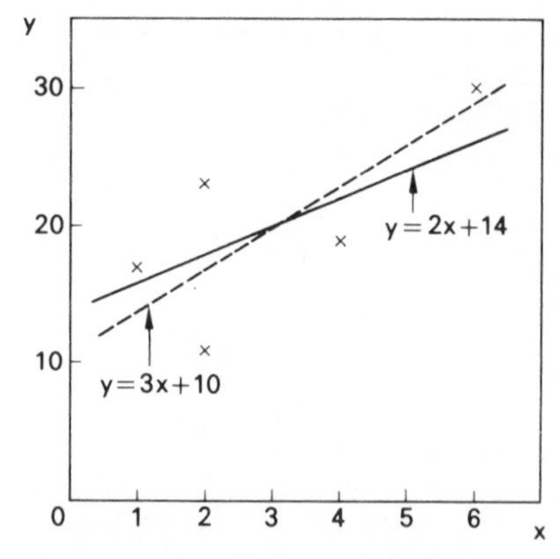

Figure 14.4B Two Equations for the Data in Figure 14.2A

Table 14.3 shows that an equation like $y = 2x + 14$ gives almost the same degree of fit. The residuals $(y - 2x - 14)$ have a standard deviation of 5.0. Figure 14.4B illustrates these equations. They look rather different from each other and the problem is which such equation to choose.

TABLE 14.3 The Fit of the Equation y = 2x + 14

(Data from Table 14.1)

						Average
x	1	2	2	4	6	3
y	17	11	23	19	30	20
2x + 14	16	18	18	22	26	20
y - 2x - 14	1	-7	5	-3	4	0

One possibility is to pick the equation that has the lowest residual variance. Because the variance is the average of all the squared deviations from the fitted line, the criterion involved here is called the "least-squares" principle.

The equation giving the minimum residual variance in the y-direction is called the regression of y on x. If we are fitting a linear equation, this must be of the form $(y - \bar{y}) = a(x - \bar{x})$, since the line has to go through the means $(\bar{x}, \bar{y})$ of the data. It can then be shown mathematically that the slope-coefficient a must equal the ratio of the covariance (xy) to the variance of x, or

$$a = \frac{\text{cov } xy}{\text{var } x}.$$

For our numerical example, the means are 20 and 3, the covariance is 10.25, and the variance of x is 4. The regression equation of y on x is therefore

$$(y - 20) = \frac{10.25}{4}(x - 3).$$

This reduces to

$$y = 2.6x + 12.2.$$

Table 14.4 shows the fit of this equation. The variance of the residuals is $94.96/4 = 23.7$ and their standard deviation is 4.9. The fit is therefore closer

TABLE 14.4 The Fit of the Regression Equation y = 2.6x + 12.2

						Average
x	1	2	2	4	6	3
y	17	11	23	19	30	20
2.6x + 12.2	14.8	17.4	17.4	22.6	27.8	20
y - 2.6x - 12.2	2.2	-6.4	5.6	-3.6	2.2	0

than for the equations mentioned earlier. In the "least-squares" sense it is the best-fitting line.

Exercise 14R gives a short-cut formula for calculating the slope-coefficient of a regression equation. Regression equations for extensive data are now usually calculated on a computer.

The Regression of x *on* y

The regression equation just discussed provides a "best fit" to the data in the y-direction. It is the equation which minimizes the sum of the squared deviations $(y - ax - b)^2$. But this is a somewhat arbitrary criterion. For example, Figure 14.5A shows one could also consider the deviations of any point $(\hat{x}, \hat{y})$ *perpendicular* to the line, or deviations in the x-direction.

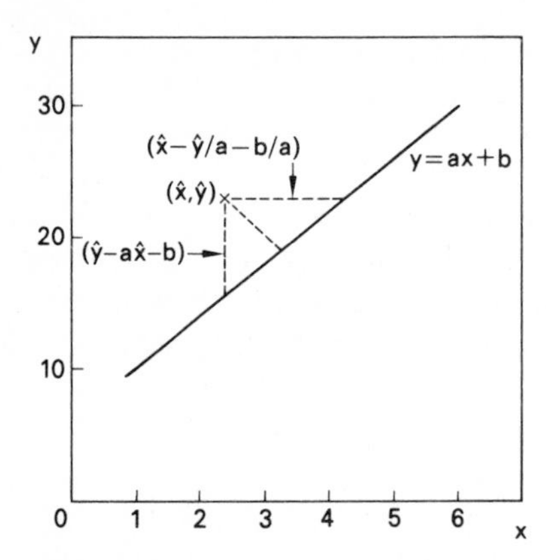

Figure 14.5A Vertical, Horizontal, and Perpendicular Deviations

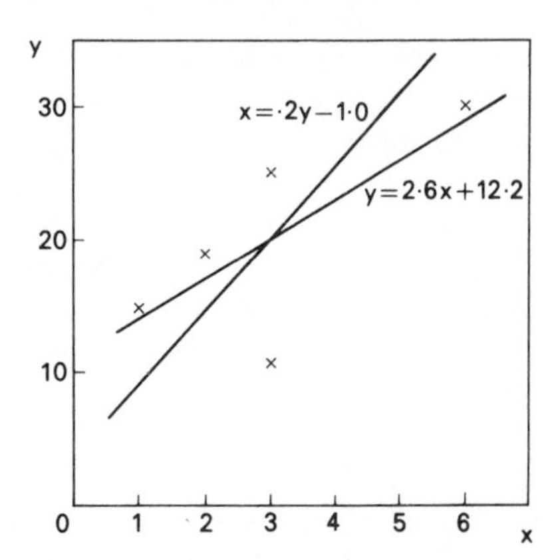

Figure 14.5B The Regressions of y on x and of x on y

Fitting a line by minimising the *perpendicular* deviations might seem an intuitively attractive approach, but is has a crippling disadvantage. Changing the units of one variable (as from inches to feet) results in a completely different equation. This approach is therefore not used in practice.

But if minimising the deviations in the *y-direction* seemed attractive, minimising the deviations in the *x-direction* must also be. The two things are different, and so are the results. Thus corresponding to the regression equation of *y on x*

$$(y - \bar{y}) = \frac{\text{cov}\,(xy)}{\text{var}\,(x)}(x - \bar{x}),$$

the regression of *x on y* is

$$(x - \bar{x}) = \frac{\text{cov}\,(xy)}{\text{var}\,(y)}(y - \bar{y}).$$

This equation is the best "least-squares" fit in the sense of minimising the deviations in the x-direction.

For our numerical example, this formula gives

$$x = .2y - 1.0.$$

Multiplying by 5 makes this equation read $5x = y - 5$, or $y = 5x + 5$. This is numerically quite different from the regression equation of y on x, $y = 2.6x + 12.2$ which we calculated earlier. The two lines are shown in Figure 14.5B. For x on y, a unit change in x corresponds to a 5-unit change in y; for y on x, a unit change in x corresponds to only a 2.6-unit change in y.

In general, any given set of data has two different regression lines. For the user this poses problems. Most statistical textbooks state that the regression of y on x is the best equation for predicting y from x; and that the regression of x on y is the best equation for predicting x from y. This idea is usually not explained any further and may seem confusing. We now examine it.

14.4 The Non-comparability of Regression Equations

If an equation is to give a correct prediction, it must hold for the new data about which the prediction is being made. But under what circumstances do the regression equations for one set of data also hold for another set of data?

Consider the height and weight data for Birmingham children referred to in Part II. These children had been classified into boys and girls, nine yearly age-groups from 5 to 13, and three different social classes. There were therefore 27 different sets of data for boys and 27 different sets for girls, a total of 54.

Each set of data has two regression equations. Figure 14.6A illustrates them for the middle social-class boys aged 5, 9 and 13 years. Table 14.5

TABLE 14.5 The Regressions of Weight on Height and of Height on Weight for Each Age-Group of Birmingham Boys in Social Class (3)

Age	Regressions of	
	Weight on height	Height on weight
5	w = 1.9h - 40	h = .35w + 28
6	w = 2.2h - 58	h = .31w + 32
7	w = 2.3h - 59	h = .29w + 33
8	w = 2.7h - 78	h = .23w + 37
9	w = 3.0h - 94	h = .19w + 40
10	w = 3.2h - 105	h = .21w + 40
11	w = 3.6h - 125	h = .16w + 43
12	w = 3.8h - 138	h = .19w + 42
13	w = 4.0h - 148	h = .16w + 45

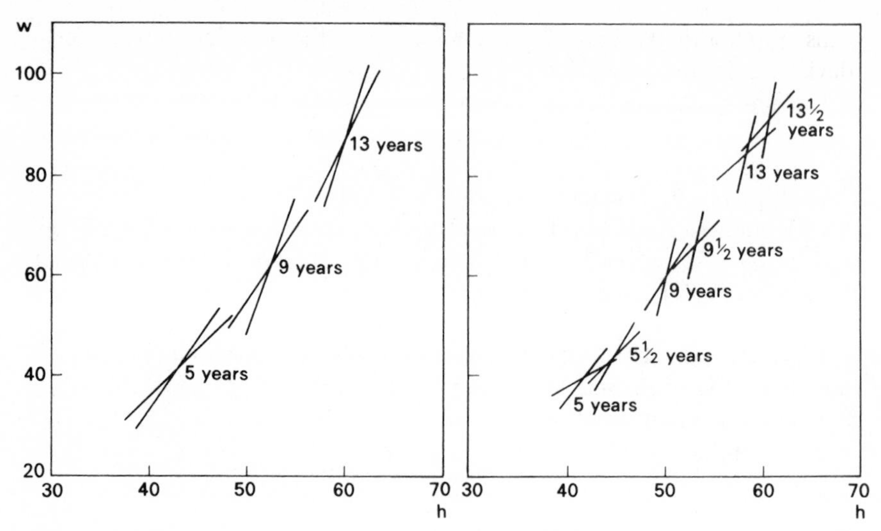

Figure 14.6A h on w and w on h in Yearly Age-Groups

Figure 14.6B The Two Regressions for Half-Yearly Age-Groups

gives both regression equations for *each* age-group (based on equations reported in kilogrammes and millimetres by Healy, 1952, and on data provided by Clements, 1954).

We can see that the equations generally differ from each other. (For ease of comparability Table 14.5a shows the regressions of height on weight also in the form $w = ah + b$, i.e. having divided through by the initial slope-coefficients.) Thus for our 54 sets of data we have a total of 108 different regression equations. Had we analysed the data for half-yearly instead of

TABLE 14.5a The Equations for Social Class (3) with the Regressions of Height on Weight written as w = ah + b

Age	Regressions of Weight on height	Regressions of Height on weight
5	w = 1.9h - 40	w = 2.9h - 8
6	w = 2.2h - 58	w = 3.2h - 10
7	w = 2.3h - 59	w = 3.4h - 11
8	w = 2.7h - 78	w = 4.3h - 16
9	w = 3.0h - 94	w = 5.3h - 21
10	w = 3.2h = 105	w = 4.7h - 19
11	w = 3.6h - 125	w = 6.2h - 27
12	w = 3.8h - 138	w = 5.3h - 22
13	w = 4.0h - 148	w = 6.2h - 28

yearly age-groups, we would have had 216 different regression equations, all even more different from each other, as Figure 14.6B illustrates.

There is nothing exceptional about these results. The height/weight data are not unusually complex. Indeed, we saw in Part II that a single generalisable relationship exists. The reason why regression analysis yields such complex results lies in the technique of analysis.

The two regression equations for a given set of readings go through the means of that data. Another set of data will usually have different means. Hence it will have two different regression equations, as Figure 14.7A shows. The two regression equations for the *first* set of data cannot both go through the means of the second set of data, because two different straight lines can only have *one* point in common. (Some special cases are discussed in Exercises 14C and D.)

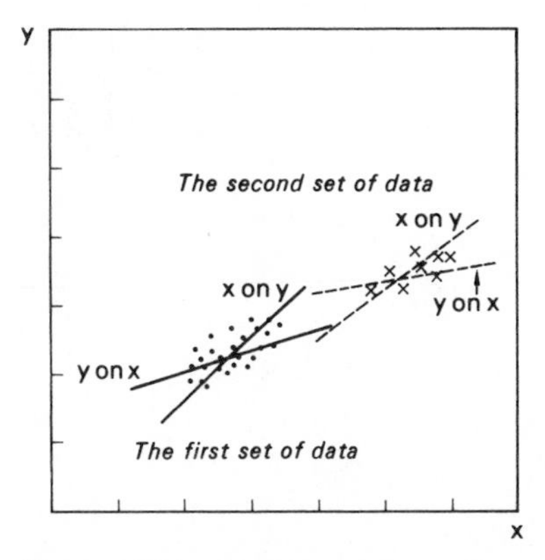

Figure 14.7A The Regressions of y on x and x on y in Two Different Sets of Data

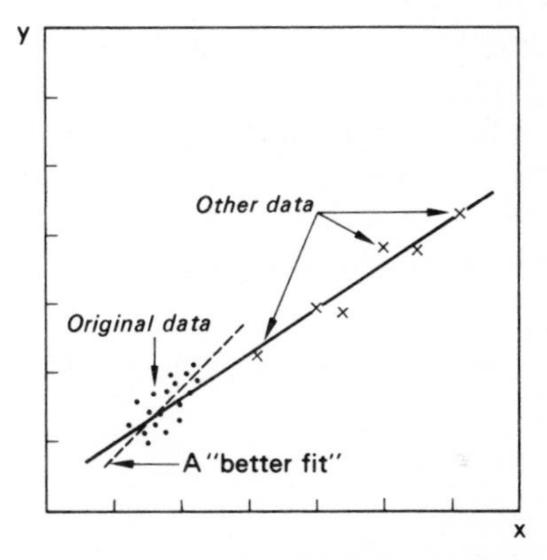

Figure 14.7B Generalization versus Better Fit

We therefore have the theorem that, in general, a regression equation fitted to one set of data cannot hold again for any other data. This is what theory says. And in practice no one has claimed anything different. There appear to be no cases quoted in any textbook where a regression equation fitted to one set of readings *has* held for another, different, set of data.

There is nothing in the theory of regression analysis that says it *should* lead to generalisable results. For an equation to be a best fit to one set of data does not say that it also needs to hold for other data. And with *two* possible regression lines for one set of data (y on x and x on y), it is generally impossible for *both* lines to hold again. They are the wrong type of equations for this purpose and no success has ever been claimed in this direction.

Residual Scatter Versus Empirical Generalisation

The basic problem is how to choose one equation from the various alternatives which give reasonable fits to a given set of readings. Regression analysis chooses an equation which gives a certain "best fit" for that particular set of data. But if one wants to use the equation for prediction to another set of data, then one wants the equation to hold also for this other set of data. That is quite a different criterion. In the one case we aim at the equation which fits one set of data *best*; in the other we aim at the equation, if any, which fits two (or more) sets of data *at all.*

There is no major conflict between fit and generalisation, if generalisation is put first. Faced with a choice among the many different equations that fit a set of data, we can choose the equation which generalises to *other* data. Figure 14.7B shows how an equation can give a good fit for the original set of data and for other data. It does not necessarily provide quite the "best" fit for either set in least-squares regression terms, but the fit is *good.* Although the regression equation gives the "best" fit, there are many other equations which are almost equally good. (That is precisely why there is a problem of choosing among such alternatives!)

As an example, we return to our five pairs of readings. The regression equation of y on x was

$$y = 2.6x + 12 \pm 4.9.$$

But at the beginning of Section 14.2 we also considered other equations like

$$y = 2x + 14 \pm 5.0,$$

$$y = 3x + 10 \pm 5.1.$$

The last term in each equation is the standard deviation of the residual scatter, given to two digits to point up the small differences that exist. Although the regression equation gives the "best fit", the residual scatters of the other equations are barely larger. We also considered a fourth equation (in Figure 14.4B) with a substantially higher slope-coefficient,

$$y = 4x + 8 \pm 5.7,$$

but even here the residual scatter is not *that* much larger than for the regression equation. Given that y varies almost 20 units from 11 to 30, could one say that the equation $y = 4x + 8 \pm 5.7$ is "wrong" whereas $y = 2.6x - 12 \pm 4.9$ is "right", just because of a 0.8 unit difference in the residual scatter?

The important conclusion is that different equations can give a very similar degree of fit. The criterion of "fit" is very inefficient in differentiating between equations. Therefore when faced with choosing between alternative equations we can select the equation that generalises to other data. This need not greatly affect the fit to the original set of data. This is the approach we discussed in Part II.

14.5 Regression with One Variable Controlled

Regression analysis is also used for data where one variable, say x, is controlled. This means that readings of y are observed only for specific values of x.

For example, we might select children aged 5, 9, 11, 14, and 15 years and measure their heights. Age A would then be a controlled variable. The data would consist of vertical arrays of the different h-values for each age selected, as in Figure 14.8A. Sometimes the controlled variable can be directly manipulated, e.g. with certain kinds of physical apparatus, in clinical trials testing different dosage levels of a drug, and in many other experimental situations.

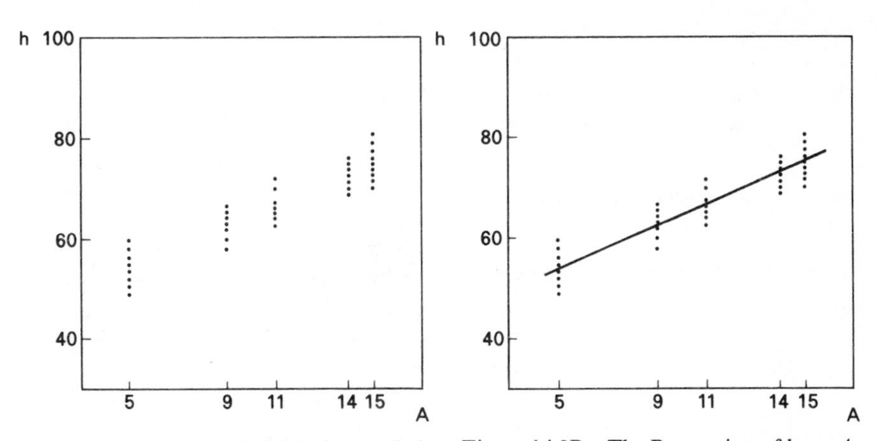

Figure 14.8A One Variable Controlled   Figure 14.8B The Regression of h on A

Here the least-squares regression principle can only be applied in one direction, in our example minimising the variance of the h-values about the line to be fitted. Thus regression analysis gives only *one* equation in this case. But the principle of minimising scatter is really not needed to fix this line. For linear data the straight line can be determined without any extraneous principle because there is no choice between different possible equations. The line must go through the mean values of h for each controlled value of A, as Figure 14.8B shows.

The situation is therefore similar to that discussed in Part II. We have more than one distinct set of data and the fitted equation is an empirical generalisation that has to hold for each set of data. The only special factor here is that each set of readings has one variable that does not vary (so all the values of A in each set are equal to their mean, $\bar{A}$).

If the means $(\bar{h}, \bar{A})$ of the different sets of data lie on a straight line, then *that* is the relationship between h and A. The scatter of the individual h-values does not determine which line to fit. (Least-squares regression would give this answer also, but virtually any "reasonable" method of fitting a line gives the same result in this case.)

The problem of choosing one equation from various alternatives only arises if the mean values of the different sets of data do not lie exactly on a straight line. But then the problem is one of fitting a straight line as a deliberate approximation to non-linear data. An initial working-solution can be determined as before, calculating the slope from the two extreme sets of data and putting the line through the overall means of x and y. (As an alternative, the regression of y on x could be calculated as a first working-solution, if it is easy to compute. But there is nothing "best" or especially attractive about this line. In fact the statistical requirements of linear regression theory are not even fulfilled, because the data are strictly non-linear.)

An initial working-solution fitted in such a situation is usually of no lasting importance. It may well need to be adjusted when further data become available, as we have seen earlier.

14.6 Errors in the Variables

A further problem with regression analysis arises if there are unsystematic errors of measurement in the data. The coefficients of the regression of y on x are affected by errors of measurement in x. In this situation instead of just two regression equations, the theory effectively distinguishes *four* different equations for any one set of data.

If x is subject to errors of measurement, which usually happens, then the regression of y on the observed values of x is generally not the same line as the regression of y on the "true" (error-free) values of x. Such problems do not arise with the lawlike relationships discussed in Part II. Using that type of analysis, the line is fitted to the mean values of each set of readings and therefore remains the same despite different kinds of error in the data. Only the residual scatter is affected by errors, which is as it should be.

With a *controlled* variable a special form of measurement error can arise. For example, if a drug dosage of 5 cubic centimetres is prescribed for a certain type of patient in a clinical trial, the "observed" values will all appear equal. But the actual amounts administered to different patients, the "true" dosages, may vary somewhat about this value.

The same applies to the height and *age* data. Figure 14.9A shows the mean heights of the five controlled age-groups. But the 5-year olds are not all exactly 5 years, that is only their *nominal* age when rounded to the nearest year. In practice, the ages will range between 5 and 6 years, or between $4\frac{1}{2}$ and $5\frac{1}{2}$, depending on how the rounding is done. (There may also be some "real" error of measurement if a 6-year-old child is wrongly classified as being 5.) Figure 14.9B shows the true picture, which brings us back again to the general case of two variables discussed in Part II. Errors in the controlled variable are therefore only a special case of this. As long as the errors are irregular and unbiased (i.e. average out at zero), they do not affect the relationship fitted to the mean values.

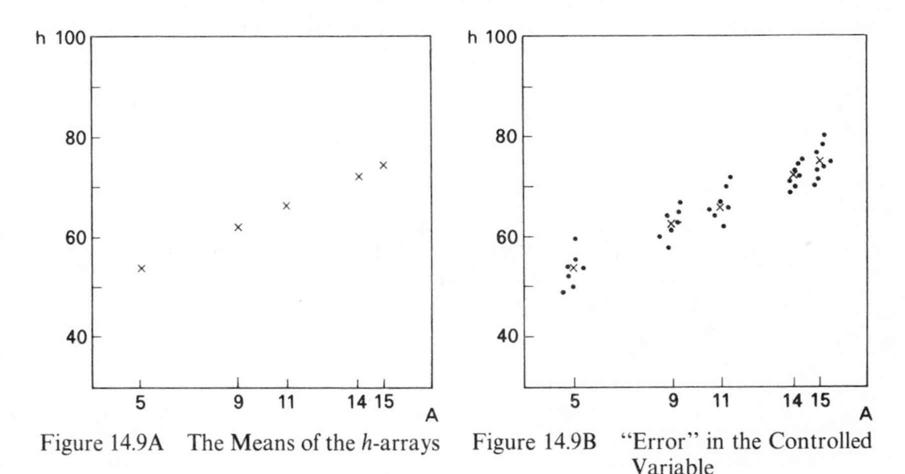

Figure 14.9A The Means of the *h*-arrays

Figure 14.9B "Error" in the Controlled Variable

14.7 Summary

Correlation coefficients and regression equations are two statistical techniques that are widely used in analysing the relationship between two variables.

The correlation coefficient, denoted by the symbol r, is an index that measures how closely the two variables are related in a given set of data. But it only describes this covariation in relation to the total variation in y or x in that particular data. Two sets of data can therefore have the same correlation coefficients, but the scatter about the fitted relationships (the residual variances) can be of different sizes and the relationships themselves can also be different.

A regression equation is the equation that gives the "best fit" to the particular set of data being analysed. But there is more than one "best" equation. The regression of y on x minimises the residual variations in the y direction, and the regression of x on y minimises the residual variations in the x direction. These two regression equations are different for any set of data. Most statistical textbooks claim the regression of y on x is best for predicting y from x, and the regression of x on y is best for predicting x from y. But it is not made clear what this means in practice, nor is there any reason why it should be so.

The problem with regression equations is that the regressions for one set of data generally no not fit any *other* set of data. The reason is that the two linear regression equations for one set of data cannot both go through the means of the other set of data. Regression equations are therefore not useful when building empirical generalisations, nor is this usually claimed for them.

Personally, I have not found either correlation or regression analysis of practical value. But they are widely used and therefore need to be described and evaluated.

CHAPTER 14 EXERCISES

Exercise 14A. Comparing Two Correlation Coefficients

A correlation of .8 has been reported for one set of data in the variables x and y. The same value has subsequently been reported for another set of data in x and y. What can one say about the x/y relationships in the two sets of data?

Discussion.

In both cases, high values of x tend to go with high values of y since each correlation is positive. However, the form of the relationships is not necessarily the same, it could be $y = 38x + 5$ in one case and $y = 0.1x - 10$ in the other. Nor is the size of the scatter about the relationships necessarily the same.

Alternatively, the relationships and/or the size of the scatter in the two sets of data *could* be the same. We cannot tell simply from the correlations.

Exercise 14B. Two Different Correlations

The correlation between the heights and weights of some 8-year-old children is .3 and that for children ranging from 5 to 10 years is .9. What does this difference mean?

Discussion.

The correlation for the 8-year olds *might* be lower because the scatter of their individual readings about a fitted height/weight equation is much larger than for the 5- to 10-year olds. But from previous knowledge it is clear that boys in a six-year age-group (from 5 to 10) differ far more from each other in their heights and weights than boys in a one-year age-group. Even if the two sets of data had the same size residual scatter, we would therefore expect a much higher correlation for the 5- to 10-year olds. (The correlations do not of course tell us whether the relationships between the two variables are the same in the two sets of data.)

Exercise 14C. Comparing Different Regressions

In Section 14.3 we fitted the two regressions

$$y = 2.6x + 12,$$

$$x = 0.2y - 1.0,$$

to the 5 pairs of readings in the numerical example (e.g. Tables 14.1 and 14.4). Table 14.6 gives a second set of data. Compare the regression lines of the two sets of data.

TABLE 14.6 A Second Set of Data

						Av.
x	11	12	12	14	16	13
y	7	41	53	44	60	50

Discussion.

The means of x and y in the new data are 13 and 50. The variances and covariance are 4, 50, and 10.25, the same as for the previous data.

The two regressions for the new data are therefore

$$(y - 50) = \frac{10.25}{4}(x - 13),$$

$$(x - 13) = \frac{10.25}{50}(y - 50),$$

giving

$$y = 2.6x + 16,$$

$$x = 0.2y - 2.8.$$

Both regressions differ from the corresponding earlier ones, although only in their intercept-coefficients.

Whenever the correlations of two sets of data are equal and the two pairs of variances are either equal or in the same ratio, the two pairs of regressions will be *parallel* to each other, as here. *The two pairs of regressions will be the same only if the two sets of means* (x, y) *are also the same*, as Figure 14.10 illustrates.

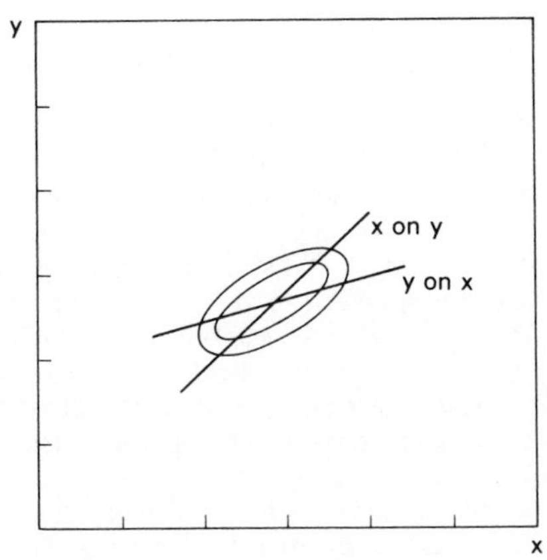

Figure 14.10 Two Pairs of Regressions the Same

The simplest case is of course when both sets of data are identical, i.e. with the same means, same correlations, and same standard deviations. The regressions fitted to the two sets of data will then be the same. But so would *any* line that was fitted by any method.

However, if the means are different in two sets of data, then at least one of the regression lines (y on x or x on y) has to be different for the two sets.

The conditions just outlined for equal or parallel regressions do not form a part of predicting a relationship. For example, if we use Boyle's Law, $PV = C$, to predict Pressure P for a certain value of V, we are only predicting that the relationship $PV \doteqdot C$ will hold again, *not* that the new data will also have the same means as the initial data and its standard deviations in the same ratio.

Exercise 14D. One Regression the Same

Can at least *one* regression be the same in different sets of data?

Discussion.

Figure 14.11 illustrates that it is technically feasible for one regression (say y on x) to be the same in sets of data with different means.

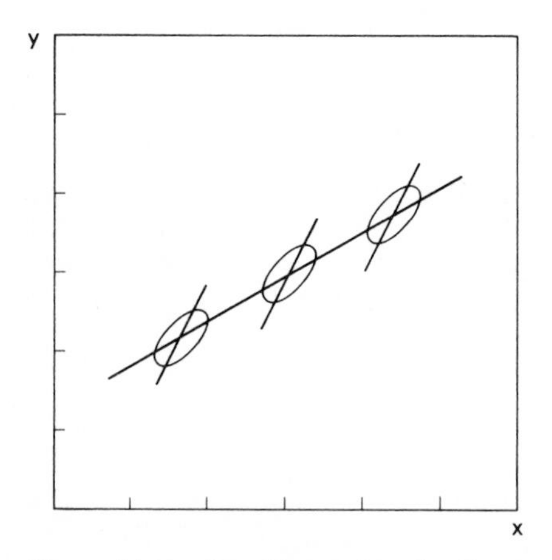

Figure 14.11 One Regression the Same

This case is not discussed in statistical literature. This seems right, because it appears to be technically "trivial", i.e. it could occur only accidentally. For example, suppose each set of the data in Figure 14.11 consisted of yearly age-groups. If each were divided into two sub-groups, such as boys and girls or *half-yearly* age-groups, with different means, then *both* regressions would differ from one sub-set to another. (Figures 14.6A and B provided relevant illustrations.)

Therefore the general conclusion remains. If two sets of data have different means, then the regressions fitted to one set of data will virtually never hold for the other set. To reach that conclusion one does not even have to analyse the data.

Exercise 14E. Two or More Sets of Data

How can one fit a regression equation to two or more sets of data?

Discussion.

The problem of fitting a regression equation to more than one set of data is not discussed in the general statistical literature.

Fitting separate regressions to each set of data generally gives the non-comparable results already illustrated.

Pooling the data into one group does not improve the situation. The result would be influenced by the arbitrary number of readings in the initial groups. Also, pooled data would not generally meet the theoretical requirements of statistical regression analysis (e.g. Normal Distributions, see also Exercise 14Q).

In any case, pooling is unnecessary. Any line fitting more than one set of data must go through the means of each, or at least approximately so. Knowing that is enough to determine a straight line if the data are in fact linear. No special principle of "least squares" or the like is needed.

Exercise 14F. Fitting a Regression to Mean Values

Can regression analysis be used to fit an equation to the mean values of different sets of data?

Discussion.

If the mean values lie exactly on a straight line, there is no problem of finding the "best fit". Only one equation is possible.

If the means do not lie on a straight line, as in Figure 14.12, then the theoretical requirements of regression analysis would not be satisfied and, strictly speaking, fitting a linear equation would be wrong (Exercise 14Q). Regression is concerned with statistical analysis of *irregular* variations.

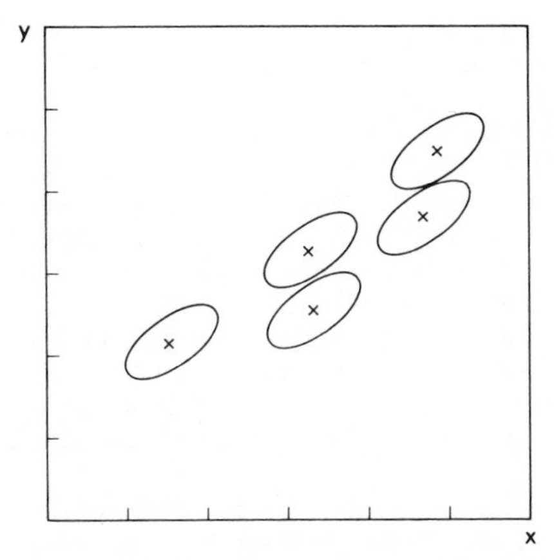

Figure 14.12 Means Not on a Straight Line

Fitting a straight line in this case would be a deliberate oversimplification and approximation of *systematic* deviations. Statistical literature does not discuss forcing a linear regression equation to fit data which are significantly non-linear.

Exercise 14G. Using a Regression Equation

Is it common to use a regression equation fitted to one set of data in the analysis of another set of data?

Discussion.

There is an apparent lack of such cases discussed in the general statistical literature. A seeming exception to prove the rule is a recent paper by Scott (1973), relating test-tank data for certain ship models to their subsequent performance on trial. (There must be other such cases, but they are not common.)

In Scott's paper, regression equations fitted to data from one test-tank laboratory held for data from another laboratory. But the two sets of data were rather similar (see Exercise 14C), so *any* type of line fitted to one would also fit the other. (The deeper methodological problem is that the "best fit" regression lines for the first set of data were not the *best* fit lines for the second set. It is therefore not strictly a case of *regression* lines holding again.)

Exercise 14H. Predictions for a Single Child

Could the height/weight regression equation for a specific age-group be used to assess whether a particular child of that age has the correct weight for its height, rather than using the more general height/weight relationship?

Discussion.

This question was answered some time ago as follows (Ehrenberg, 1968, p. 235):

"I have a daughter aged 7 who weighs 58.4 lbs. and stands 51.2 inches high.

"The regression of weight on height for 7-year-olds (in Table 14.5) is $w = 2.3h - 59$, so that my daughter is .3 lbs. overweight. The regression of height on weight for 7-year-olds is $h = .29w + 33$, so that at 51.2 inches my daughter is 1.3 inches too tall for her weight.

"Being both too heavy for her height *and* too tall for her weight may be less confusing to statisticians than to my daughter, but she will be 8 years old in a day or two and then everything will be different anyway. Eight-year-olds are generally taller and heavier than 7-year-olds, and the regressions necessarily differ.

"For 8-year-olds, the regression of weight on height is $w = 2.7h - 78$, so that in a day or two my daughter will be 1.6 lbs. *underweight* for her height, compared with being .3 lbs. overweight now. (This of course does not happen with the lawlike relationship $\log w = .02h + .76$, which holds for both the 7- and the 8-year-olds.) All this leaves me with a conviction that statisticians never actually *use* the regression equations which they calculate."

A fundamental question is why should one use the 7-year-old data for Birmingham boys in 1947 for a comparison with one's 7-year-old London daughter in 1968? Given that the regression equations differed for all the age-groups in 1947, why should a comparison with the same *age-group* but different time, place, and sex be relevant? Yet a comparison with the regression equations for any other age would have led to different results, too.

Only a generalisable relationship that has been found to hold under a wide range of conditions, such as time, place, sex, age, etc., can be used as a yardstick or norm. Only for such an equation do we have evidence that the differences between then and now do not matter.

Exercise 14I. The Least-squares Principle

What justifications are there for adopting the least-squares principle?

Discussion.

In terms of obtaining simple results, there would appear to be no justification.

On *a priori* grounds, Lindley (1947) in a major review paper said that the method of least-squares "is not easily justified except in certain cases" without saying what they are. Kendall (1951) in another review paper quotes Lindley to the effect that "there is something to be said for accepting the principle of least-squares in its own right" but he does not say what this something is. Anscombe (1967) tried more recently to justify some modifications of the least-squares approach and was reduced to ambiguous phrases like "it seems desirable", "a satisfactory solution", "there is reason to think", "it is natural to eliminate", and "we need not be abashed". Anscombe also said that those who disapprove of least-squares regression methods altogether have "undoubtedly much good reason on their side".

No one seems to have given a clear reason for adopting the regression method. In the past, the practical man has accepted the theoreticians' judgment that the least-squares approach will give the "best" solution; and the theoretician has believed that this kind of "best fit" is what the practical man wants. But although a physicist, for example, will want to know the scatter of observed readings about Boyle's Law, he never needed the scatter-minded statistician to tell him what the law actually is.

Exercise 14J. Collecting the Data

Does the way data in two variables are collected match the way regression analysis then deals with the data?

Discussion.

No. Collecting data in two variables usually consists of collecting a number of undifferentiated "items" (e.g. all boys aged 11 in a certain

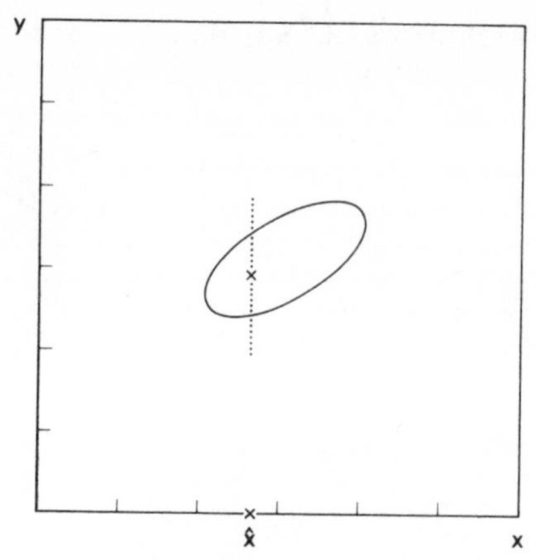

Figure 14.13 Values of y for a Selected Value of $\hat{x}$

school) and then measuring or recording their x and y values, as is illustrated by the ellipse in Figure 14.13.

The procedure differs from that used in the *analysis*, if regression is used. There we first select all items with x-values equal to some specific value $\hat{x}$. Only *then* do we look at their y-values. Figure 14.13 illustrates this selective approach in the analysis. (The case where the data are *collected* by first controlling one variable, x say, at certain specific values, has been discussed in Section 14.5.)

Exercise 14K. Significant Relationships

How do we judge whether there is a significant correlation or regression?

Discussion.

When fitting an equation to two variables x and y, whether by regression or other methods, the basic question is whether the observed y-values deviate less from the fitted line than they do from their own overall mean.

If the difference between the two types of deviation is small, then the relationship is weak. (As noted earlier, a correlation of 0.5 means that the residual standard deviation is only about 13% smaller than the standard deviation of y from its mean.)

If we are dealing with *sample* data, a correlation may appear in the data due to errors of random sampling, when no such relationship exists in the population as a whole. This can be tested by statistical means, as discussed in Chapter 18, Section 18.6.

But there is no special merit in finding a relationship in one's data, especially if it means overinterpreting minor fluctuations in the data. It could be just as effective, and a far simpler result, to establish that y is *not* related to x, or that there is only a very weak relationship with a great deal of scatter.

Exercise 14L. Perpendicular Deviations

Illustrate the statement in Section 14.3 that the linear equation obtained by minimising perpendicular deviations is influenced by the choice of units.

Discussion.

Figures 14.14A and B show a simple illustration provided by G. J. Goodhardt. Suppose we have 4 readings of height and weight, as in Figure 14.14A. The line with the least perpendicular deviations is the vertical line shown, $h = 0.5$ yards. (This is visually obvious and can also be proved mathematically.)

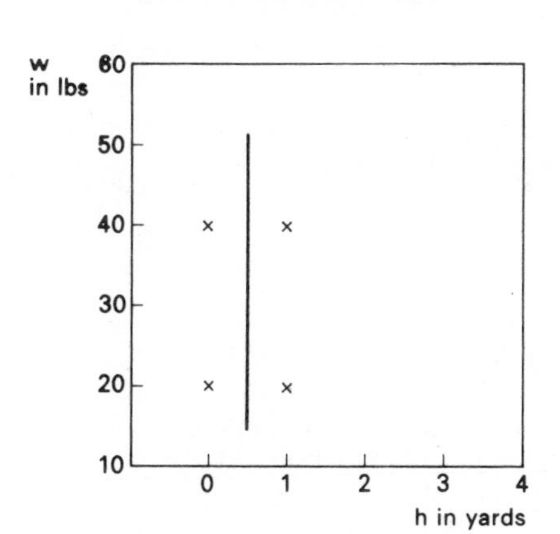

Figure 14.14A $h = \frac{1}{2}$ yard

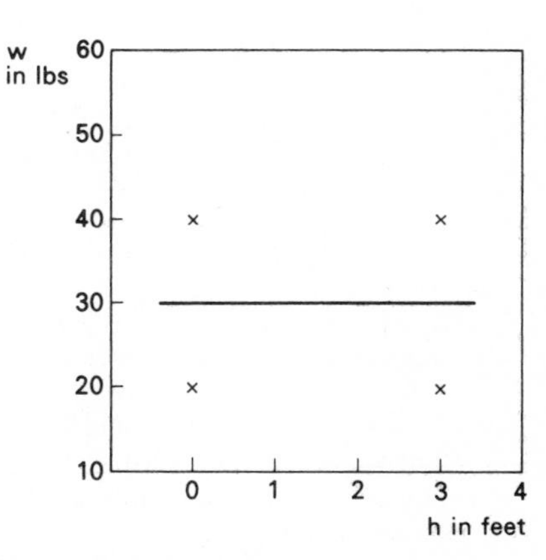

Figure 14.14B $w = 30$ lbs

In Figure 14.14B the scale of measurement for height has been changed from yards to feet. The best-fitting "perpendicular" line here is $w = 30$ lbs, quite different from the previous line. This example is an artificial one, but the same kind of effect occurs with any type of data.

Exercise 14M. Why are There Two Regression Lines?

Why does the regression of y on x not give the best-fitting equation in terms of deviations in both the x- and y-directions?

Discussion.

The regression of y on x is the straight line that gives the minimum variance of deviations in the y-direction. Similarly, the regression of x on y gives the minimum variance of deviations in the x-direction.

These two lines cannot be the same (unless all the points lie exactly on a straight line anyway). Figure 14.15 illustrates the problem. For a single point $(\hat{x}, \hat{y})$ and two particular lines, A and B, the figure shows that line A is closer to the point than line B in the vertical y-direction. In contrast, line B is closer to the point than A in the horizontal x-direction.

It follows that for a large number of observed points, the line minimising the (squared) deviations in one direction need not be the same as that

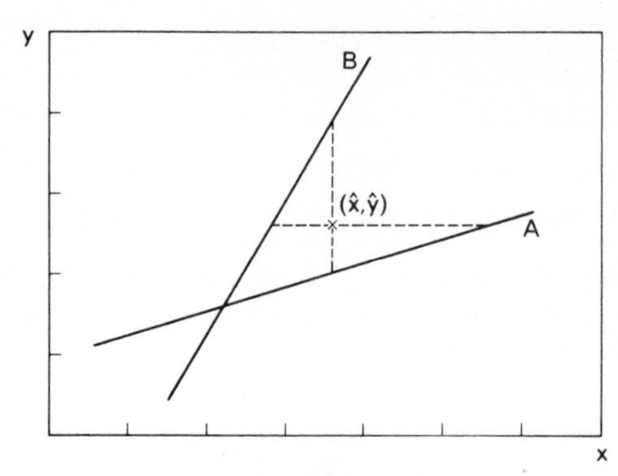

Figure 14.15 Closer to Line *A* in the *y*-direction and to Line *B* in the *x*-direction

minimising the deviations in the other direction. This can be proved rigorously.

Mathematically there is no doubt that there have to be two different regression lines, if a regression is defined as minimising the residual variance of $(y' - ax - b)$ or $(x - cy - d)$. The question is whether this criterion is useful in summarising empirical data. It has been said that anyone who finds least-squares regression analysis easy cannot have understood it at all. But fitting a straight line to some data should not be that difficult.

Exercise 14N. A Special Case?

In Section 14.5, data with one variable controlled were said to be special cases of the situation considered in Part II, where equations were fitted to the means of different sets of data. But isn't this also a special case of regression analysis?

Discussion.

It is a special case of lawlike relationships because nothing is different. The fact that the x readings do not vary in each set of data affects nothing else.

In contrast, general regression analysis deals with one undifferentiated set of readings and yields *two* regression equations, whereas here we have two or more sets of readings and these yield only *one* equation. This is therefore not what is usually meant by something being merely "a special case" of a more general form of analysis.

Exercise 14O. Best for Prediction

In Section 14.3 it was said that the regression of y on x is usually claimed to be best for predicting y from x, and the regression of x on y best for predicting x from y. What does the word "prediction" mean here?

Discussion.

Lindley (1947) defines the situation carefully. (No other writer seems to discuss the question so explicitly.) Lindley says in effect that in regression theory the prediction in question is for a new reading from the population which has already been observed and analysed.

But if we know that the reading comes from a particular population whose characteristcs have already been established, there is no problem of prediction. (The new observation must fit, within the usual errors of sampling.) Conventional statistical theory apparently uses the word "prediction" differently from the ordinary scientist or man in the street, who generally uses the word to refer to new data about which nothing is known *directly* when the prediction is being made.

Exercise 14P. Statistical Conditions

Are there technical restrictions on the kinds of data for which correlations and regressions can be calculated?

Discussion.

Formal statistical theory generally stipulates that for correlation and regression analysis the data should follow a Bivariate Normal Distribution. This means that the distributions of x readings and y readings should each be Normal, and that the distribution of the y-values corresponding to any particular value of x should also be Normal. If plotted on a graph the data should look roughly like that shown in Figure 14.1B (an egg-shaped ellipse).

This restriction is partly because of statistical sampling theory, and partly so that the standard deviations used in regression analysis have a descriptive meaning in terms of Normal Distributions. The restriction also guarantees that the mean values of y for a given value of x lie on the regression line (except for error with sample data). This is another way of defining the regression equation.

Statistical theory does not provide alternative analytic procedures for *non-Normal* data. However, correlation coefficients and regression equations tend to be calculated even when the data do not strictly satisfy the requirements, like the numerical example in the main text.

Exercise 14Q. A Computing Formula for the Covariance

Devise a short-cut formula for calculating the *covariance.*

Discussion.

The covariance of x and y is defined as the average cross-product of the x and y deviations from their means, $\bar{x}$ and $\bar{y}$; i.e.

$$\text{Covariance}\,(x, y) = \frac{\text{Sum}\,(x - \bar{x})(y - \bar{y})}{n - 1}$$

where n is the sample size. This formula entails first working out all the n individual deviations $(x - \bar{x})$ and $(y - \bar{y})$, and then cross-multiplying them and adding. As a short-cut, Sum $(x - \bar{x})(y - \bar{y})$ can be calculated from the sum of the cross-products of the individual readings, Sum (xy), minus a simple "correction term" for the means, $n\bar{x}\bar{y}$. Thus,

$$\text{Covariance}\,(x, y) = \{\text{Sum}\,(xy) - n\bar{x}\bar{y}\}/(n - 1).$$

(The mathematical proof is the same as for the short-cut formula for the variance in Exercise 11E.) For the numerical example in Table 14.1 this short-cut formula gives Sum $(xy) = 341$, $n\bar{x}\bar{y} = 300$. Hence the covariance is $41/4 = 10.25$, as before.

From this simple expression and the corresponding ones for the variances of x and y, the values of the correlation and regression coefficients can be calculated more easily. (Deviations of readings from their means, of the form $(x - \bar{x})$ and $(y - \bar{y})$, are the basis of factors called "moments" in mathematical statistics. Hence this particular type of correlation coefficient is named the "product-moment" correlation, which distinguishes it from many others invented in the earlier part of the century that are no longer used.)

CHAPTER 15

Multivariate Techniques

The most commonly used statistical techniques for analysing data in more than two variables are multiple regression and component or factor analysis. These are extensions of the regression and correlation methods used to analyse a *pair* of variables.

A multiple regression equation is derived by applying the least-squares principle between one variable and several others on which it might depend. The method has been most highly developed in econometrics, a statistically oriented branch of economics. Component or factor analysis on the other hand aims to group a large number of undifferentiated variables into sub-patterns or factors.

A common feature of these procedures is that the results depend on the particular version of the analysis technique adopted. The various arithmetical procedures need not be described in detail here because there are simple-to-use computer packages. Instead, we can concentrate on a broad description and an evaluation of the results.

The methods are often used when seeking to establish interrelationships among large numbers of variables in a single set of previously undigested data. They do not appear capable of providing generalisable results, nor is this usually claimed for them. I therefore do not find either the methods or the results useful, but that is of course a personal view. Since the methods are widely used in certain branches of statistics they require description and evaluation.

15.1 Multiple Regression

Suppose we want to analyse the yield levels of a certain crop over a number of years. Yield could depend on many factors: rainfall, temperature, the amount of fertiliser applied, the amount of sunshine, etc. So we might seek a multivariate equation. For illustrative purposes here we shall only consider two explanatory or "independent" variables, fertiliser f and temperature t, in relation to yield y. Table 15.1 gives readings for five successive

TABLE 15.1 Crop-Yield, Temperature and Amount of Fertiliser in 5 Successive Years

	Year					
	'61	'62	'63	'64	'65	Average
Yield, y	150	170	100	150	180	150
Temperature, t	11	12	5	7	15	10
Fertiliser, f	19	16	15	17	23	18

years, from 1961 to 1965. (Arranging the data in ascending order of y, say, would show clearly that there is strong correlation between all three variables. The problem is to *describe* the relationships in question.)

In general, no reasonably simple equation will estimate the crop-yields exactly. But the regression method will minimise the residual deviations, i.e. give the "best" fit for this particular set of data. Using this method, we want to determine a linear equation of the form

$$y = at + bf + c.$$

The coefficients a, b, and c have to be determined at those values which minimise the variance of the deviations $(y - at - bf - c)$. The coefficients a and b can be calculated from the correlation coefficients r_{yt}, r_{yf}, r_{ft}, and the standard deviations s_y, s_t, s_f of the variables. Thus, the coefficient a in our example, called the partial regression of y on t, is

$$a = \frac{r_{yt} - r_{yf}r_{ft}}{1 - r_{ft}^2} \cdot \frac{s_y}{s_t}.$$

This expression looks complex but is straightforward to use. (The calculations are now often done automatically on a computer, but Exercise 15A outlines the arithmetic.) If f and t are not correlated, the coefficients a and b simplify to the ordinary regression coefficients of y on t and y on f. The intercept-coefficient, here c, is determined by making the equation go through the overall mean values of all three variables (i.e. $c = \bar{y} = a\bar{t} - b\bar{f}$).

Applied to the data in Table 15.1 these formulae give the multiple regression equation

$$y = 8t - f + 88.$$

Table 15.2 shows the fit of this equation. The residual standard deviation over the five years is 14, which is fairly small compared with the 80-unit range in crop-yields. Because it is a regression equation, this is the smallest residual standard deviation that could be achieved by any linear equation among these three variables in this set of data.

TABLE 15.2 The Fit of the Multiple Regression Equation
y = 8t - f + 88

	Year					Average
	'61	'62	'63	'64	'65	
Yield						
Observed, y	150	170	100	150	180	150
8t - f + 88	157	168	113	127	185	150
y - 8t + f - 88	-7	2	-13	23	-5	0*

* Standard deviation: 14

The fit of a multiple regression equation is often reported in a different manner, namely in terms of the multiple correlation coefficient, R. This measure is defined as the correlation between the observed values of y and the estimated values, $(8t - f + 88)$. It can be interpreted like the ordinary product-moment correlation coefficient in Section 14.2 as

$$R^2 = 1 - \frac{\text{residual variance}}{\text{variance of } y}.$$

In Table 15.2 the variance of the residuals $(y - 8t + f - 88)$ is 194 and the variance of y is 950. Thus the value of R^2 is $1 - 194/950 = .8$, and R is about .9. The fitted equation therefore accounts for about 80% of the variation among the variables ("variation" being conventionally measured in terms of the variance).

Interpreting the Regression Equation

In a purely "mathematical" sense, the interpretation of the equation $y = 8t - f + 88$ is straightforward. If f does not vary, then a 1-degree increase in temperature will increase yield by 8 units. If t does not vary, then a 1-unit increase in f will decrease y by 1 unit. If *both* f and t increase by 1 unit each, the total effect will be the sum of these separate effects: an increase in y of 7 units.

But the equation does not necessarily mean this in real life. The data analysed gives no evidence of how y varies with t when f is stable. Nor does it give any reason why the equation should state correctly the effects of *joint* variations in f and t, namely that an increase in fertiliser and a drop in temperature act *independently* of each other (e.g. that a 5-unit increase in f and a 2-degree drop in t will decrease yield by 21). The only *observed* increases in fertiliser occurred when the temperature also increased.

The negative coefficient of -1 for f in the equation appears to say that increased fertiliser will decrease yield, which might seem like nonsense. But if the five sets of readings are the only data we have, this objection is not tenable. The chemical in question could have some deleterious effects, either intrinsically or because of some impurity, or because the applications are at too high a level and "burn" the plants. Or the fertiliser may encourage lush growth but diminish the actual crop that is harvested (e.g. a lot of straw but little wheat). Or the negative coefficient may be due to other factors altogether, quite unconnected with the effects of the fertiliser as such.

We need more information about the effects of various levels of fertiliser on this crop at various temperatures before we can begin to know how to interpret these phenomena, let along any equation that may be fitted. The particular equation fitted here is a multiple regression one, the "best" fit for this particular set of data, but there is no reason why the same coefficients should hold for any other data we may analyse next.

Alternative Equations

Any doubts one may have about the interpretation of the regression equation are confirmed by the existence of alternative equations which also give fairly good fits to the data but not in the minimum residual-variance sense of regression analysis.

One example is the equation

$$y = 5t + 2f + 64,$$

as shown in Table 15.3. It does not matter how this was derived. The question is how well it works for the present and, ultimately, for other data.

TABLE 15.3 **The Fit of the Alternative Equation $y = 5t + 2f + 64$**

	Year					
	'61	'62	'63	'64	'65	Average
Yield:						
Observed, y	150	170	100	150	180	150
5t + 2f + 64	157	156	119	133	185	150
y - 5t - 2f - 64	-7	14	-19	17	-5	0*

* Standard deviation: 15

The difference in the degree of fit of the two equations is small—a residual standard deviation of 15 here and 14 for the regression, compared with the 80-unit range of y-values in the data (and an R still of .9). It is especially small considering the fact that the regression equation implies an 8 unit increase in crop-yield per degree rise in temperature and the other only a 5-unit increase, and that the regression implies a 1 unit *decrease* in yield for each unit increase of fertiliser and the other a 2-unit *increase*. (This has nothing to do with the small number of readings in the example. The same variability in coefficients would occur for similar data which extend over many more years or come from other sources.)

So we again see that markedly different equations can give approximately the same degree of fit. This occurs especially if the "independent variables", here fertiliser level and temperature, are themselves correlated, as in the present instance. The lowest values of both f and t occur in 1963, and the highest in 1965. (In econometrics this is referred to as "colinearity" of the variables.)

Problems arise over which equation to fit and how to interpret its coefficients because multiple regression analysis seeks to find the "best" answer to a complex problem by analysing an isolated set of data. Irrespective of how limited or incomplete the data, the solution is always "best" with scant regard to whether it is any good. (A danger with high-powered salesmanship is that it often deludes the salesman himself.) These difficulties tend to be eliminated by building generalisable relationships from an increasingly wide range of different sets of data, as was described in Part II, where the additional data determine the equation to be fitted.

15.2 Component and Factor Analysis

Component and factor analysis are related techniques based on the correlation coefficient. They are used when there are a large number of different types of measurements for a given set of items (e.g. height H, weight W, girth G, leg length L, etc. for a sample of people). Because factor analysis started with the analysis of intelligence tests in psychology, the different measurements are often called "test variables".

These techniques aim to structure the data by reducing the numerous test variables to a smaller number of variables (components or factors) which account for most of the variation in the given data. A component is simply a linear equation of the test variables

$$\text{Component} = aH + bW + cG + dL + \text{etc.},$$

i.e. a weighted average. The numerical coefficients a, b, c, etc. are any numbers one cares to choose (subject only to the technical condition that the component must have a unit variance). A "factor" is a somewhat more complex version of the same model, with an "error term" added. For simplicity we talk here mostly of *component* analysis.

The difference between these techniques and multiple regression is that there are no direct observations of the component or factor variable on the left of the equation. Instead component and factor analysis *create* new variables, usually by using some special analytic techniques. Having created these new variables, the analyst then has to find out what they are and what they mean.

An Illustrative Example

Suppose we have a sample of children and data giving six body measurements for each. The input for the analysis is a table of the correlation coefficients between all pairs of test variables, usually called a correlation matrix, as in Table 15.4. Each correlation between different variables necessarily occurs twice in the table, e.g. L with H and H with L. The diagonal entries are necessarily unity, e.g. the correlation of H with H. In practice the input may be 20 or more variables, with several hundred correlation coefficients. The purpose of the analysis is to find some sort of grouping or underlying structure for the test variables.

TABLE 15.4 The Correlations Between Six Body Measurements for a Group of Children

	H	L	A	C	G	W
Height	(1)	.7	.9	.3	.2	.8
Leg Length	.7	(1)	.8	.4	.3	.6
Arm Length	.9	.8	(1)	.4	.5	.7
Chest Circ.	.3	.4	.4	(1)	.8	.3
Girth	.2	.3	.5	.8	(1)	.4
Weight	.8	.6	.7	.3	.4	(1)

To help in this process a number of new variables, the components or factors, are created. The first step is to derive the correlations of each new component with the original test variables. These correlations are called the "loadings" of the test variables on the component.

Factor and component analysis each has many different versions and the results depend on which version is being used. We shall first illustrate the general nature of the approach in terms of a component extracted by a method akin to "centroid analysis", which for many years was the most popular form of factor analysis. Suppose we calculate the average of each row of the correlation matrix in Table 15.4. This gives a series of six numbers, .65, .63, .72, etc., shown to one digit (.7, .6, .7, etc.) in Table 15.5. These new numbers must each be less than or equal to 1 and it is therefore possible for each to be correlation coefficients.

We now define a new variable, the component M say, as that linear function of the six test variables which has these six numbers as its correlation coefficients with the six test variables. The new variable M therefore has correlations of about .7 with height, .6 with leg length, etc.

To understand this new variable better we note that all the test variables are "standardised" in this analysis. This means that the scales of measurement are changed so that each test variable has a mean of 0 and a standard

TABLE 15.5 The "Loadings" of the First Component M

	The First Component M
Height	.7
Leg Length	.6
Arm Length	.7
Chest	.5
Girth	.6
Weight	.6

deviation of 1. This is the same as changing from inches to centimetres, except that the new unit of measurement is chosen on the basis of the specific data. For example, if h is the height in inches for the sample of children with a mean of 45 and a standard deviation of 3, then the standardised height variable H used in the analysis is

$$H = \frac{h - \text{mean}}{\text{stand. dev.}} = \frac{h - 45}{3}.$$

A child 39 inches high will then have a standardised score of $H = -2$, i.e. it will be 2 standard deviations below the mean height. This standardising applies to all the test variables and also, as a matter of definition, to all the components.

We can now form the regression equations of the test variables on the component M. Since all the standard deviations equal 1, the slope-coefficient a in a regression equation is equal to the correlation coefficient. Thus

$$a = \frac{\text{cov}\,(x, y)}{\text{var}\,x} \quad \text{and} \quad r = \frac{\text{cov}\,(x, y)}{\sqrt{\text{var}\,x}\sqrt{\text{var}\,y}},$$

and since for standardised variables $\text{var}\,x = \sqrt{\text{var}\,x} = \sqrt{\text{var}\,y} = 1$,

$$r = a.$$

Because all the means are now zero, we simply have

$$H = .7M$$
$$L = .6M$$
$$A = .7M, \quad \text{etc.}$$

From the general formula for a regression equation that the residual variance = variance $y(1 - r^2)$, we know that r^2 is the proportion of the variance of y that is accounted for by the relationship with x, leaving $(1 - r^2)$ unaccounted for. Therefore the component M accounts for $.7^2 = .5$ of the observed variance of H, for $.6^2 = .36$ of the variance of L, for $.7^2$ of the variance of A, and so on. This one component accounts for a total of $\text{Sum}\,(r^2) = (.7^2 + .6^2 + .7^2 + .5^2 + .6^2 + .6^2) = 2.3$ or almost 40% of the total observed variance of all six standardised test variables (which is 6 since each has a variance of 1).

To account for the remaining 60% of the observed variation we need further components. It is generally found that a small number of components can account for the greater part of the variance. In effect it is one of the main considerations of component (and factor) analysis to find the least possible number of components to account for the generally large number of test variables.

A common method for doing this is to extract the *Principal Components.* The way this works is that the first principal component accounts for more of the total variance of the test variables than any other linear equation could. From the remaining variation the second principal component is extracted, and again accounts for more of that variation than any other component could. And so on. (By definition, the principal components are uncorrelated.)

The component M in Table 15.5 accounted for 40% of the total variance but Principal Component analysis can do better than that. Table 15.6 gives the loadings of the first three principal components, labelled I, II, and III.

TABLE 15.6 The Loadings of the First Three Principal Components I, II, and III

	Principal Components		
	I	II	III
Height	.9	-.4	.1
Leg Length	.8	-.2	.4
Arm Length	.9	-.2	.1
Chest	.6	.7	.1
Girth	.6	.7	.2
Weight	.8	-.2	.4
Sum of $(\text{Loadings})^2$	3.6	1.3	.4

Principal component I accounts for as much as 3.6/6 or 60% of the total observed variance. Component II accounts for about 22% and component III accounts for about 7%. There must be six components for six variables, and because the first components here were chosen to be so big, the remaining three must be small. Each can only account for an average of about 4% of the total variance (making a total of 100% for all six). Such small components are then ignored. It is in this way that component analysis finishes up with fewer components than test variables. In practice, where the number of test variables is usually 20 or more, only 3, 4, or 5 components tend to be extracted (or considered further) and hence there is a marked reduction in the number of variables.

Interpretation and Rotation

Now that we have extracted some components, e.g. the principal components, or "centroid" components like M, the next step in the analysis must be to interpret them. Principal component I accounts for 60% of the total observed variation (80% of the variance of H, 64% of the variance of L, etc.). This might seem fine, but we still do not know what this new variable actually is.

The interpretation of a component (or factor) is usually done by looking at the loadings and noting which test variables are highly correlated with the component. The component is then named accordingly. *This is usually the end of the analysis.*

For example, the correlations in Table 15.6 with component I are all large and positive, so that it is highly correlated with all six body measurements: if the values of the six measures for a particular person are large, the value of Component I will be large. Therefore it can be regarded as a measure of "Size". Component II is positively correlated with Girth and Chest Circumference but negatively with Height, etc., so it may be considered a measure of "Shape". Component III is more difficult to interpret in this way.

Difficulties in interpreting components are not uncommon. When they occur, alternative solutions to those first extracted (e.g. the first three principal components) are often sought. It is usual to adopt some linear functions of the initial components, e.g.

$$\text{New Component} = \hat{a}\text{I} + \hat{b}\text{II} + \hat{c}\text{III},$$

where $\hat{a}$, $\hat{b}$, and $\hat{c}$ are any numbers one cares to choose, subject to the technical restriction that the new component is again a standardised variable with unit variance. (Given that the initial components were uncorrelated with each other, or "orthogonal", the required condition is that $\hat{a}^2 + \hat{b}^2 + \hat{c}^2 = 1$.)

A change from one set of components to another is called a "rotation" because the technical procedures involved were initially developed by graphical or geometrical means. Rotation is a major part of modern component (or factor) analysis.

Alternative Solutions

There is in fact no unique solution in component (or factor) analysis. Other components can account for the observed correlations equally well. (One does not even have to start with principal components.) We shall now consider some alternative solutions for the data on the six body measurements.

Table 15.7 illustrates two other components, P and Q. (At this stage it does not matter how these new variables were derived.) Following the usual procedure of interpreting components by inspecting their loadings, we see that P is highly correlated with all the variables. It therefore looks like another measure of "Size" (but not the same as the "Size" component I in Table 15.6). Component Q is correlated .6 with Height, only .2 with Leg Length, and *negatively*, $-.6$ and $-.5$, with Chest Circumference and Girth. The bigger the chest and waist, the lower the score on Q. It may therefore be called a measure of "Thinness".

TABLE 15.7 The Loadings on Components P and Q

	Components	
	P "Size"	Q "Thinness"
Height	.8	.6
Leg Length	.7	.2
Arm Length	.8	.4
Chest	.8	-.6
Girth	.6	-.5
Weight	.7	.4

These two components do not look very different from the principal components I and II in Table 15.6 except for a change of signs in Q, and they account for almost as much of the observed variation, namely 4.6, as against 4.9 for I and II. But their make-up is not the same, as we shall see later.

Table 15.8 gives another pair of components, labelled T and S. They are "rotations" or linear functions of P and Q, namely

$$T = .8P + .6Q,$$

$$S = .8P - .6Q.$$

TABLE 15.8 The Loadings of Components T and S

	Components	
	T "Tallness"	S "Chestiness"
Height	1.0	.3
Leg Length	.7	.4
Arm Length	.9	.4
Chest	.3	1.0
Girth	.2	.8
Weight	.8	.3

To interpret T and S we note that T is very highly loaded in Height (a correlation of 1, to the nearest first decimal). It may therefore be regarded as a factor of "Tallness". Correspondingly, component S may be regarded as a measure of "Chestiness".

Table 15.9 gives still more examples, X, Y, and Z, but these are not functions or "rotations" of an earlier solution. Component X is perhaps another measure of general "Size", or maybe better called "Tallness". Component Y, unlike any of the previous components, differentiates between Height and Leg Length (a positive correlation of .4 with L and a negative one of $-.4$ with H). It is therefore a measure of "Legginess" (being long in the leg relative to one's height). Similarly, component Z differentiates between Chest Circumference and Waist Girth, so it can be interpreted as reflecting an "Hour-Glass" shape (big chest and small waist or vice versa). However, Components Y and Z only account for 7% and 6% of the total variance.

As noted earlier, the analysis usually stops when the components or factors have been *named* and some kind of structure has been imposed on the initial data.

TABLE 15.9 A Three-Factor Solution

	Component		
	X "Size"	Y "Legginess"	Z "Hour-Glass"
Height	.9	-.4	.2
Leg Length	.9	.4	.2
Arm Length	.9	-.1	-.2
Chest	.4	.1	.3
Girth	.4	.1	-.3
Weight	.4	-.3	-.2

Components and Factors

Instead of concluding the formal analysis when the components have been "named", each individual in the sample could now be given a score or value on each of these new variables. These scores can be calculated from the linear equations in the initial test variables that define the components, as stated at the very beginning of this section. But in practice this is seldom done.

Factor analysis differs in this respect in that the new variables cannot actually be calculated. This is because of differences in the definition of the techniques. But since in component analysis the values of the new variables are seldom calculated, this difference between the two techniques may not be crucial. In other respects the two techniques are rather similar. Factor analysis used to be practised more in the first half of the century, but component analysis has become far more popular now, perhaps because the computational difficulties connected with principal component analysis have been by-passed by the computer.

As already stated, component scores are seldom calculated (or used) in any subsequent work. But it is helpful to illustrate their nature. The principal components in Table 15.6 are given in terms of the six body measurements by the equations

$$\mathrm{I} = .12H + .11L + .13A + .09C + .09G + .11W,$$

$$\mathrm{II} = .27H + .14L + .12A - .46C - .47G + .15W,$$

$$\mathrm{III} = .23H - 1.47L - .27A - .44C + .54G + 1.47W.$$

(The coefficients are derived by matrix algebra from the initial correlation matrix in Table 15.4 and the loadings matrix in Table 15.6.) Each principal component is therefore a function of all or most of the variables. This is typical of most components that are reported in practice (and implicitly, of many factors).

Some of the alternative solutions given above were, however, deliberately chosen to differ in this respect. For example, components P and Q in Table 15.7 are functions of only two variables, Height H and Chest Circumference C,

$$P = .6(H + C),$$

$$Q = .8(H - C).$$

The components S and T in Table 15.8 are "rotations" of P and Q, i.e. linear functions of the form

$$S = .8P + .6Q,$$

$$T = .8P - .6Q,$$

so that

$$S = .8 \times .6(H + C) + .6 \times .8(H - C) = H,$$

$$T = .8 \times .6(H + C) - .6 \times .8(H - C) = C,$$

to a close degree of approximation. Thus any of the original variables can itself be a "component".

All of the alternative solutions discussed earlier were derived by deliberately choosing some function of the initial variables. For example, the "Legginess" component Y in Table 15.9 is simply $1.3(H - L)$, and the "Hour-Glass" component Z is $1.6(C - G)$. It is therefore not surprising that these two components correlated with the test variables in the way they did.

There is nothing wrong with such "arbitrary" methods of constructing components. A component is merely *any* linear function of the test variables. The justification and meaning of any particular component lies in what it is empirically found to measure.

Problems of Interpretation and Use

Component analysis and factor analysis are highly quantitative techniques, but the interpretation of their results tends to be *qualitative*, some new variables are "named". Whether it is useful to have an analysis which asserts that body measurements are mainly made up of three "factors", "Size", "Shape", and "Legginess", must at best remain a matter of personal judgment.

In general, even factor analysts make little *numerical* use of their results in further work. (Instead, they carry out new factor analyses.) There are several reasons for this.

One is that the "parsimony" achieved in reducing a large number of test variables to a few components generally does not result in any of the original variables actually being discarded, but just in adding additional ones, the new components. (As we have seen, components I, II, and III typically are functions of *all* the initial variables.) Moreover, before the analysis these new variables were altogether unknown quantities, not even their existence was known. Therefore no previous results exist with which these components can be directly compared. This is what occurs in every such analysis.

Next, the concern with the proportion of the total variance that each component accounts for leads to difficulties. In an analysis such as that of Table 15.9, the "Legginess" component would generally be discarded because it accounted for only a small proportion of the total observed variation. But this depends entirely on which other test variables were included in the analysis, a more or less arbitrary decision. If other measures of leg length (e.g. length of shin and length of thigh bone) had been included, legginess would be a numerically "important" factor. (In this particular respect, component and factor analysis have been likened to a sewer: "What you get out of it depends on what you put into it.".)

Factor and component analysis are often said to be useful at the early stages of analysing new kinds of data, when nothing much is known. But such a degree of ignorance cannot occur often. Furthermore, the techniques do not facilitate comparisons across other sets of data and the building up of generalizable quantitative knowledge. Both the input and the output (the "loadings") are correlation coefficients. As seen in the last chapter, if the correlations or loadings in two studies are the same, it does not follow that the relationships in question are the same. The process of standardising the test variables makes comparisons even more difficult. In one set of data a variable might need to be divided by a standard deviation of 3, in the next by a standard deviation of 5, and so on. Thus one cannot readily compare the different results, nor is this generally claimed for the techniques.

An alternative approach to multivariate data is to start with more structured data. For instance one could study the systematic interrelationships between different body measurements for different groups of children (differing by race, age, sex, etc.). Chapter 9 gave examples where the aims were broadly the same as here, to reduce a large number of variables to simpler patterns. Only the methods and the results differed.

The results of a factor or component analysis are usually not altogether "obvious" even after the analysis, especially with regard to the quantitative details. One is therefore relying on the particular version of the analytic technique used to "justify" one's findings (e.g. "as shown by factor analysis with 'Maximax' type of rotation"). While component and factor analysis are often regarded as *objective* ways of reducing large numbers of variables, in fact they are not. The choice of variables to put into the analysis, whether to use factor analysis or component analysis, the choice of a particular version of these techniques, the choice of a particular type of "rotation", and the interpretation of the results are all *subjective*. There is nothing inherently wrong with subjectivity since judgment has to be exercised in many problems in data analysis. But the adoption of some arbitrary analytic technique should not be regarded as objective justification for one's results.

15.3 Other Multivariate Techniques

Other statistical techniques of multivariate analysis include discriminant analysis, cluster analysis and multi-dimensional or non-metric scaling. These techniques are as yet less widely used than multiple regression and component or factor analysis.

Cluster analysis and multi-dimensional scaling are currently modish and have received considerable impetus through the increasing availability of computer programmes which take the drudgery out of the arithmetical calculations. Where factor analysis seeks to establish groups of test variables which measure the same thing (i.e. are "highly loaded" in the same factor), cluster analysis aims to find groups of items (e.g. people) that are relatively similar to each other. (An analogy from biological taxonomy would be the classifying of plants or animals into species, but this particular set of results was not achieved by any method of *statistical* cluster analysis.) Multi-dimensional scaling also seeks to establish groupings in data, but it uses less information (or fewer assumptions) than the other techniques.

However, when facing many variables, the usual analysis problem is not that we have little information, but that we do not know how to interpret and integrate the information that we already possess. None of these techniques seem to have shown themselves capable of building a growing body of coherent knowledge.

Discriminant analysis is an older technique with a rather different purpose. The question posed is whether two sets of objects differ from each other. A typical example from anthropology is two sets of skulls found in different locations. Are they different or do they stem from the same kind of people? For any one measurement (e.g. depth of cranium) it may not be clear whether the two sets of skulls differ "significantly". The two observed *means* differ, but there is a good deal of overlapping scatter in the readings from individual skulls. Discriminant analysis then aims to construct a linear function of a number of such measurements so that the mean of this new function discriminates "best" between the two sets of data, relative to the scatter of values for individual items.

If the difference between the means is "significant" when subjected to a statistical test of significance, then the two sets are regarded as coming from different populations. However, since the original data do not usually derive from any explicit random sampling from a larger population, it is not altogether clear what this kind of procedure really signifies.

15.4 Summary

Multiple regression and component or factor analysis are relatively advanced statistical techniques used to structure the relationships between large numbers of variables in a single set of data.

Multiple regression gives the "best fit" to a single set of multivariate data, and component and factor analyses create a small number of completely *new* variables to explain the large number that have been observed.

Both types of technique suffer from difficulties in interpreting the results. Since the analyst can choose among many variations in the techniques, they are less "objective" than might appear. They also lack the discipline of having to obtain reproducible and generalisable results, since the techniques are not designed to lead to such findings.

CHAPTER 15 EXERCISES

Exercise 15A. Calculating the Multiple Regression Equation

Using the data in Table 15.1 and the formula for the partial regression coefficient in Section 15.1, calculate the multiple regression equation $y = 8t - f + 88$. Also calculate the multiple correlation coefficient $R = 0.9$.

Discussion.

The partial regression coefficient of y on t, with f "partialled out" or "held constant", was defined as

$$\frac{r_{yt} - r_{yf}r_{tf}}{1 - r_{ft}^2} \times \frac{s_y}{s_t}.$$

We therefore need to calculate the pair-wise correlation coefficients r_{yt}, r_{yf}, and r_{ft} and the standard deviations s_y and s_t. Applying the formula of Exercise 14Q to the data in Table 15.1, the reader should obtain the values

correlation between y and t, $r_{yt} = .92$,

correlation between y and f, $r_{yf} = .67$,

correlation between f and t, $r_{ft} = .79$,

and $s_y = 30.8$, $s_t = 4{\cdot}0$. The partial regression coefficient of y on t is therefore

$$= \frac{.92 - .67 \times .79}{1 - .79^2} \times \frac{30.8}{4.0}$$

$$= 8.0.$$

Similarly, the partial regression coefficient of y on f with t partialled out is

$$\frac{.67 - .92 \times .79}{1 - .79^2} \times \frac{30.8}{3.2} = -1.5,$$

$$= -1 \text{ to the nearest whole number for simplicity.}$$

(Rounding this coefficient to -2 makes virtually no difference to the fit of the equation.) The intercept-coefficient is calculated by substituting the means of three variables in the equation with these two coefficients, i.e.

$$150 = 8 \times 10 - 1 \times 18 + \text{intercept-coefficient},$$

so that its value is 88.

The multiple correlation R can also be calculated from the pair-wise correlation coefficients. The formula given in specialist texts for the multiple correlation of y with f and t is

$$R^2 = \frac{r_{yt}^2 + r_{yf}^2 - 2r_{yt}r_{yf}t_{ft}}{1 - r_{tf}^2}$$

$$= \frac{.92^2 + .67^2 - 2 \times .92 \times .67 \times .79}{1 - .79^2}$$

$$= .84,$$

so that R is about .9. (In Section 15.1 we calculated the same value of R from an alternative formula in terms of the variance of the residual deviations.)

Exercise 15B. The Linear Multiple Regression Model

What are the main assumptions in the linear multiple regression model?

Discussion.

First, that the observed relationships are (approximately) linear. If they are not, this can often be overcome by transforming one or more of the variables.

Second, that there is no interaction among the independent variables. In our example this means that the effect of a unit change in fertiliser f does not depend on the temperature t. But there is no reason why this should be the case. Indeed, at normal temperatures, application of a suitable fertiliser will have a positive effect on yield, but at extreme temperatures nothing grows so the level of fertiliser will have no effect at all. In less extreme cases the two factors may still interact. If there is interaction, it will show up in systematic patterns of the deviations from the fitted line. Sometimes this can be overcome by non-linear transformations of scale.

Third, that one has selected the main variables that matter. If not, this will show up in any other set of data because the relationship between y, f, and t will be very different. This should mean that some other factor or factors has not been allowed for.

Fourth, that the deviations or "errors" from the fitted line fulfil certain statistical requirements, without which the least-squares regression method is not technically valid. For example, there has to be "homoscedasticity" (i.e. the average size of the "errors" has to be constant all along the line). Next, the "independent" variables, t and f, have to be free from any errors of measurement or the like (which is seldom true). The errors in the *dependent* have to be normally distributed and independent of each other, or "serially uncorrelated", which is often not the case in time-series, for example. (More complex regression procedures have been developed, especially in econometrics, for situations where some of these assumptions are not true.)

Finally, that each variable covers a good deal of its relevant range of variation for various *different* values of each of the other variables. Otherwise the equation mostly consists of extrapolations beyond the observed data.

Exercise 15C. The Best-fitting Line for Other Data

The multiple regression equation is the "best-fitting" line for the given data in the least-squares sense. Will it also provide the best fit for other data?

Discussion.

Suppose the equation in question also holds for another set of data, in that it goes through the new mean values and has irregular deviations of about the same size as in the original data. Then there will almost certainly be other linear equations which can fit the new data better (such as the multiple regression equation for that set). There is no reason why the original line should be the best-fitting one for the new data, even when it fits at all. This implies that mechanical application of a "best-fit" criterion gives results which generally cannot hold for other data. There is nothing in the literature of statistics, either in the theory or in the results, to indicate the contrary. As Corlett (1963) has put it in his *Ballade of Multiple Regression*:

"Your optimum only is bonum
for the data you've fitted it to!"

Exercise 15D. A Text-book Example

In his *Principles of Econometrics*, Henri Theil (1971, p. 101) gives for his basic example of multiple regression some time-series data on textile consumption (C), real per capita income (I), and the relative price of textiles (P) in the Netherlands from 1923 to 1939 (indexed on 1925 and transformed to logarithms, as shown in Table 15.10).

The question Theil poses is that since classical demand theory indicates real income and relative prices are the variables that determine the consumption of various commodities, to what extent does statistical data show these variables account for the variation in textile consumption over time? Answer his question.

TABLE 15.10 Time-Series of Dutch Textile Consumption

Year	Volume of Textile Consumption per Capita * (1)	Real Income per Capita * (2)	Relative Price of Textiles * (3)	Logarithms of Columns (1)–(3) $\log_{10}(1)$ (4)	$\log_{10}(2)$ (5)	$\log_{10}(3)$ (6)
1923	99.2	96.7	101.0	1.99651	1.98543	2.00432
1924	99.0	98.1	100.1	1.99564	1.99167	2.00043
1925	100.0	100.0	100.0	2.00000	2.00000	2.00000
1926	111.6	104.9	90.6	2.04766	2.02078	1.95713
1927	122.2	104.9	86.5	2.08707	2.02078	1.93702
1928	117.6	109.5	89.7	2.07041	2.03941	1.95279
1929	121.1	110.8	90.6	2.08314	2.04454	1.95713
1930	136.0	112.3	82.8	2.13354	2.05038	1.91803
1931	154.2	109.3	70.1	2.18808	2.03862	1.84572
1932	153.6	105.3	65.4	2.18639	2.02243	1.81558
1933	158.5	101.7	61.3	2.20003	2.00732	1.78746
1934	140.6	95.4	62.5	2.14799	1.97955	1.79588
1935	136.2	96.4	63.6	2.13418	1.98408	1.80346
1936	168.0	97.6	52.6	2.22531	1.98945	1.72099
1937	154.3	102.4	59.7	2.18837	2.01030	1.77597
1938	149.0	101.6	59.5	2.17319	2.00689	1.77452
1939	165.5	103.8	61.3	2.21880	2.01620	1.78746

*Index base 1925 = 100.

Discussion.

The multiple regression equation for the data (in terms of the log variables) is

$$\log C = 1.14 \log I - .83 \log P + 1.37.$$

The reader can calculate this from the data, using the formulae set out in Exercise 15A. The equation fits to within a residual standard deviation of about .022 in $\log C$ units, which is small compared with the range of variation of $\log C$ from 2.00 to 2.22.

Exercise 15E. Alternative Equations

The multiple regression equation gives the "best" fit to the textile data, but there must be other equations that fit almost as well. Derive one.

Discussion.

Suppose we look at the last three columns of Table 15.10:

Column (4) increases fairly steadily from 2.00 to 2.22.
Column (5) varies little and irregularly between 1.99 and 2.02,
Column (6) decreases almost steadily from 2.00 to 1.79.

Columns (4) and (6) therefore complement each other. They run from 2.0 to about 2.2 and from 2.0 to about 1.8, and the two figures generally add to about 4. So we have approximately that

$$\log C + \log P \doteqdot 4.$$

Table 15.11 shows this more clearly (in the form $\log CP \doteqdot 4$).

TABLE 15.11 Textile Consumption, Price, and Income

(Netherlands, 1923-1939)

Year	log C	log P	log CP*	log I	log CP/I**
'23	2.00	2.00	4.00	1.99	2.01
'24	2.00	2.00	4.00	1.99	2.02
'25	2.00	2.00	4.00	2.00	2.00
'26	2.05	1.96	4.01	2.02	1.99
'27	2.09	1.94	4.03	2.02	2.01
'28	2.07	1.95	4.02	2.04	1.98
'29	2.08	1.96	4.04	2.04	2.00
'30	2.13	1.92	4.05	2.05	2.00
'31	2.19	1.85	4.04	2.04	2.00
'32	2.19	1.82	4.01	2.02	1.99
'33	2.20	1.74	3.99	2.01	1.98
'34	2.15	1.80	3.95	1.98	1.97
'35	2.13	1.80	3.93	1.98	1.95
'36	2.23	1.72	3.95	1.99	1.96
'37	2.19	1.78	3.97	2.01	1.96
'38	2.17	1.77	3.94	2.01	1.93
'39	2.22	1.79	4.01	2.02	1.99
Av.	2.12	1.87	3.99	2.01	1.98

* log CP = log C + log P. ** log CP/I = log C + log P - log I.

We also note from Table 15.11 that the small variations that exist in this sum tend to be in line with the small variations in the income variable, $\log I$. If we subtract $\log I$ from $\log C + \log P$, we have a quantity that varies even less. It is more or less constant at about 1.98. Hence we have the equation

$$\log C + \log P - \log I = 1.98, \quad \text{or} \quad \log C = \log I - \log P + 1.98.$$

The residual standard deviation of this equation is .025, compared with .014 for the multiple regression.

The new equation is similar to the multiple regression shown in the last exercise, but much simpler. The intercept-coefficient, 1.98, is the only coefficient that is not unity. If rounded this coefficient is 2.0, which is the log of 100. It is primarily a scale-factor caused by all the variables being indexed as "1925 = 100". (If the indices had been expressed as "1925 = 1.00", this coefficient would effectively be unity as well.)

Exercise 15F. The New Equation $\log C = \log I - \log P + 1.98$
Consider the meaning of this new relationship.

Discussion.
Eliminating the logarithms, the relationship reads $CP = 96I$ in the indexed variables C, P, and I ("1925 = 100"). Expressed in terms of real data, the relationship will therefore be

$$CP = kI,$$

where k is some numerical coefficient which depends on the units of measurement of C, P, and I. (The multiple regression would correspondingly read $CP^{.83} = \hat{k}I^{1.14}$, where $\hat{k}$ is a numerical coefficient reflecting the antilog of the intercept-coefficient 1.37 and the change to the original units of measurement.)

The product CP stands for the per capita volume of textile purchases times their relative price, and thus equals per capita *expenditure* on textiles. The relationship $CP = kI$ therefore says that per capita expenditure on textiles was an approximately constant proportion of consumer income.

However, real income varied very little over the 17 years in Table 15.11 (the fluctuations average at about 5%). We therefore have as a simpler approximate result that CP, per capita expenditure on textiles, was approximately *constant*.

(One further question is to what extent the adjustments of incomes in turning actual incomes into "real" incomes reflect the variation in price levels used to adjust textile prices to "relative" prices? An examination of this point may also throw light on the small decreasing trend in the quantity ($\log C + \log P - \log I$), in the last column of Table 15.11.)

Because of the simplicity of the $CP = kI$ relationship, it is relatively easy to move away from the "1924 = 100" type of index and "adjusted" figures to examine the data more openly in terms of simple systems of relationships. We may also be moving from curve-fitting to economics.

The data here may seem exceptionally simple. But if multiple regression in skilled hands produces such results in a *simple* case, what can it do in more complex ones?

Exercise 15G. Other Things Being Equal
Under what conditions can the fitted equations be applied?

Discussion.
Theil (1971, p. 116) interprets the multiple regression equation

$$\log C = 1.14 \log I - .83 \log P + 1.37$$

as meaning that when real incomes go up by 1 per cent and price goes down by 2 per cent, textile consumption will go up by about $1.14–2(-.83) =$ 2.8 per cent, if all other determining factors remain the same (the common "ceteris paribus" assumption of economics). Since he does not establish what these other factors are, this assumption can only mean that the equation will hold in all those other situations where it will hold.

There is no reason for such a situation ever to arise. For example, if the initial analysis had been from 1923 to 1938 instead of to 1939, multiple regression analysis would have given an equation with different coefficients. Such *different* equations cannot all hold for other data.

In contrast, the coefficients of the equation $CP = kI$ do not depend on the almost haphazard selection of readings. Therefore it is possible for this equation to generalise. Whether it does is a matter for empirical research. No doubt there will be cases where it does not hold, where the relationship between the variables is different. This one also has to determine and explain.

Its simplicity makes it easy to compare the equation $CP = kI$ with other cases, e.g. for Dutch textiles in other years, for textiles in other countries, for other products in Holland or in other countries, when volume goes *down* (as well as for the present case when volume went up), when (real) income levels change, and so on. In this way, we can build empirically based knowledge of the conditions under which the result can be applied and of those where it cannot.

Exercise 15H. Prior Knowledge

How can any other data on consumption, price, and income now be analysed?

Discussion.

In introducing the Dutch textile example, Theil stated that the classical demand theory of economics indicates real income and relative prices as the variables which determine the consumption of a commodity (see Exercise 15D). Now we have more specific theoretical knowledge: for the Dutch textile data, $CP \doteqdot kI$. The question is whether this equation holds for the new data. If it does not, the question is what the discrepancies are like, what an alternative relationship would look like, and how far would this new equation generalise? There is no longer any need for a "blind search" procedure (such as looking for a "best fit") when faced with another set of data.

Exercise 15I. Factor and Component Analysis

What is the main difference between factor analysis and component analysis?

Discussion.

A component is a multivariate linear equation (or weighted average) of all the standardised observed variables, i.e.

$$\text{Component} = aH + bW + cG + \cdots,$$

where $a, b,$ and c are numerical coefficients. Different methods of component analysis will determine different numerical coefficients.

In contrast, a factor is a new variable that cannot be directly expressed in terms of the observed data. This makes it more difficult to deal with.

Exercise 15J. Creating New Variables

Are there any constraints on the kinds of components or factors one may construct?

Discussion.

By definition any factor or component is a new variable. Any linear combination of the test variables may be put forward as a new component. There is not even any restriction to the given test variables, which are only an arbitrary collection anyway.

Exercise 15K. The Comparison of Different Component Analyses

Table 15.12 reproduces the loadings of two components, P ("Size") and Q ("Thinness"), on the six test variables.

TABLE 15.12 The Loadings on Components P and Q

	Components	
	P "Size"	Q "Thinness"
Height	.8	.6
Leg Length	.7	.2
Arm Length	.8	.4
Chest	.8	-.6
Girth	.6	-.5
Weight	.7	.4

Suppose that the same set of loadings on these six variables were found in another study, for two components provisionally labelled D and E. Are P and D the same, and Q and E?

Discussion.

Each component analysis says that a variable exists, D in one case, P in the other, which has correlations .8 with H, .7 with L, etc. It does not follow that D and P are the same since correlations cannot determine that sort of thing. Many different variables can have correlations of .8 with H, .7 with L, etc. These variables may be correlated to some extent, but they need not (and generally will not) be the same—we cannot tell. (It is possible for two variables, X and Y, each to have a correlation of .7 with a third variable, Z, but to be completely uncorrelated with each other.)

The comparison problem is made worse because all the variables in component or factor analysis are standardised. Thus the heights in one study may have a mean of 45 inches and a standard deviation of 3 inches. The standardised height variable H in the resulting factor analysis is treated as

$$H = (\text{height in inches} - 45)/3.$$

In the second study the heights may have a mean of 60 inches and a standard deviation of 2 inches. The standardised height variable $\hat{H}$ in that component analysis is then

$$\hat{H} = (\text{height in inches} - 60)/2.$$

The two standardised height variables, H and $\hat{H}$, are therefore not the same. A child 51 inches tall will have a score of $(51 - 45)/3 = 2$ for H and of $(51 - 60)/2 = -4\frac{1}{2}$ for $\hat{H}$. If component P has a correlation of .8 with H, and component D has the same correlation of .8 with $\hat{H}$, it certainly is no evidence that D and P are the same (i.e. that any particular individual would have the same score on both D and P).

In principle, when interpreting different component analyses it might be possible to make allowances for the different means and standard deviations of the test variables in each set of data. But in practice these values are usually not even reported. Furthermore, a component is generally a linear function of *all* the test variables, but sometimes one or more test variables are changed from one study to another. It is then generally impossible for any component in the first study to be the same as a component in the next study.

The literature of factor and component analysis does not usually claim either that there are any procedures for comparing results from different studies, or that the quantitative results of different analyses are the same.

Exercise 15L. Useful for an Initial Look?

It is frequently suggested that factor analysis is useful for sorting out data where there is no prior knowledge of the relationships between the variables.

Discussion.

If there is no prior knowledge, there can be no way of judging whether one type of factor solution is more "meaningful" than another. If such an interpretive judgment of the results is nonetheless made, e.g. that it is "reasonable" for a factor to be heavily loaded in both chest circumference and girth, then the implied knowledge could have been used to analyse the given data in the first place. ("The correlation between C and G in Table 15.4 is high, now that we look!")

Exercise 15M. The Analysis of Structured Data

How can the techniques of multivariate analysis discussed in this chapter be applied to the kinds of multivariate data discussed in Chapters 9 and 10?

Discussion.

Techniques like multiple regression and factor analysis apply to a single *unstructured* set of data. There is no way of using the techniques on two or more sets of (structured) data. In Chapters 9 and 10 we had readings on the different variables for different *groups* of items, e.g.

(i) body measurements for children of different ages, races, etc.,
(ii) size measurements for trees from different rootstocks,
(iii) attitudinal responses to different brands, among users and non-users of each, etc.

The only way to apply the present techniques to such data is to *pool* the different sets of readings, thus losing one's prior knowledge of the differences between them.

If the experimenter or analyst has used his prior understanding of the phenomena to obtain data in a structured form, he does not need the techniques of multivariate statistical analysis discussed in this chapter. The procedures in Part II will suffice.

Exercise 15N. The Naming of Factors or Components

What is the point of a form of analysis which merely creates new variables?

Discussion.

It would be unfair to dismiss component or factor analysis outright just because it merely creates new variables and calls them names. For example, it *could* be valuable to establish that human intelligence is made up of "general ability", "verbal ability", "spatial ability", "arithmetical ability", and other such factors. Again, an art critic might distinguish paintings according to the extent to which they have "a strong line", "a sense of composition", "a sensitive use of colour", together with more specific factors like "a Rembrandesque handling of light", "a Rubenesque feel for flesh tones", etc.

Such concepts are of practical use if they achieve a certain concensus. In particular, in scientific work one does not accept a single analyst's results or opinions. Instead, detailed numerical agreement of the results of different investigators' empirical results is needed, and it is this which factor and component analysis have failed to supply.

PART IV: SAMPLING

A sample is a selection from a given set or "population" of items. The purpose of sampling is generally to save time, effort, and money by dealing with a sub-group instead of the whole population.

To be useful, a sample must be more or less representative of the population from which it was selected. There is always *some* loss in accuracy by dealing with a sample, but this may be acceptable if the amount of error is known.

A major contribution of modern statistics has been the theory and practice of random sampling. Procedures of taking a random sample are outlined in Chapter 16. The major theoretical concept is the *sampling distribution.* This describes how different samples vary, as discussed in Chapter 17. Making a statistical inference from a specific sample to the population is discussed in Chapter 18. In general, the larger the sample the smaller the possible sampling errors. Problems of statistical inference therefore matter most when dealing with small samples.

CHAPTER 16

Taking a Sample

In this chapter we discuss the physical process of selecting a sample from a population. The detailed work is not necessarily difficult, but tends to be carried out either by specialists or by people with previous experience. However, the general reader needs some appreciation of sampling operations in order to understand the nature of the results.

16.1 The Purpose of Sampling

In Part III we discussed the incidence of boys and girls in different families. The results were based on a total of 5,017,632 children: all the legitimate births recorded in Saxony in the years 1876–85 (Giessler, 1889). It seems obvious that virtually the same conclusions would have been obtained from a tenth of the children, still half a million readings, or even from less, if the right kind of sample had been taken.

A good sample of a few thousand is accurate enough for almost any purpose, and in many cases a few hundred or less will do. Some studies involve much larger numbers, and complete "censuses" of every item are still not uncommon. This is often because adequate information is required about many separate sub-groups of the total population, and so the total number of readings builds up.

Little is lost by taking a sample and cutting down the time, effort, and money required to collect and analyse extensive data. Another advantage is that more care can usually be devoted to the individual measurements in a small sample study. Sometimes sampling is more than a convenience, as when the act of measurement seriously interferes with the object being measured, e.g. in testing matches to see that they will burn. Sampling then becomes essential.

But just any sub-set of the population will not do. A sample must be more or less representative of the population being studied. But even with the best forms of sampling some degree of accuracy is usually lost. Thus sampling is largely a matter of economics, balancing the size and cost-implications of the possible sampling errors against the costs of full data collection and analysis.

The statistically safe way of selecting a sample is to pick items "at random". But in many situations no one would dream of "putting all the items in a hat, shuffling them, and pulling out the required number". Most sampling situations are not statistical in this sense at all.

For example, we usually just take a sip from a cup of tea to see if it is hot. A doctor takes a blood sample by deliberately selecting a vein near the surface. In seeing whether a goose is cooked, we usually check with a small cut in one of its legs or its breast. In starting up a new barrel of beer, the barman runs off a couple of pints and then checks the *third* pint to see that the beer is not cloudy.

In none of these cases is the sample "statistically" representative. Yet the information remains valid. This is because we have a great deal of prior knowledge and understanding of the situation. We either know that the material is more or less homogeneous, so that it does not matter greatly which sample item one selects, or we know how different parts of the system are related.

Statistical sampling is required when there is *substantial and unpredictable variability*. An example is measuring the proportion of households which own freezers. We know that some households own one and some do not: there cannot, in a sense, be greater heterogeneity than that. We also know that the better-off are more likely to own one, but otherwise we are rather ignorant about who owns what. This is the kind of situation where statistical sampling is usually required to obtain a representative result.

16.2 Types of Statistical Sampling

A sample is a selection of objects or measurements taken from a specific *population* of such items. The aim is to make the results from the sample tell us more or less what we would have found by measuring the whole population. There are various ways of selecting a sample, but only with random sampling is it possible to know how representative the sample results are likely to be.

Two popular forms of selecting a sample are "convenience sampling", e.g. selecting the first 100 on a list because that is easiest, and "plausibility sampling", e.g. selecting the middle child in each family because it should be "average". Both types of procedure can give consistently wrong results. Furthermore, neither the nature nor the size of the sampling errors are usually known. These methods are not acceptable as possible forms of *statistical* sampling, i.e. selection procedures which on their own provide representative results.

An alternative is to select a sample *systematically*, e.g. by going down a list and taking every other or every tenth item or whatever, depending on how large a sample is needed. This might appear to provide a good cross-section of the population, but it can still lead to biased results.

For example, if we aim to sample half the houses in a street, we could systematically select every other house. This might result in all the houses with even numbers. These could all be on one side of the street and would generally have their front doors facing in the same direction; they could also be the older and larger houses, say. The results could well be very different from the other houses in the street, and hence be unrepresentative. Furthermore, we would not generally know how big a bias and even what *kind* of bias is involved. We would therefore not know how well, or how badly, the sample represents the particular population, here the whole street.

However, there are many situations where systematic sampling is used successfully. This is when enough is already known about the population to be sampled to suppose that it has no obvious or dramatic regularities. Certain forms of "purposive" sampling can also give representative results. An example is "quota sampling" in social and market surveys involving human populations. Here each interviewer is set certain quotas of different types of people to interview (e.g. five white-collar workers aged 45 to 65, three non-working housewives aged under 35 with husbands in a manual occupation, etc.). Each interviewer personally selects the individuals within each "quota". This is where bias can enter despite additional rules and regulations; e.g. not more than two persons to be selected in one street, and the interviews restricted to specified districts (usually selected on a random or probability basis).

It has been shown empirically, by checks against known data, that well-controlled quota sampling gives representative results for quite a wide range of topics (e.g. certain forms of mass-behaviour which are not closely related to those demographic factors where quota samples might still be biased, such as for people in particular occupations). Quota sampling is generally cheaper than the main alternative, random sampling. But the crucial point is that quota sampling can only carry much conviction in those limited situations where previous experience has shown it to work.

Random Sampling

Sample bias is generally avoided by using a random or probabilistic procedure for selecting the sample. This is a method that works whatever the nature of the population sampled. Leaving things strictly to chance involves very precise procedures, it is not to be confused with being haphazard. It certainly does not eliminate sampling errors, but their likelihood can be calculated theoretically. An error which is *known* is no longer simply an error, it can be allowed for. In particular, there are many situations in which random sampling eliminates the risk of any consistent bias in the results.

Suppose again that we have to sample half the houses in a certain street and that 60% of the houses have a garage. We want to avoid samples which are heavily biased in this or any other respect. This can be done by tossing an unbiased coin for each house in the population and selecting it for the sample if the coin comes down heads. There is still a chance that *all* the houses in the sample will be ones with a garage, but the probability will be small.

The chance that the first house picked will have a garage is .6 or 60%. The chance that the next house selected for the sample will have a garage is also .6. But this selection is independent of the first (independent tosses of a coin), so that the chance of *both* houses having a garage is only $.6 \times .6 = .36$. The chance of three houses all having a garage is about .2. The probability that a sample of n houses would be made up entirely of those with a garage would be $.6^n$ (the first term in a Binomial Distribution). Less than one in every 100 possible random samples of 10 houses would consist only of houses with garages. The probability would decrease to less than 1 sample in a million (.0001%) for samples of 25 houses. Similar results apply to the chances of a sample being greatly atypical in any other respect.

We have here three basic properties of random sampling:

(i) the chance of an unrepresentative sample is relatively small;
(ii) this chance decreases as the size of the sample increases;
(iii) the chance can be calculated.

16.3 Random or Probability Sampling

Selection of a random sample proceeds with something like the following sequence of steps:

(i) all members of the specific population to be sampled are individually identified (e.g. on a list);
(ii) each item is numbered or otherwise coded;
(iii) the numbers are put on separate slips of paper;
(iv) these slips are put in a hat and thoroughly shuffled;
(v) the required number of slips is then selected blindly.

This identifies the members of the population who are to make up the sample.

The crucial step in this sequence is shuffling all the slips in the hat. This aims to produce a completely irregular or effectively random mixture. All the slips should have the same probability of being selected, and it should not matter how the slips are picked.

But, in practice, when selecting slips from a hat, most people do not simply select one after the other in a systematic manner. We usually follow some irregular practice, to be fair, like one from near the top, then one from near the bottom, and so on. We do this because we do not completely trust the shuffling of the slips. We realise that complete irregularity or randomness is difficult to achieve and add an additional stage of irregular selection.

The concept of randomness is in fact a *theoretical* one. There is no physical process which is exactly random, or if it were, we would have no means of telling. But as with other forms of applied mathematics, certain observable phenomena can be modelled by the theoretical concept *to a close degree of approximation.*

Selections from numbered slips shuffled in a hat can be tested for systematic patterns. If no noticeable ones emerge, the process can be regarded *as if* random. But in practice, such shuffling is seldom very good (slips close together when put into the hat tend somewhat to stay together). In any case, this particular procedure is clumsy when large samples are required, as is tossing a coin.

Therefore most random samples are drawn using specially prepared tables of so-called random numbers or computerised procedures for producing random number sequences. Tables of "random numbers" are prepared from processes where experience has shown that the results appear predictably irregular (e.g. certain kinds of electronic phenomena, or the results of certain mathematical calculations, like the last digits in successive square roots of a given number). But none of these processes are "really" random, and all have to be tested empirically to establish that they adequately approximate *theoretical* randomness.

The reason for all this paraphernalia is that if the physical selection process can be successfully approximated by the theoretical concepts of randomness, then the theory of probability can be used to make useful inferences about the results (as will be discussed in Chapters 17 and 18).

16.4 Technicalities of Random Sampling

In practice, random samples are usually selected using a published table of random numbers (e.g. Fisher and Yates, 1957; Lindley and Miller, 1966). Table 16.1 illustrates an extract from such a table.

TABLE 16.1 A Small Extract of "Random Numbers"

41	03	59	24	78	54	14	48	27	05
53	26	08	33	10	98	62	46	16	94
96	17	25	92	41	17	55	13	73	59
43	61	20	39	65	62	18	15	70	66
65	04	96	78	37	13	98	90	62	28

Starting at any haphazardly chosen point, one reads off successive numbers in some direction, horizontally or vertically, until n numbers are accumulated for a sample of size n. For example, suppose we want to select a sample of 5 from a list of 400 items coded in 3-digit figures running from 20 (i.e. 020 to 419). We might start with the 08 in the third column and second row, and read off groups of three digits, going from left to right. This gives

083 310 (986) 246 169 (496) 172.

The numbers in brackets are simply ignored because there is no corresponding item in the population.

Multi-stage Sampling. Various refinements to this kind of simple random sampling are used in many situations. For example, in sampling human populations it is common first to select a sample of towns (or districts in towns), and then to select individuals in the chosen towns. This is *multi-stage* sampling: first towns, then individuals. It avoids having to use a list of all individuals in the population. Instead, only a list of all towns is needed, and then lists of all individuals in the selected towns. Random sampling would be used at each stage, i.e. in selecting towns as well as individuals.

Cluster Sampling. If more than one individual is sampled per town, we have a form of *cluster* sampling, each town providing a "cluster" of individuals. Cluster sampling is often less expensive than simple random sampling. For example, in surveys involving personal interviews it is often cheaper to interview several people in the same district. But in general, cluster sampling is somewhat less efficient *statistically*. The chances of sampling errors are not reduced as much as the sheer size of the sample might imply, because people in the same district may tend to resemble each other. The cost advantages therefore have to be balanced against the estimated loss of statistical accuracy.

Stratified Sampling. The population may in the first instance be divided into sub-groups or "strata" and an appropriate number sampled at random from each stratum. For example, instead of selecting a simple random sample of 10 children from a population where it is known that boys and girls occur in a 50:50 mixture, two samples of exactly 5 boys and 5 girls may be randomly selected. With simple random sampling, only 25% of all possible samples of 10 would be split exactly into 5 boys and 5 girls. Stratified sampling ensures that every sample is strictly in the right proportion of boys and girls. The representation of any factor related to sex would also be improved.

But to select stratified samples one needs prior information on the appropriate split of the population. It must also be physically possible to assign each member of the population to his stratum *before* the sample is selected. In stratified sampling our prior knowledge about the population can be used without risk of bias, unlike the earlier case of judgment sampling. Returning to the example of sampling houses in a street, we can group all the houses by sides of the street and by whether they are corner houses, and then sample each stratum separately.

Weighting. If an unstratified sample of 10 has given 6 boys and 4 girls and we *know* that boys and girls are 50:50 in the population, the results in each stratum can be "weighted" to bring the sample into line with the population proportion (e.g. by multiplying all the girls' readings by 1.5). This kind of "posterior" stratification is usually less effective than prior stratification. The weighted portion of the sample has an undue effect on sampling errors (e.g. an untypical girl would count for 50% more than an untypical boy).

Weighting of sample results is also widely used when the sample is out of proportion for other reasons than the errors inherent in random sampling. A sample of adults may have too low a proportion of very young and very old people, who for different reasons are difficult to interview in a survey. In such cases the beneficial effect of corrective weighting on sampling errors is often less striking than seems to be thought. Furthermore, while the weighted sample will have the right proportion of old and young people it will not necessarily be representative of the right *kind* of old or young people.

Probability Sampling. The situation where each item in the population is given the same probability of selection is called simple random sampling. The possible advantages of random sampling (unbiased results with calculable chances of error) also arise in more general *probability sampling*, where items have unequal probabilities of selection. As long as these different probabilities are *known*, then the sample results can be appropriately weighted.

For example, if individuals are selected by randomly choosing one person from a sample of households, people from large households will be under-represented. But this occurs to a known degree if the size of each household is recorded. Multiplying the results for each sampled individual by the number of people in his household will then redress the imbalance. Alternatively, households can first be selected with probabilities proportional to their *size*. This is often the more efficient procedure in multi-stage sampling (for example in selecting towns and then clusters of people within them).

Variable Sampling Fractions. Instead of sampling the same proportion of readings from each stratum or cluster of the population, the proportion can be made to vary. Selecting one individual per household, whether large or small, was a case in point. In other cases a higher proportion is taken in strata which are of particular interest or where the variability of the data is known to be exceptionally great, in order to achieve greater statistical accuracy there. This occurs particularly if some important strata are numerically small in the population. For example, there may be far more patients suffering from a relatively mild attack of a certain illness than from a severe one. In studying the illness one might then sample a much higher proportion of those severely affected in order to achieve an adequate sample size of this group.

Generally the *overall* statistical accuracy of the *total* sample will then be less than that of a straight sample of the same size. Before the numbers from the different strata can be added and analysed they have to be brought back into line with the proportions in the population by weighting. The advantages of increased accuracy for certain sub-groups therefore have to be balanced against the likely reduction in accuracy in the overall results.

Sampling with Replacement. In the normal process of simple random sampling from a population of N members, exemplified by putting N numbered slips into a hat and successively selecting a sequence of n slips, the slips do not all have the same probability of being selected.

With the first selection each slip has a chance of $1/N$, but with the second selection each of the remaining $(N - 1)$ slips now has a chance of $1/(N - 1)$, and so on. This kind of "sampling without replacement" is therefore a form of "probability sampling", i.e. sampling items with different but known probabilities. This can be allowed for in the analysis, but it makes parts of it considerably more complex.

In contrast, each sampled slip could be replaced in the population before the next slip is selected. Then at each stage the probability of any slip being selected is $1/N$. This is called "sampling with replacement". (An item in the population may then be picked more than once, but it would not actually have to be *measured* more than once, as long as the relevant numerical result is counted the appropriate number of times.) Because all the probabilities of selection are equal, the theoretical mathematics of sampling *with* replacement is much easier than that of sampling without replacement.

In practice, however, most sampling is carried out without replacement because it is usually physically more convenient. This can generally be ignored in the analysis and the simpler forms of analysis appropriate to sampling *with* replacement used instead. This "fudging" is possible because the numerical results of the two types of sampling are almost the same unless n, the size of the sample, is large compared with N, the size of the population. If not, fairly simple correction formulae can often be used.

Randomised Experiments. Random sampling also plays a major role in statistically controlled experiments. To assess the effectiveness of some drug for an illness, or the effect of some fertiliser on a crop-yield, experimental control can be introduced by taking a group of patients or some plots of ground and dividing them into two sub-groups *at random.* One sub-group is treated and the other is left alone as a control.

This is an important procedure when the material under examination is very uneven. Some patients are inherently more likely to improve than others, and some plots of ground are more fertile than others. This could greatly affect the apparent results of the study. But in a randomised experiment the chance of the control group having substantially fewer spontaneous recoveries or fewer fertile plots than the treated group is relatively small and can be calculated. The theory and practical applications of the design of statistical experiments have been highly developed and are further discussed in Chapter 19.

Further Reading. The techniques of probability sampling are described more fully in various specialised texts (e.g. Kish, 1965; Moser and Kalton, 1971; Yates, 1960; Cochran, 1963: Deming, 1960; Hansen, Hurwitz, and Madow, 1953).

16.5 The Results of Taking a Sample

We now consider the results of taking a single random sample. The population sampled are the 491 households whose half-yearly purchases of Corn Flakes were analysed in Chapters 12 and 13. Suppose we have taken a simple random sample of 10 of these 491 households and measured their half-yearly Corn Flakes purchases, as set out in Table 16.2. This is now all the direct information we have about the population.

TABLE 16.2 Results for a Random Sample of 10 Households from the Population of 491

(Purchases set out in order of their size)

Half-yearly purchases of Corn Flakes	Av.
0, 0, 0, 0, 1, 1, 2, 3, 5, 7	1.9

The sample made an average of 1.9 purchases. Although we hardly expect this result to represent the population exactly, it does tell us *something.* For example, the average number of purchases in the population is unlikely to be 19 or 190, even a sample of 10 seems to tell us that. It is the role of statistical sampling theory to set more precise limits on the information that a sample gives us.

In interpreting such data we might have some background information. For example, we might know that in the previous year annual sales of Corn

Flakes in Great Britain amounted to almost 150 million packets. With about 20 million households in the country, this gave an average rate of 3.7 packets per household per half-year.

Last year's national rate of sale was therefore twice as high as the average rate of 1.9 packets in our sample data now. But we also know that there are no really dramatic seasonal trends in the consumption of breakfast cereals, and that the dominant brand's sales do not usually drop by 50% from one year to the next. That kind of thing does not happen. It therefore looks as if the specific population of 491 households did not behave like the country at large.

If we had measured all 491 households we would know. But having measured only a sample of 10 of these households, another possibility is sampling error. Perhaps the population of 491 *did* make an average purchase of about 3 or 4 packs, like the country in general, and it was only our small sample which was different. This is a typical question for sampling theory to deal with: how likely is a result of 1.9 packs for a random sample of 10 if in fact the rate was 3.7 for the population of 491?

Sampling theory can never tell us more than if we had measured the whole population, in this case all 491 households. Its role is to bridge the gap between a sample and what we want to know about the population. This we discuss in the next two chapters.

16.6 Summary

A sample can never give us more information than would the population from which it was selected. Samples are usually taken to save time, effort, and cost in data collection. With a random sample one can calculate the chance of an unrepresentative result and reduce it by choosing an adequately large sample-size.

Various elaborations, such as multi-stage stratified sampling, can be used to reduce costs for any given level of accuracy. Random sampling also plays a fundamental role in controlled statistical experiments.

CHAPTER 16 EXERCISES

Exercise 16A. Judgment Sampling

Replying on "judgment" to avoid bias in selecting a sample can be difficult. Here are 6 equally-spaced points (e.g. houses in a street:

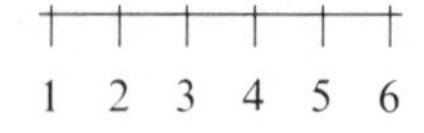

Consider how people would set about selecting a "representative sample" of two of these points.

Discussion.

Most people would first select one of the two "middle" points, say number 3. Next they would generally choose a more extreme point, over to the right like number 5, to "balance things out":

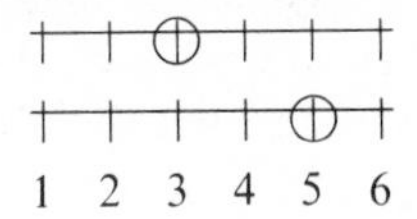

We now have a sample of two readings with a mean of 4, compared with the "population" mean of 3.5. This is close. But the *scatter* of the sample readings is smaller than that in the population. In particular, the end values 1 and 6 have been omitted completely. Most people would not pick these, they are "too extreme" or "not typical", although they account for a third of the population (and corner houses tend to be more expensive).

One could try to allow for this particular kind of bias, but generally it is found that systematic biases in judgment sampling are difficult to eliminate completely and with certainty.

Exercise 16B. "Drawn at Random"

"A set of 10 cards numbered 1 to 10 is shuffled, and three cards are drawn at random." (From "Experiment I", Brookes and Dick, 1969, p. 8.) Comment.

Discussion.

The three cards cannot have been drawn *at random.* Instead, they must have been selected in some more or less deliberate or haphazard way. The approximate randomness of the process arose from the *shuffling.* (For the three cards to have been selected literally at random, ten slips of paper numbered 1 to 10 would have had to be put into a hat, shuffled, and three selected haphazardly, giving three numbers. These would then identify a random sample of three cards. These three cards could then be selected by looking at the numbers on all *ten* cards and picking out the relevant three!) Statisticians who confuse explicit randomness with haphazard selection make a difficult concept almost impossible.

As another example, a recent examination question had a housewife go to a deep-freeze cabinet in a supermarket and "select three packets at random". But a housewife certainly does not pull out all the packets, number them, put correspondingly numbered slips of paper in a hat, shuffle, etc. Apart from the insult to housewives, the phrasing of the question ignores the whole nature of the problem, that the selection of packets is generally *biased.* Packets at the bottom of the cabinet are picked less frequently.

Exercise 16C. A Random Sample

"Table X gives the weight of hearts and the weight of kidneys in a random sample of twelve adult males between the ages of 25 and 55." (From "Example 16.1", Moore, 1969, p. 253.) Comment.

Discussion.

It seems unlikely. One can take a random sample of 12 men from the population, but how does one weight their hearts and kidneys unless they are *dead*? The sample must therefore have been from a population of dead men (or from a set of appropriate records), so it will not be representative of men generally. One also wonders if the population was aged from 25 to 55, or if that was merely the age-range for the 12 men.

A further question is why a sample of only 12 was taken instead of some more "adequate" sample like 50 or 100. It looks as though this was not a proper sample at all, random or otherwise, but just all the readings available. It is a common practice to upgrade some haphazard collection of readings by calling it "a sample".

(In point of fact, the 12 sets of readings here *were* a random sample from the records for male patients who had died at a certain hospital. But some scepticism seemed called for.)

Exercise 16D. Defining the Population

What is a *population* in statistical sampling?

Discussion.

A population has to be defined in terms of the operational conditions of observing it, and not merely by specifying a particular group of living individuals or material items.

For example, a nation's population would have to be defined *operationally* not merely as all persons "living" there, but as all those who can be observed to be alive at a given point in time, using a specified method of observation. This might then exclude nationals who were abroad, and include foreigners, whether resident or visitors. Is that what was intended?

For the practical purpose of measuring every individual in such a population or of taking a sample from it, the population must be defined in terms of a listing of all individuals (or some equivalent operating procedure). When examined closely, such a list might not include the army, prisoners, and people in hospitals, whilst other people might be listed more than once (e.g. students at home *and* at their university). A list will inevitably be out of date (some members of the list will have died, newborns will not be included, and some people will have changed addresses).

Some members of the population, although explicitly included in the list, will be difficult or impossible to observe or measure. In measuring human populations (e.g. in a sociological survey), people abroad or ill, travelling salesmen, the old and the rich, are often difficult to contact. Others may *refuse* to be measured or interviewed.

None of these problems has anything to do with sampling as such. But it emphasises that the population to be sampled is not necessarily the one we would like to sample. Instead, care has to be taken in (i) explicitly defining the population actually sampled, (ii) how this might differ from the population one had in mind, and (iii) how one can reduce or allow for any biases involved. These are matters for the technical expert experienced in dealing with the particular type of population in question.

There will always be some fuzziness in defining a population. For example, we may aim to define it as all people alive and in the country at

12 noon on a certain day. But for some people we will not know whether they were alive, or in the country, *exactly* at 12 noon. A test of whether we have a reasonably well-defined population is whether we could physically list them all and draw a *proper random sample*. If not, we need to reconsider what we are talking about.

Exercise 16E. Sampling in Time

How can we take a representative sample of events occurring over time?

Discussion.

Time is rarely sampled statistically. We would have to be able to take a proper random sample of all the relevant moments in time, the "population". This is only possible with deliberate planning. For example, suppose we wanted to make traffic counts in a certain street for a random selection of 100 one-minute intervals tomorrow. We would have to select 100 random numbers between 1 and 1,440 and then make the corresponding observations.

Again, the number of deaths per year due to measles from 1964 to 1973 is not a *sample* of such readings. No population of years has been explicitly defined from which these 10 years were selected, randomly or otherwise. Instead, the 10 years amount to a "mini-population" of their own.

Exercise 16F. Sampling the Same Population Again

Having measured the heights of a sample of boys from a certain school, can we measure another sample the next day?

Discussion.

No. Suppose that the first sample were taken by appropriately selecting a number of boys from the school list, calling them to a certain room, and measuring their heights.

Following the same procedure the next day will not necessarily give the same results. This is not because of the sampling (the samples could be large) but because the *populations* on the different days are not the same. This may seem odd because by and large we would expect the measurements of heights to be very similar on different days. But insisting that the populations differ is not splitting hairs.

We only know the measurements of boys' heights will be much the same from one day to the next because we know that people's heights do not vary in the short run. This knowledge has nothing to do either with sampling or with the definition of the populations, but with the particular measurement: heights.

If our observations concerned what the boys wore, or what they did, results could differ markedly from day-to-day, and especially between a Friday and a Saturday, say. In general, the lessons the boys took, their ages, or anything of any kind would be different. To suppose that different populations are the same is to beg the entire question. Two populations can have the same properties in some respects (e.g. the same average height, etc.), but this does not mean that they are operationally the same population.

Exercise 16G. Taking More than One Sample

Is it in fact possible to take more than one sample from the same population?

Discussion.

This can only be done in rather artificial and limited ways. For example, if from some population we take a random sample of 500 boys, numbered consecutively from 1 to 500, then the even-numbered and the odd-numbered boys selected each constitute a (smaller) random sample from the same population.

If the even-numbered boys were measured in the morning and the odd ones in the afternoon, we would be dealing with two samples from *two distinct populations*, the boys in the morning and the boys in the afternoon. (Their heights might be the same, but their activities, the state of their metabolism, etc., would not be.)

This implies that if the 500 boys were measured at one-minute intervals at 9.00, 9.01, 9.02 and so on, one would be dealing with samples of 1 from 500 populations. But this is only the case if the times when each boy is measured are recorded, and even then one may choose deliberately to oversimplify by ignoring this information and regarding the 500 boys as a random sample from "boys that day", which they are.

Exercise 16H. An Indefinite Population

"If an observation, such as a simple measurement, is repeated indefinitely, the aggregate of the results is a population of measurements." (Fisher, 1950, p. 2). Comment.

Discussion.

The concept of an indefinite population is ill-defined. How do we know that the measurements will not change in the course of time? Is the experimenter allowed to rest every now and then in carrying out the indefinite sequence of measurements? Does he have to work at night and at weekends? What happens when he dies? Who is he anyway, or is more than one experimenter involved?

It is not clear what the population is, nor how any proper (random) sample selection can take place. The items in an indefinitely large population cannot be listed and sampled. This view of a population appears to be meaningless for any practical purpose.

The alternative is to define explicitly some *limited* populations of observations (and possible samples therefrom). One can see whether such different populations have the same observed properties.

Exercise 16I. Games of Chance

Are 10 successive throws of a coin a random sample?

Discussion.

No, not according to the definitions used so far. Firstly, there is no formal random sampling procedure involved in selecting the ten throws from some larger number of throws. Secondly, there is no firmly defined population of throws.

In games of chance the outcome of each event (e.g. throwing a coin, throwing dice, etc.) can be at best a "quasi-random" event, under suitable empirical conditions (see Section 13.2). A sequence of throws can therefore behave *like* a random sample (and indeed be used in selecting "random" samples from a real population, e.g. half the houses in a street). But this is because of something in the inherent nature of each physical observation, not because any explicit random sampling procedure has been used in selecting a sequence of throws.

CHAPTER 17

Sampling Distributions

A sampling distribution summarises the variation of all possible random samples of a given size from a population. In practice, one rarely takes more than one random sample from a population, so the sampling distribution mostly remains a theoretical concept. It is used in estimating the accuracy of a single random sample from the population, as will be discussed in Chapter 18.

There are a variety of sampling distributions for different summary measures like the mean, the variance, the correlation or regression coefficients, and so on. In this chapter we concentrate on sampling distributions of the mean because they are the most important and the simplest.

17.1 Some Empirical Sampling Distributions

In an earlier chapter (Table 12.9) we looked at the half-yearly purchases of Corn Flakes for a population of 491 households and found that their mean purchase was 3.4 packs. Here we show what happens when we take a number of random samples from this population. Table 17.1 illustrates three empirical sampling distributions of the mean, for a selection of random samples of size $n = 1$, 2, and 10 from this population, and Table 17.1a illustrates the sampling distribution of the mean for samples of $n = 40$.

TABLE 17.1 Empirical Distributions of Sample Means for Random Samples of Different Sizes

(Average half-yearly Corn Flake purchases of 200 samples of 1, 100 samples of 2, and 20 of 10 households)

	Value of Sample Mean												Ave-rage*	Stand. Dev.*
	0-	1-	2-	3-	4-	5-	6-	7-	8-	9-	10-	11+		
Samples of 1 %	40	17	11	4	4	5	3	1	4	0	2	9	3.4	4.7
Samples of 2 %	29	21	12	8	5	4	5	4	1	1	2	8	3.4	3.9
Samples of 10 %	5	5	30	15	20	15	10	-	-	-	-	-	3.4	1.5

* Calculated from the ungrouped values

TABLE 17.1a The Distribution of the Means of Samples of 40 Households
(20 samples of 40 households)

	Value of Sample Mean									Ave-rage*	Stand. Dev.*
	<1.8	1.8-	2.2-	2.6-	3.0-	3.4-	3.8-	4.2-	4.6-		
Samples of 40 %	-	10	10	15	25	10	20	10	-	3.4	.8

* Calculated from the ungrouped values

Few of the individual smaller samples have means anywhere near the population mean of 3.4. For example, 40% of the samples of $n = 1$ have mean values of zero. A single sample of one reading cannot provide very accurate information about the population. But for larger samples, such as with $n = 40$, most of the sample means lie between about 2.6 and 4.5, only a unit or so from the population mean. Therefore in most cases any single sample of 40 will indicate *fairly* accurately the actual population mean of 3.4.

To describe the empirical variability of the various sample results, the sampling distributions must be specified, i.e. their shapes and summary measures like their own means and standard deviations. The illustrations in the tables show three important features.

(i) The *average* value of each sampling distribution is 3.4, which is equal to the population mean.

(ii) The standard deviations of the sampling distributions decrease markedly as the sample sizes increase, but less than proportionately; e.g. doubling the sample size does not halve the scatter.

(iii) The shape of the distributions changes as the sample sizes increase. For $n = 1$ or 2, the distribution is highly skew, but for $n = 40$ the distribution is fairly close to Normal; 70% of the readings lie less than $\pm$ standard deviation of the average mean value of 3.4.

These illustrative empirical results generalise and can be developed theoretically.

17.2 The Sampling Distribution of the Mean

Suppose we have a population with a mean μ ("mu", the Greek m) and a standard deviation σ ("sigma", the Greek s), to use the conventional statistical notation of Greek letters for population values.

Then it can be shown that the distribution of the means of all possible random samples of size n from that population has the following properties:

(i) its mean is equal to the population mean, μ;

(ii) its standard deviation is $\sigma/\sqrt{n}$;

(iii) its shape is approximately Normal, except for small samples if the population is not Normal.

These results are fundamental in statistical sampling and have a firm theoretical basis. They follow mathematically from the fact that each reading in a random sample is selected according to the rules of probability.

But in practice we do not know the values of the population parameters μ and σ. It is therefore of little immediate use to know that the sampling distribution of the mean is approximately Normal with (unknown) mean μ and standard deviation $\sigma/\sqrt{n}$.

However, we have our sample of n readings with mean m and standard deviation s. We can use these sample values to estimate the mean and standard deviation of the sampling distribution. But unfortunately it is then not true that this distribution is Normal with standard deviation of $s/\sqrt{n}$ (its standard deviation is the unknown $\sigma/\sqrt{n}$). How then can we describe the sampling distribution of the mean?

Publishing under the pseudonym "Student", W. S. Gossett (a Brewer of Arthur Guinness Ltd.) presented a solution to this problem in 1908. This marked the first breakthrough in the exact statistical treatment of small samples. He took the ratio of the sample mean m to the observed value $s/\sqrt{n}$, a quantity known as Student's "t"

$$t = \frac{m}{s/\sqrt{n}}$$

and established the mathematical nature of its sampling distribution for samples from a Normal population.

The t-distribution has a larger scatter than a Normal Distribution. For example, with samples of size $n = 5$, about 95% of the readings of a t-distribution lie within ± 2.6 standard deviations of the average. When $n = 10$ and $n = 20$, 95% of the readings lie within ± 2.2 and 2.1 standard deviations of the average. In a *Normal* Distribution, 95% of the readings lie within only ± 2 standard deviations of the average. The t-distribution therefore depends on the sample size n (or on the *degrees of freedom* of the data, which here are $(n - 1)$, as described for the variance in Exercise 11H).

The important point illustrated by the above results is that the t-distributions do not differ radically from the Normal Distribution, except for very small samples (n less than 10). Thus whether 95% of the readings lie within ± 2.2 or ± 2.0 standard deviations is only a small difference. Even for rather small samples of 10, 20 or so, having to estimate the population σ from the sample values therefore hardly affects the sampling distribution of the mean. (For large samples, say $n = 1000$ or more, this would not be surprising since s must then be a close estimate of σ anyway.)

The practical implication of Student's achievement is therefore not so much that he provided the exact answer to the sampling distribution of the mean for very small samples (since these seldom occur), but that it follows from his findings that the problem he tackled does not greatly matter. The Normal Distribution will serve in most cases as a close approximation even when using the *estimated* standard deviation $s/\sqrt{n}$.

The Standard Error of the Mean

The quantity $s/\sqrt{n}$ is commonly referred to as the *standard error of the mean* and tends to be regarded as a property of the one sample actually observed, implying a standard or "average" level of sampling error. Nonetheless, it has to be remembered that it is by definition an estimate of the true standard deviation $\sigma/\sqrt{n}$ of the distribution of the means of all possible random samples of size n from the same population.

It is a remarkable achievement to be able to estimate from a single sample how different samples might vary. We can do it because of the nature of random sampling, where each item in a sample is selected independently of all previous items. Thus, a single random sample of size n is effectively composed of n separate random samples of size 1, or $n/2$ random samples of size 2, etc. Therefore, one single sample *does* tell us about the variability of different samples, but smaller ones. The formula $s/\sqrt{n}$ then provides the link for samples of different sizes n.

The standard error formula $s/\sqrt{n}$ fits in with common sense. As the sample size n increases, the scatter of the sample means decreases, more of the sample means lie close to the population mean, μ. But the scatter does not decrease in direct proportion to the increase in sample size, since additional readings provide only marginal extra information. As we saw from the illustration in Section 17.1, the sampling distribution for $n = 1$ had a standard deviation of 4.7, but for $n = 40$ the standard deviation was .8. To estimate this figure from the first, we have to divide 4.7 by $\sqrt{40}$, which gives as an estimate $s = .75$ or .8. Thus the standard error formula uses as the divisor the *square-root* of the sample size, $\sqrt{n}$. To *halve* the standard error the sample size has to be *quadrupled*.

One can use the standard error formula to determine the sample size required for a particular degree of accuracy. For example, if we want to sample mean with a standard error of about .5 and know from previous results that the standard deviation of the data is about 4, then we would have

$$\frac{4}{\sqrt{n}} = .5.$$

A sample size of about 64 would be required. However, since the formula involves the squre-root of n, the precise sample size is not critical. For $n = 100$ the standard error of the mean would be .4, and for $n = 50$ it would be .6, neither of which is very different from .5.

17.3 The Difference Between Two Sample Means

The sampling distribution of the difference between the mean values of two independently drawn samples is of great practical importance. For example, one could have an experimental group and a control group, each consisting of a random sample from a larger population. The sampling distribution of the difference of two means can then be used to assess the accuracy of the observed difference between the two samples.

If the sizes of the two independent samples are n_x and n_y, then the sampling distribution of the differences in their means $(m_x - m_y)$ will have a standard error of

$$\sqrt{\left(\frac{\sigma_x^2}{n_x} + \frac{\sigma_y^2}{n_y}\right)},$$

where σ_x^2 and σ_y^2 are the variances of the x and y variables. This formula contains the sum and not the difference of the two separate variances because the distribution of $(m_x - m_y)$ is subject to the sampling errors of both x and y. Therefore there will generally be *more* error than for either m_x or m_y alone.

The numerical value of the standard error has again to be estimated, by using the observed standard deviations s_x and s_y to replace σ_x and σ_y. The sampling distribution follows a t-distribution with $(n_x + n_y - 2)$ degrees of freedom but, as already noted, unless the degrees of freedom fall well below 10 or the population is highly non-Normal, the t-distributions are almost identical to a Normal Distribution.

17.4 Other Sampling Distributions

Any summary measure of sample data varies from sample to sample and this variation is described by a sampling distribution. Examples are the variance, standard deviation or range of a sample, the proportion of zeros in the data, or the correlation and regression coefficients for two variables. When n is very large these sampling distributions also tend to be Normal, but otherwise they are generally more complicated than that of the mean. The subject can get very technical, but for the practical user the broad principles already discussed remain the same. For specific applications one can usually refer to suitable numerical tables showing the proportion of samples with values exceeding some particular level. We now discuss a particular example.

The χ^2-Distribution

The χ^2-distribution (called chi-squared) is an example of a more complex sampling distribution. One practical use is in assessing how well theoretical models fit sample data. This is discussed in Chapter 18; here we consider the nature of the relevant sampling distribution.

Suppose we have a sample of 200 readings with a mean of 9 and a standard deviation of 3, as shown grouped in the top line of Table 17.2. Suppose further that we *know* that the sample comes from a Normal population (or it could be from some other type of distribution, e.g. an NBD or a Poisson, etc.). The theoretical values for a Normal Distribution are shown in the bottom line (i.e. 68% of the 200 readings should lie between 6 and 12). The observed values clearly differ somewhat from the population ones. The question here is how much will the results of *different* random samples vary from these theoretical norms?

TABLE 17.2 **Observed and Theoretical Numbers for a Sample of 200 Readings from a Normal Distribution**
(Mean 9, Standard Deviation 3)

	Values						Total No. of Readings
	<3	3-	6-	9-	12-	15-	
No. of readings							
Observed	7	30	61	73	25	4	200
Theoretical	5	27	68	68	27	5	200

To answer this question the usual approach is to calculate an overall measure of agreement of fit between the observed and theoretical values and then to consider how the value of this measure might vary from sample to sample. One possible measure to use is the mean deviation between the observed and theoretical values, which here is 20/6 = 3.3. But the sampling distribution of this quantity is technically difficult to establish.

A more tractable measure is Sum {(Observed − Expected)²/Expected}, i.e. the sum of the squared differences between the observed and theoretical frequency in each group, divided by the theoretical value. In our example, this would give us

$$\frac{(7-5)^2}{5} + \frac{(30-27)^2}{27} + \frac{(61-68)^2}{68} + \frac{(73-68)^2}{68} + \frac{(25-27)^2}{25}$$

$$+ \frac{(4-5)^2}{5} = 2.6.$$

(The smaller the value, the closer the theoretical model fits the data.)

This measure is useful because its distribution for different samples can be calculated, a result due to Karl Pearson in 1900. It approximates to a χ^2-distribution, whose mathematical properties are already well-known. (The χ^2-distribution is a special case of the Gamma-distribution mentioned earlier, which occurs in several distinct ways in statistical theory.)

For our 200 readings the appropriate χ^2-distribution is one with three "degrees of freedom". This is the difference between the six categories into which the data have been grouped and the three parameters used in fitting the theoretical model (the standard size n, the mean, and the standard deviation). The theoretical χ^2-distribution is shown in Table 17.3 (from Elderton, 1902). Data with different degrees of freedom have different χ^2-distributions.

TABLE 17.3 The χ^2-distribution with 3 Degrees of Freedom

3 degrees of freedom	Values of χ^2												Total
	<1	1-	2-	3-	4-	5-	6-	7-	8-	9-	10-	11-	
No. of readings %	20	23	18	13	9	6	4	2	2	1	1	1	100%

This distribution is very skew, with a long positive tail. But it shows that 18% of samples of 200 from a Normal Distribution would give a χ^2-value of between 2 and 3. Thus the data in Table 17.3 seem quite typical: many other random samples would give similar χ^2-values.

The shape of the χ^2-distribution depends only on the number of degrees of freedom, not on the sample size. The number of different χ^2-distributions that need to be tabulated for reference is therefore limited to the possible levels of degrees of freedom. A further simplifying feature is that for relatively large degrees of freedom the distribution becomes increasingly humpbacked and symmetrical and tends to an approximately Normal form. This tendency can be accelerated by mathematically transforming χ^2 to $\sqrt{(2\chi^2)}$. Then the skewness of the distribution is reduced and the distribution is almost Normal even for relatively *low* degrees of freedom (i.e. 20 or 30). The distribution of $\sqrt{(2\chi^2)}$ has a mean of $\sqrt{(2\nu - 1)}$ and a unit standard deviation, where ν (Greek n, or "nu") is the conventional symbol for the number of degrees of freedom. The tendency of most sampling distributions to approximate a Normal shape for reasonably large sample sizes (or numbers of degrees of freedom as here) is a major simplifying feature in the general statistical theory of sampling.

17.5 Sampling Theory for Other than Simple Random Sampling

Until now our discussion has been concerned with sample data obtained by simple random sampling, but in practice most sampling is not done this way. As discussed in the last chapter, sampling is almost always done *without* replacement, and in sample surveys some form of stratified multi-stage cluster sampling is usually employed. This affects the sampling errors that occur.

The formula $\sigma/\sqrt{n}$ applies to sampling *with* replacement; sampling without replacement decreases the standard error. Dividing the population into strata before sampling should also reduce the size of sampling errors (but the effect is small if the stratifications are not very discriminating). In contrast, multi-stage cluster sampling (e.g. selecting a number of towns and then a "cluster" of individuals in each town) generally *increases* the size of sampling error, compared with a simple random sample of the same size.

Unfortunately, statistical theory has not yet told us very much about the numerical effects of most of these types of sampling. Instead, the formulae for simple random sampling tend to be used, in the hope that they will adequately approximate the true answers. Sometimes estimates are made of the "design factor", the extent to which the sample design has affected the size of the sampling errors. However, delving more deeply into these matters is largely a subject for the professional statistician and the specialist textbook.

One case that is clear is sampling without replacement. The standard error formula $\sigma/\sqrt{n}$ implies that the size of the sampling error depends on the sample size n and not on that of the population. But many people instinctively feel that a large population can only be adequately represented by a large sample. This is in fact correct for sampling without replacement: the larger the population, the larger the sampling error, the sample need therefore to be larger to give the same degree of accuracy. However, the *numerical* effect of the population size on the sampling error is trivial if the sample is a small portion of the population, as is almost always the case.

The statistician's usual assertion is that sampling error only depends on sample size and not on population size. For sampling with replacement (which nobody practises) this is true; for sampling without replacement it is in principle false, but in practice a close and simple approximation to the truth.

17.6 Summary

Statistical sampling theory deals with the way a particular summary measure of a sample, such as its mean, varies from sample to sample.

For simple random samples of size n from the same population, the sampling distribution of the mean has three simple properties. Its own mean equals the population mean μ; its form tends to be Normal except for very small samples; and its standard deviation is $\sigma/\sqrt{n}$, where σ is the standard deviation of the individual readings in the population.

The value of σ can be estimated by the standard deviation s of the observed sample. The sampling distribution then strictly requires use of Student's t-distribution, but except for very small sample sizes it differs little from a Normal Distribution with standard deviation $s/\sqrt{n}$.

The quantity $s/\sqrt{n}$ measures the scatter of the means of different samples of size n and is usually called the standard error of the mean. It is the basic formula in statistical sampling theory.

The difference between the means of two independent samples of size n_x and n_y, drawn from the same population, will have an estimated standard error

$$\left(\frac{s_x^2}{n_x} + \frac{s_y^2}{n_y}\right),$$

where s_x and s_y are the standard deviations of the x and y readings.

Sampling distributions for other summary measures are usually more complicated than those for the mean. However, for large sample sizes they generally tend to be Normal. This fact is the major simplifying feature in the statistical theory of sampling.

CHAPTER 17 EXERCISES

Exercise 17A. Empirical Samples

To illustrate the nature of random samples, draw a series of such samples from a specified population and consider the results.

Discussion.

As an example, consider taking samples of 10 from the population of 491 households whose half-yearly purchases of Corn Flakes were described in Table 12.9. The households can be arbitrarily labelled from 1 to 491, then sets of 10 numbers between 1 and 491 inclusive can be read off from a table of random numbers or the like. Doing this for 20 samples gave the results in Table 17.4.

Each sample broadly resembles the population, where most households bought either no Corn Flakes or only one or two packs, and few households bought many packs. But the resemblance is not very precise, since the results for the different samples are so variable. For instance, the numbers of non-buyers vary from 2 to 6 out of 10, and the means vary from .6 to 5.8.

This kind of sampling generates a sampling distribution for any aspect of the data one may wish to consider, e.g.

the sample means, 2.2, 3.1, 2.3, 1.7, etc.;

the percentage of zeros or non-buyers, 30%, 20%, 50%, 40%, 50%, etc.;

the standard deviations of the samples, 2.2, 4.8, 3.5, 2.1, 4.9, etc.;

the k-parameter of the Negative Binomial Distribution, 1.8, .5, .5, 1.1, .7, etc. (using the formula $k = m^2/(s^2 - m)$ in terms of the sample mean m and variance s^2) if an NBD is fitted to the data.

The sampling distributions of these other statistics are in many cases highly skew. For example, the k-values for the 20 samples in Table 17.6 are, in order of size,

.2, .3, .3, .4, .4, .4, .4, .5, .5, .7,
.7, .8, .9, .9, 1.0, 1.1, 1.4, 1.5, 1.8, 9.0.

TABLE 17.4 Twenty Random Samples of 10 from the Given Population of 491 Households in Table 12.9

(Half-yearly Purchases of Corn Flakes)

Samples of 10	Number of purchases made												Average
	0	1	2	3	4	5	6	7	8	9	10	11+*	
1st sample	3	2	1	2	-	-	2	-	-	-	-	-	2.2
2nd "	2	4	1	-	-	-	-	-	-	-	-	(17)	3.1
3rd "	5	-	2	1	1	-	-	-	-	-	-	(12)	2.3
4th "	4	3	-	1	-	1	1	-	-	-	-	-	1.7
5th "	5	-	1	1	-	-	-	-	1	-	-	(14)	3.7
6th "	6	1	1	1	-	-	-	-	-	-	-	(17)	2.3
7th "	3	3	2	-	-	1	-	-	1	-	-	-	2.0
8th "	6	1	1	-	1	-	-	-	-	-	-	(13)	2.0
9th "	5	1	1	-	1	-	-	-	-	-	-	(14, 22)	4.3
10th "	2	3	-	1	-	1	1	-	1	-	-	(16)	4.1
11th "	3	1	1	-	-	-	-	-	-	2	-	(12, 13, 14)	5.8
12th "	6	2	2	-	-	-	-	-	-	-	-	-	.6
13th "	4	1	1	-	-	2	-	-	-	1	-	(29)	5.0
14th "	2	3	2	-	1	1	-	-	-	-	-	(36)	5.2
15th "	3	1	1	-	1	2	-	-	-	1	1	-	3.6
16th "	3	2	1	-	1	1	-	-	1	-	-	29	5.0
17th "	6	1	-	-	-	-	-	1	-	-	-	(20, 24)	5.7
18th "	6	1	-	-	-	-	-	1	-	-	-	(13, 13)	4.1
19th "	2	3	-	-	-	-	-	-	-	-	1	(11)	3.0
20th "	5	1	1	-	-	2	-	1	-	-	-	-	2.7

* The actual values of the 11+ readings

The range of these sample values is large, from .2 to 9.0, and the average is 1.2. This compares badly with the k-value of .50 for the whole population (see Section 12.3). Thus using the sample value of k to estimate the population value gives biased results, i.e. the wrong answer on average.

Often the size of this type of bias can be established by theoretical analysis, so that a correction factor can be devised. This, and other methods of deriving better estimators, is part of the more advanced theory of statistics and typifies some of the more complex problems in sampling. The sample mean is a somewhat exceptional measure because it gives an unbiased estimate of the population mean and hence its sampling theory is particularly simple.

Exercise 17B. The Expected Value of the Sample Mean

Outline a proof that the average value of the sampling distribution of the mean, m, of a sample is equal to the population mean, μ. Is the equivalent result true for the range?

Discussion.

Consider random samples of two readings, selected with replacement. For a particular sample, the readings are x_1 and x_2. The sample mean is $(x_1 + x_2)/2$. The sampling distribution of the means of such samples therefore has a mean equal to the average, or statistically "expected", value of $(x_1 + x_2)/2$ over all possible samples of two readings. (The expression "expected value" is a useful way of describing the average value of a reading across all possible samples. It is what one "expects" to obtain on average.)

Because the two readings x_1 and x_2 were sampled independently of each other, the expected value of $(x_1 + x_2)/2$ equals the expected value of $x_1/2$ across all possible samples, plus the expected value of $x_2/2$ across all possible samples. Now the expected value of a single reading x_1 across all possible samples is μ, by definition the average value of all the readings in the population. Similarly, the expected value of x_2 is μ. Hence the expected value of the sample mean $(x_1 + x_2)/2$ is $2\mu/2 = \mu$, the population mean, which is the required result. The argument generalises readily to samples greater than 2.

This argument may seem almost simple-minded. But it is only possible because in random sampling with replacement the two items, x_1 and x_2, are selected independently of each other. Without this independence we could not consider the expected value of $x_1/2$ separately from that of $x_2/2$.

For instance, this kind of proof is not possible for sampling without replacement (although for the sample mean the same *result* still holds). More generally, the equivalent result will not hold for a statistical measure that is not a "linear function" of the n sample readings, i.e. a function of the form $a_1x_1 + a_2x_2 + a_3x_3 + \cdots + a_nx_n$, where the coefficients $a_1, a_2, a_3, \ldots$ are fixed *a priori*. (For the sample mean, all these coefficients are equal to $1/n$.) The variance, for example, is not a linear function of the readings in this sense. Nor is the range, because the coefficients of all but the highest and lowest readings are zero, but are not determined *a priori*.

The expected value of the range for a sample size n is always biased. For any particular sample it is always either the same or smaller than the range of all the readings in the population, it cannot be *larger*. Hence the average value of the range of all possible samples of size n cannot be equal to the population value of the range.

Exercise 17C. The Standard Error of the Mean

Outline a proof that the standard error of the mean of a single random sample of n readings is $\sigma/\sqrt{n}$, where the population standard deviation is σ.

Discussion.

The theoretical formula for the standard error of the mean holds because the items in a simple random sample are selected independently. Consider again a sample of two readings x_1 and x_2, with a mean of $(x_1 + x_2)/2$. The variance of the sampling distribution of such means is then by definition the expected value of

$$\left(\frac{x_1 + x_2}{2} - \mu\right)^2.$$

This expression can be written as

$$\left(\frac{x_1 - \mu}{2} + \frac{x_2 - \mu}{2}\right)\left(\frac{x_1 - \mu}{2} + \frac{x_2 - \mu}{2}\right)$$

$$= \frac{(x_1 - \mu)^2}{4} + \frac{2(x_1 - \mu)}{2}\frac{(x_2 - \mu)}{2} + \frac{(x_2 - \mu)^2}{4}.$$

The average or "expected" value of $(x_1 - \mu)^2$ over all possible samples is simply the population variance σ^2: that is how the variance is defined. The same holds for $(x_2 - \mu)^2$. Thus the expected value of $(x - \mu)^2/4$ is $\sigma^2/4$.

Since the values of x_1 and x_2 are selected independently, we can start considering the middle term with one particular value for x_1, say $\hat{x}_1$. Then the expected value of $(\hat{x}_1 - \mu)(x_2 - \mu)$ across all possible values of x_2 must be zero, since the average of x_2 is μ. Similarly for any other value of x_1. Hence the expected value of the middle term is zero.

It follows that the expected value of the above expression is $\sigma^2/4 + 0 + \sigma^2/4 = 2\sigma^2/4 = \sigma^2/2$. The same kind of argument can be used to show that the variance of the sampling distribution of the mean for samples of size n is σ^2/n. The standard error of the mean, i.e. the standard deviation of its sampling distribution, is therefore $\sigma/\sqrt{n}$.

Exercise 17D. The Distribution of the Sample Mean

Explain why the sampling distribution of the mean tends to be approximately Normal for large enough samples.

Discussion.

The mean of a random sample of n readings is a variable made up of the average of n independent random variables. This will tend to follow a Normal Distribution for large n, because of the Central Limit Theorem (see Section 13.2 and Exercise 13K). The detailed mathematics required to prove the Central Limit Theorem are complex, but at least it is easy to illustrate that in taking reasonably large samples from a highly skew distribution, the sampling distribution of the mean must tend to be humpbacked.

Consider the population of 491 households and their purchases of Corn Flakes, as set out in Table 12.9a. This distribution was very skew: 39% of the households made 0 purchase, 14% bought 1 pack, and only 11% bought 10 packs or more.

A sample with a large mean, say 10 or 20, would have to include large numbers of heavy purchasers. Because individual households are sampled at random and independently, the chance of this happening is very small. For example, since only 11% (0.11) of all households buy 10 or more packs, the chance of getting 10 such heavy buyers in a random sample of 10 households is $(0.11)^{10}$ or roughly 1 in a thousand million. By a similar argument, very small sample means will be rare. The sampling distribution of the mean will thus tend to be humpbacked.

Exercise 17E. The Standard Error of the Difference of Two Means

Outline a proof of the formula

$$\sqrt{\left\{\frac{\sigma_x^2}{n_x} + \frac{\sigma_y^2}{n_y}\right\}}$$

for the standard error of the difference of the means, $(m_x - m_y)$, for independent samples of size n_x and n_y from populations with means μ_x and μ_y and standard deviations σ_x and σ_y.

Discussion.

The variance of the sampling distribution of the quantity $(m_x - m_y)$ is the expected value of the squared deviations

$$\{(m_x - m_y) - (\mu_x - \mu_y)\}^2$$

across all possible samples of n_x and n_y readings. We can write this expression as

$$\{(m_x - \mu_x) - (m_y - \mu_y)\}\{(m_x - \mu_x) - (m_y - \mu_y)\}$$
$$= (m_x - \mu_x)^2 - 2(m_x - \mu_x)(m_y - \mu_y) + (m_y - \mu_y)^2.$$

The expected value of the middle term is zero because the x and y samples are selected independently. (For any given sample value of m_x, the expected value of $(m_y - \mu_y)$ is $(\mu_y - \mu_y) = 0$, etc.). Therefore the expected value of the above expression reduces to the expected value of $(m_x - \mu_x)^2 + (m_y - \mu_y)^2$, which is $\sigma_x^2/n_x + \sigma_y^2/n_y$.

Exercise 17F. Sampling Without Replacement

Discuss the effect of sampling without replacement on the sampling error of the mean. As an example consider sampling from a very small population.

Discussion.

We are sampling n items from a population of N items. If each selected item is replaced before the next item is picked, each item in the population has an equal chance of $1/N$ of being selected. But if a selected item is *not* replaced, the chances for successive items are $1/N$, $1/(N - 1)$, $1/(N - 2)$, etc. The sample mean will still be equal to the population mean when averaged across all possible samples of n, but the standard error of the mean will be smaller than in sampling with replacement.

To illustrate, consider a very simple population which consists of only three items with values 1, 3, and 5. We aim to sample two of these. Then in sampling without replacement, the first item selected might be "1", and the next item would have to be either "3" or "5". There are in fact six possible samples of two: 1 & 3, and 1 & 5, 3 & 1 and 3 & 5, 5 & 1 and 5 & 3. The sample means are 2, 3, 2, 4, 3, and 4. These six values differ on average by .67 from the population mean of 3.

In sampling *with* replacement, however, after selecting the "1" for the first item, there are still three equally likely possibilities for the second item, namely 1 again, or 3, or 5. There are therefore nine equally likely samples:

the six earlier ones plus 1 & 1, 3 & 3, and 5 & 5. The sample means are 1, 2, 2, 3, 3, 3, 4, 4 and 5. These nine values differ on average by .89 units from the population mean, which is greater than the value .67 when sampling with replacement. Thus the "unlucky" chance of hitting the same item twice increases the average size of the sampling error. However, this effect is only sizeable if the sample is a large proportion of the total population.

It can be shown by simple but relatively lengthy mathematics that the standard error of the mean for sampling without replacement is

$$\frac{\sigma}{\sqrt{n}}\sqrt{\left\{\frac{(N-n)}{(N-1)}\right\}}.$$

When the population size N is very large compared with the sample size n, the quantities $(N - n)$ and $(N - 1)$ are both virtually equal to N, and the value of their ratio is very close to 1. The standard error formula is then almost equal to the value $\sigma/\sqrt{n}$ for sampling *with* replacement.

For example, taking a sample of 100 from a population of 1,000 leads to a factor of .95. Thus the *correct* standard error formula for sampling without replacement is only 5% smaller than the value given by the simpler $\sigma/\sqrt{n}$ formula. The difference is not large. If one is sampling 1,000 items from a population of one million, the correction factor would be about .9995, i.e. a difference of less than 1‰, which is clearly negligible.

These results account for the fact that while it is common practice to sample without replacement (usually the easier operation physically), this is ignored in the analysis (which is then easier).

Exercise 17G. Binomial Sampling

Discuss the sampling distribution that arises in industrial quality control when items are randomly selected from a large batch of manufactured items of which a proportion p are defective.

Discussion.

Random sampling from a population in which each observation can take one of two values (e.g. Defective or Non-defective, or boy or girl) is a common way for the Binomial Distribution of Section 12.4 to arise.

Consider the first item randomly selected. It will have a chance p of being defective and $(1 - p) = q$ of not being defective. The next item selected will have an equal chance p of being defective, if sampling is with replacement or if the population is large. If the two items are selected independently, the probability of both being defective is $p \times p = p^2$, that of one being defective is $p \times q + q \times p = 2pq$, and that of neither being defective is q^2. This is the (positive) Binomial Distribution with $n = 2$.

Thus it can be shown that the sampling distribution in such a case takes the form of the Binomial Distribution for any sample size n.

CHAPTER 18

Statistical Inference

When dealing with sample data one needs to do three things: (1) estimate the characteristics of the population sampled; (2) assess the accuracy of these estimates; and (3) check one's prior hypotheses about the data to determine whether any deviation in the sample result is due only to the likely errors of sampling.

The concept of the sampling distribution is used to infer from a single random sample the characteristics of the population from which it has come. In the discussion in this chapter we shall assume that the data have been selected from a specified population by simple random sampling. Calculations for other forms of probability sampling are more complicated, but the general principles remain the same.

18.1 Estimation

To estimate a population characteristic from a sample of readings commonsense suggests that we might simply calculate the corresponding value for the sample. For instance, the mean of the sample gives a good estimate of the population mean.

But for other kinds of statistics it is not always as easy. For example, if the variance of a sample of n readings is defined as $\Sigma (x - \bar{x})^2/n$, the average or "expected" value of this quantity across all possible samples will be fractionally smaller than the population value of the variance. This systematic error or "bias" in the sample estimate can be eliminated by defining the variance as $\Sigma (x - \bar{x})^2/(n - 1)$, using the divisor $(n - 1)$ as described in Chapter 11. Such problems typically occur when trying to argue from a sample to the population from which it came.

Absence of statistical bias, i.e. giving the right answer *on average* across all possible samples, is one possible criterion in judging a good estimator. An alternative is aiming at the most *accurate* estimate from the single sample we have. Another widely used estimating principle, especially in complex situations, is "maximum likelihood". This means picking an estimate of the population value which, if it were true, would give the highest probability or "maximum likelihood" that one would have observed the particular sample data actually observed. These three criteria do not always lead to the same answer. But despite these theoretical difficulties, in many common situations there are no important problems of how to estimate the population value. The corresponding sample values often serves as a fairly adequate estimator.

18.2 Confidence Limits

Having chosen a sample estimate of the population value we now have to consider how accurate it is. Because we are dealing with random samples, the answer will be in the form of probabilities, i.e. how *likely* we are to be wrong.

For example, suppose we want to estimate the population means from a random sample of 100 readings with a mean of 15, a standard deviation of 4, and an estimated standard error of $4/\sqrt{100} = .4$. We saw in Chapter 17 that we can make rather precise statements about the variability of the results obtained from different samples of this kind. For instance, because the sampling distribution of the mean would be approximately Normal, about 95% of the sample means would lie within twice the standard error ($2 \times .4 = .8$) of the population mean μ. But since we do not know the value of μ, this does not tell us how close μ is to our observed sample mean of 15.

To get a better answer, suppose that our sample were actually one of the 95% of all possible samples whose means lie within the two standard error limits $(\mu - .8)$ and $(\mu + .8)$. The difference between the unknown μ and our observed value 15 must then be less than .8. We can therefore turn this statement around and say that μ must in this case be less than .8 away from the observed value 15, i.e. that μ must lie between 14.2 and 15.8. Since this case occurs in 95% of all samples, we can say that μ must lie within the 2 standard error limits (here $\pm .8$) of the observed sample mean for 95% of all samples.

If we make this statement, we will be correct 95% of the time, i.e. the probability of being right is .95. This is commonly referred to as one's "confidence" in being right in saying that the population mean lies between 14.2 and 15.8, and these two-standard error limits are referred to as the "95% confidence limits".

This kind of result is more complex than it appears on the surface. We are *not* saying that the unknown value μ has a probability of .95 of lying between the two-standard error limits. The value of μ cannot have a probability distribution, it has one particular value (with probability 1, if one likes). Instead, we have to make the more convoluted statement that μ will lie between the two-standard error limits for 95% of all possible samples. Even this is quite an achievement. From a single sample we can tell how accurate it probably is!

The Level of Confidence

By the same form of argument, we can determine other confidence limits. Since the sampling distribution of the mean is approximately Normal for sample sizes of 30 or more, we can use results such as those in Table 18.1. For example, we can say that the population mean will lie between ± 3.3 times the standard error and expect to be right in 99.9% of all possible samples and wrong in only 1 in a 1,000 cases.

TABLE 18.1 Descriptive Characteristics of the Normal Distribution

Distance from the mean in terms of the standard deviation	The proportion of readings lying within the stated limits
± 2.0 s.d.	95%
± 2.6 s.d.	99%
± 3.0 s.d.	99.7%
± 3.3 s.d.	99.9%

In our numerical example with a standard error of 0.4, we can therefore say that we expect the population mean to lie

between 14.2 and 15.8 with 95% confidence,
between 14.0 and 16.0 with 99% confidence,
between 13.7 and 16.3 with 99.9% confidence.

The notable feature of these results is that the risk of being wrong decreases sharply, but the size of the confidence limits increases relatively little. Thus although the population mean will lie outside the limits 14.2 and 15.8 in 5% of all samples, in all but 1 in 1,000 cases it will lie only just outside these limits, by up to 0.5 units. Even in that one case μ will mostly be *just* beyond the 13.7 and 16.3 limits. This means that even if our original sample had been that 1 in 1,000 case, our sample estimate of μ would not be much more inaccurate. We can therefore be pretty sure that the population mean lies roughly between 14 and 16, or fractionally outside these limits.

Prior Knowledge

In our numerical example we have supposed so far that we merely have a sample of 100 readings with a mean of 15 and standard deviation of 4, without saying what the data refer to. In practice, we would generally have some degree of prior knowledge of the situation.

For example, suppose that the population consists of the 50,000 employees of a large firm and we want to determine the average number of days in 1973 that each employee was absent from work. Without analysing any data we already know certain facts.

Firstly, assuming a 5-day week, we know the answer cannot lie outside the limits 0 and about 250. Secondly, we know the average rate almost certainly lies well below 100 unless there was a special circumstance, such as a 5-month strike, which someone would already have noticed and told us about. Thirdly, there will be results on absenteeism in previous years, for other firms in 1973, and so on. For example, if the firm's figures in the three previous years were 19, 17 and 18, the 1973 result should be something like 18, otherwise something exceptional must have occurred.

To get much closer to the truth, we have to measure what actually happened in 1973. We could reach the answer through the attendance records of all 50,000 employees (subject to any problems of measurement inherent in such records), but even a relatively small random sample of 100 would tell us a great deal.

If the sample gives an average of 15 days absent and the variation from employee to employee has a standard deviation of 4, the average for the total work-force is almost certainly not as high as 50 or as low as 5. This seems obvious. But how much narrower can we make the limits while still feeling "almost certain"?

The more specific we make the estimate, i.e. the narrower the limits, the more risk we run of their being wrong. For example, if we set the limits at 14.5 to 15.5, we cannot be "almost certain" since the population value may well lie outside. The important contribution of sampling theory is that it can give us a more precise measure of our risk of being wrong; e.g. 1 in 20 or 5% for the ± 2 standard error limits of 14.2 to 15.8 noted above, and 1 in 1,000 for the ± 3.3 limits 13.7 to 16.3.

In recent years, certain theoretical procedures have been developed to try to improve such inferences from sample data still further by explicitly taking into account one's prior knowledge of the situation, such as the implication of last year's results, etc. The basic step in the so-called *Bayesian* approach is to try to translate this prior information into "prior probabilities" about the likely value of the unknown 1973 rate of absenteeism. Suppose one can somehow determine a zero probability that it is 13 or less, a .01 probability that it is 14, .1 that it is 15, and so on (with a peak probability of say .6 that it is 18 as in the preceding years). The "posterior probabilities" of the likely value are then obtained by combining the prior probabilities with the information contained in the sample, using a well-known result in probability theory due to Thomas Bayes in the middle of the eighteenth century. This result essentially says that given a sample value of 15, the probability of the population mean taking some particular value, say 16, is proportional to the *prior* probability that μ would be 16, multiplied by the probability that a sample result of 15 would occur if the population mean were *in fact* 16.

This method of "adjusting" prior probabilities in the light of sample evidence seems at first to make sense, but it is not widely used in ordinary statistical inference. One difficulty is fixing on the prior probabilities themselves. There is usually no objective way of determining them and they are generally referred to as "subjective probabilities". In fact ordinary people (scientists, administrators, etc.) do not usually think explicitly in *precise* probabilistic terms. Therefore adjusting prior probabilities which no one has been explicitly thinking about is after all perhaps not a very obvious method to use.

The usual place for prior information is not in sample estimation and the fixing of probabilities, but in determining the kinds of *hypotheses* one wishes to test.

18.3 Testing a Statistical Hypothesis

In analysing data one usually has some presupposition or hypothesis to test or explore. Typically, we might expect average absenteeism in 1973 to be 18 days, as in previous years. If the actual result for 1973 is 15 days, then our expectation was wrong and the hypothesis that it would be 18 in 1973 is rejected. This is straightforward.

However, problems arise if the 1973 result of 15 is based only on a *sample*. Perhaps absenteeism in 1973 really was 18 but we were "unlucky" in that our particular random sample of 100 employees happened to give a way-out result. How likely is it that the difference between the observed sample result of 15 and our initial hypothesis of 18 is only due to random sampling error? It is this narrow type of uncertainty problem that is tackled in testing statistical hypotheses.

The specific statistical hypothesis that is tested is usually called the "null hypothesis". In our example suppose the null hypothesis is that the population mean $\mu = 18$ (with a standard deviation of 4). Then the means of samples of 100 would have a Normal sampling distribution with mean 18 and standard error $4/\sqrt{100} = .4$. It follows that our observed value of 15 differs from the mean of 18 by $7\frac{1}{2}$ times the standard error. This is well beyond the 1 in a 1,000 probability level noted in Table 18.1. In fact this value would be observed less than once in a million random samples of 100, if the samples really came from a population with mean 18.

Given that the observed sample is so very unlikely if the null hypothesis were true, we "reject" this hypothesis. After all, it was only a hypothesis to be pitted against the facts. Sampling error *might* have accounted for the discrepancy between 15 and 18, but we have now calculated that this is extremely unlikely.

To illustrate the opposite kind of result, suppose we started with a null hypothesis that the population mean $\mu = 15.5$. Then the observed sample value of 15.0 lies just over one standard error away. For a Normal Distribution, values more than one standard deviation away from the mean occur in almost 30% of all samples (about two-thirds of the readings lie *within* one standard deviation). Therefore the observed sample result is quite likely if the hypothesis that $\mu = 15.5$ is true, and we have no reason to reject this hypothesis.

The Level of Significance

In the two cases just discussed, the null hypotheses were either highly unlikely or highly likely in the light of the sample data.

We now consider a less clear-cut type of result, a sample mean which is two standard errors away from the hypothesised population value. With a Normal Distribution sample results further from the mean occur in only 5% of all possible samples. This is *fairly* unlikely: in any 20 empirical studies there would be only one such result. Therefore at the 5% probability level it is conventional to reject the null hypothesis and to call the observed result "significantly" different.

But such a cut-off point is arbitrary. One should never think of a two-standard error result as being strongly "significant", while with a 1.9 sample result the null hypothesis is "acceptable". Instead, anything like a 1 in 20, a 1 in 15, or a 1 in 25 chance should be considered rather unlikely.

The difficulty is that a 1 in 15 or a 1 in 20 result is not "impossible", therefore one might wrongly reject a null hypothesis when it is in fact true. One could cut the chances of rejecting a true null hypothesis by adopting a more stringent level of significance, say 1 in 100 or 1 in 1,000. But this would *increase* the chance of accepting a wrong null hypothesis.

These so-called "errors of the first and second kind" (either rejecting a true null hypothesis or accepting a false one), together with the notion of the "power" of a test of significance in discriminating between alternative hypotheses, are part of the theory of statistical inference which was highly developed by Jerzy Neyman and Egon Pearson around 1930, following Fisher's lead in the 1920's.

The technical arguments simplify considerably because a small change in the difference between an observed sample value and the null hypothesis has a marked effect on the probability of whether the difference occurred merely by chance. Problems of interpretation therefore only arise when sample observations differ from the null hypothesis by about $1\frac{1}{2}$ to $2\frac{1}{2}$ times the standard error. This is quite a narrow "twilight" range. More discrepant sample values almost unambiguously lead to the *rejection* of the null hypothesis, with a probability of less than 1 in 100 of its being true, while sample values differing by well under two standard errors must equally unambiguously be regarded as being *consistent* with the null hypothesis.

When sample values fall into the twilight range one usually either rejects the null hypothesis "tentatively", or "has doubts about it". There is always the possibility of taking another set of readings to reduce these doubts.

18.4 The Choice of Hypothesis

The choice of the null hypothesis is the crucial feature in tests of significance. A "significant" result means that the null hypothesis has to be rejected, it was the wrong hypothesis. The analyst was therefore wrong in choosing it, often presumably either through ignorance of his subject matter or incompetence. The results are not as he had thought. With sample data this might be caused by a rather "unlucky" random sample, but the purpose of establishing a result as "significant" is to show that this particular possibility is highly unlikely. Thus the null hypothesis was almost certainly really wrong.

Occasionally an unexpected observation can be important; that is how some discoveries are made (Fleming's discovery of Penicillin is a popularly quoted example). But it is not advisable to make a habit of collecting data that differ from one's expectations. It only shows that one is consistently incompetent in selecting the appropriate hypotheses to investigate.

Despite this, some tradition has grown up in the last few decades that "significant" results are "good" results. Findings are reported as being 5%, 1% or even 0.1% significant, and the symbols *, **, *** tend to be attached to the results (as in hotel guides for the ignorant tourist, as Sprent (1970) has put it).

But it is easy to choose a hypothesis that will amost certainly differ from an observed result. The more absurd the hypothesis the more "significant" the observed sample result will be. Instead, one's choice of hypothesis should depend on one's prior knowledge or expectations. Then there is little problem. For example, if the rate of absenteeism in previous years has been about 18 days, *that* is the relevant hypothesis, unless additional prior information leads one to expect something different now. Again, in analysing the height and weight data of some group of children, the generally appropriate hypothesis is the result $\log w = .02h + .76$, unless one has relevant additional information (such as that the children are older girls, or babies, or undernourished). However, if one's prior knowledge is not clear-cut, then *that* is part of the situation and one needs to say so. The empirical study will then be more of a fishing expedition to throw some light on the situation, rather than a rigorous test of some crucial hypothesis. All one needs to do is to attach confidence limits to the estimated values.

The problem of choosing an appropriate null hypothesis is highlighted by a particular form of null hypothesis: that the population value should be *zero* (probably the reason for the name "null hypothesis"). A typical example is that there should be no difference between two mean values, e.g. between the responses of a treated and a control group in a clinical trial.

But in most cases there is nothing objective about the choice of such a null hypothesis of zero. The analyst usually chooses it to provide "a fig-leaf of scientific respectability", not because of anything he *knows*. He does not generally expect the null hypothesis of zero difference to be true. Few clinical trials are carried out because the drug is expected to have no effect. Testing such a no-effect null hypothesis is mostly a game: the analyst wants to prove himself "scientific".

In a clinical trial of a drug, the scientist's expectation or hypothesis might be that the drug will *decrease* blood pressure (not increase it), and that it will do so by about 10 units. That is what other more or less similar studies, or theory, or his gut-feeling, lead him to expect. In dealing with a sample of treated patients, a decrease of 10 units is therefore the appropriate statistical null hypothesis to test. Only if he is checking to confirm the previous *failure* of a drug, or the absence of unwanted side-effects, would the "no-difference" type of null hypothesis be relevant.

The fear remains that the analyst might mislead the reader (or himself?) by ignoring the possibility that an apparently positive sample result might still only represent a zero situation in the population as a whole. But establishing that an observed result agrees with one's prior hypothesis within the limits of sampling error does not absolve one from taking note of this sampling error. If the sampling error is so large that zero is included in the two standard error confidence limits, then the sample evidence on its own cannot exclude the possibility that there really is no difference. The implication is that the investigation was badly designed—the samples should have been large enough to lead to a clear distinction between the expected "positive" null hypothesis and zero.

"No-difference" tests of significance are widely carried out with correlation and regression coefficients. The usual null hypothesis tested is that of a zero value (no correlation) in the population. For example, given that the observed correlation between x and y in the sample is 0.3, could it be that there is really *no* correlation in the population? But the analyst who reports fifty "significant" correlation coefficients has merely picked the wrong hypothesis fifty times. As already said, he should not make a habit of it.

One reason for the popularity of the zero null hypothesis here is that its meaning is clear: x and y are *unrelated*. If the results are established as "significant" (i.e. *not* zero), too often the meaning of the result is left unexplored. What does a correlation of 0.6 *mean*, or a regression coefficient of 2.5? How do these values compare with previous experience? Is a generalisable pattern of results building up? Unfortunately such questions tend to remain unanswered if the emphasis is merely on establishing that the coefficients are effectively non-zero.

There is nothing remarkable in finding that x is related to y in some particular set of data. But somehow reporting a "significant" correlation of 0.3 based on a sample of 100 seems to be treated as more important than finding a correlation of 0.3 in the whole population, or in a large sample of 10,000 (where anything is "significant").

The question raised by such a "significant" result is "so what?". Finding a coefficient which is significantly different from zero is at best a *starting-point* in the analysis. As Gatty (1966) has put it

> "Statistical significance of a correlation or regression coefficient merely means that there is a pretty good chance that it is in fact a number different from zero. One should not exaggerate the worthwhileness of a coefficient simply because it probably differs from zero."

18.5 Empirical Variation

Except in the earliest stages of studying a topic, there is generally a great deal of previous empirical information about the variability of the material in question. Thus the reliability or statistical significance of a sample result can be judged in other ways than just from that result itself.

For example, consider data on the heights and weights of a random sample of 100 girls from some specified larger population

Average height: 49 inches,

Average weight: 56 lbs.

Without any other information about the girls the only null hypothesis that can be used is the general relationship

$$\log w = .02h + .76 \pm .01,$$

discussed in Part II of this book. We are effectively predicting that this equation should hold again.

The logarithm of 56 is 1.75, so the new result deviates from the hypothesis by

$$(1.75 - .02 \times 49 - .76) = .01 \text{ log lb units.}$$

This deviation is very much in line with all the earlier results, where the means of various groups of boys or girls gave a fit to within average limits of about .01 log lbs. In that sense, the deviation for the new data is not different from the null hypothesis.

But if we look at the new data purely from the point of view of random sampling, we have to note that the deviations for *individual* children generally have a standard deviation of about .04 (see Section 6.4). Hence the standard error of the mean for the sample of 100 girls is $.04/\sqrt{100} = .004$. The observed discrepancy of .01 is therefore statistically significant at almost the 1% level of probability. This means that if we measured a larger sample or the whole population of these girls, we would have to expect the result to deviate from the line $\log w = .02h + .76$. Thus the deviation was not only due to sampling error.

We can resolve the apparent contradiction between these two conclusions by noting that most of the height and weight means discussed in Part II also differ "significantly" from the equation $\log w = .02h + .76$. The deviations there were also not due to sampling alone; there were other factors involved. General experience shows that observed data do not fit any model exactly. At best they fit only within some close, more or less irregular, limits. Sampling errors usually account for only a small part of these deviations.

One usually judges discrepancies in new data against the general run of discrepancies found previously. If these earlier discrepancies have not yet been explained, i.e. their causes *in addition* to sampling error, one can hardly say more about the *new* result than whether it fits in with previous experience.

But if the new result is beyond the usual limits, e.g. a height/weight discrepancy of .06 log lbs, one would have to ask if it were merely due to sampling errors (which it could be in a very small sample or for an *individual* reading) or whether some additional "real" factor were also involved. In most reasonably well-developed areas of study, tests of significance therefore perform mostly a negative function, a form of hygiene, to establish whether some unusual result is merely an unlucky sampling error.

18.6 Specific Tests of Significance

There are various commonly used procedures to test statistical significance, i.e. to test whether the difference between a hypothesis and a sample result is due to a *real* difference in the population or merely to the errors of random sampling. We shall start here with tests involving sample means and then briefly cover tests involving variances, correlation and regression coefficients, goodness-of-fit procedures, and contingency tables.

A technically useful device in many tests of significance is the number of "degrees of freedom" in the data. This often identifies the particular sampling distribution to be used. For the χ^2-distribution discussed in Chapter 17 the degrees of freedom were the number of groupings into which the data were classified, minus the number of "constraints" on the data caused by fitting a theoretical model or comparable calculations. For detailed *quantitative* data (in contrast to such grouped or *qualitative* data), the degrees of freedom are generally defined as the sample size minus the number of constraints. This accounts for the divisor $(n - 1)$ commonly used to calculate the variance of the deviations from the mean: one degree of freedom has been used to estimate the sample mean.

The Mean of a Sample

To test the statistical significance of hypotheses about the mean of a sample of n readings, we have to use the t-ratio of the sample mean m to its estimated standard error, $s/\sqrt{n}$:

$$t = \frac{m}{s/\sqrt{n}}.$$

This ratio follows Student's t-distribution with $(n - 1)$ degrees of freedom if the population sampled is Normal or near-Normal. Table 18.2 summarises the three most commonly used significance levels for t-distributions with various degrees of freedom. Thus 95% of the means of samples of $n = 6$ readings (i.e. 5 degrees of freedom) lie less than 2.6 times the standard error from the population mean. The t-distribution significance levels are very similar to those of the Normal Distribution with unit standard deviation except for very small sample sizes ($n = 10$ or less). Thus, in practice, one can generally use the Normal Distribution values, as shown in the last column of Table 18.2.

TABLE 18.2 **Common Significance Levels for the t-distribution**
(Multiples of the sample standard deviation within which 95%, 99% or 99.9% of the values lie)

	Degrees of freedom							
	1	2	5	10	15	20	30	**Large***
95%	13	4	2.6	2.2	2.1	2.1	2.0	**2.0**
99%	64	10	4.0	3.2	2.9	2.8	2.7	**2.6**
99.9%	640	32	6.9	4.6	4.1	3.8	3.6	**3.3**

* As for the Normal Distribution

If the population data are non-Normal, the distribution of sample means for small samples will not follow a t-distribution. But for large enough samples the distribution will still be approximately Normal. What is "large enough" depends on the nature of the population distribution. Even in fairly extreme cases, e.g. sampling from a skew Poisson or Negative Binomial type of distribution, the distribution of sample means for $n = 50$ tends to be close to Normal.

To test whether the mean m of an observed sample differs significantly from a hypothesised value μ, we look in Table 18.2 to find the probability with which the value

$$\frac{m - \mu}{s/\sqrt{n}}$$

will be exceeded. For example, if a sample of 6 workers has an average absentee level of 14 days with a standard deviation of 4, the t-value against the hypothesised rate of 18 days is $(14 - 18)/(4/\sqrt{6})$, or approximately 2.5. This virtually reaches the 5% significance level of 2.6 for $(n - 1) = 5$ degrees of freedom. Thus if the population mean were 18, values at least as different as 14 would occur in only 1 out of 20 samples of $n = 6$. The probability that this would occur by chance errors in the sampling is therefore sufficiently unlikely that one would generally reject the null hypothesis that $\mu = 18$.

An alternative hypothesis with μ somewhat less than 18 would make the observed result appear more probable. It is not necessarily clear *what* alternative hypothesis one should consider. But in the absence of other information, the most likely value would be 14, i.e. simply the observed sample mean.

The Difference Between Two Means

To determine the significance of the difference between two means, m_x and m_y, of two independent samples of n_x and n_y readings with variances s_x^2 and s_y^2, we calculate the t-statistic

$$t = \frac{(m_x - m_y) - (\mu_x - \mu_y)}{\sqrt{(s_x^2/n_x + s_y^2/n_y)}}.$$

Here $(\mu_x - \mu_y)$ is the hypothesised difference in the two population means. We then assess the numerical value of the t-statistic against a t-distribution with $(n_x + n_y - 2)$ degrees of freedom, since *two* means have been fitted. (This is again virtually identical with a Normal Distribution with unit standard deviation when the degrees of freedom are greater than 10 or 20 or so.)

If the null hypothesis is that there is no difference in the population means, then $\mu_x - \mu_y = 0$ and the numerator of the t-statistic simplifies to $(m_x - m_y)$. In such cases the hypothesis being tested usually says that the populations sampled also have the same variances and the same shape, i.e. that their properties are altogether the same. Therefore one could calculate a single "pooled" estimate of the variance for the two samples combined (on the basis that the hypothesis is true). This consists of the average squared deviations of the $n_x + n_y$ readings from their *overall* mean $(n_x m_x + n_y m_y)/(n_x + n_y)$. This leads to a t-distribution with $(n_x + n_y - 1)$ degree of freedom because only *one* mean is fitted. The t-test is then slightly more sensitive, but the gain is almost negligible, especially with relatively large samples.

More than Two Means—Simple Analysis of Variance

To test whether there are significant differences among the means of three or more samples one can use the Variance-Ratio or F-statistic (named after Sir Ronald Fisher):

$$F = \frac{\text{Variance estimate based on means}}{\text{Variance estimate based on individual readings}}.$$

Table 18.3 sets out data for random samples of four readings from three larger populations, A, B, and C.

TABLE 18.3 **Three Samples of 4 Readings Each**

	From Population		
	A	B	C
	2	3	4
	4	4	6
	4	6	6
	6	7	7
Mean	4	5	6

The variances of the three samples are 8/3, 10/3, and 8/3, which look very similar. They can therefore be averaged into an overall estimate of the variance σ^2 of the individual readings within each population, giving (26/3)/3 or 2.9, on the null hypothesis that the three populations have the same distributions (i.e. the same means, the same variances, and the same shapes). This is the denominator of the F-statistic above.

We can also estimate the variance σ^2 of the individual readings from the mean values of the three samples. The observed variance of these three means is

$$\frac{(4-5)^2 + (5-5)^2 + (6-5)^2}{2} = 1.$$

The null hypothesis says that the three populations are the same. So, if the null hypothesis is true, in effect we have three sample means from the same population. For samples of $n = 4$ the variance of the sampling distribution of these means would be $\sigma^2/4$. Therefore we can estimate σ^2 by multiplying the observed variance of the sample means by 4, giving a value of $4 \times 1 = 4.0$, the numerator in the F-statistic above.

We now have two possible estimates of the population variance, 4.0 and 2.9. The question is whether the difference is merely due to sampling error (i.e. the null hypothesis) or whether the larger value of 4.0 reflects real differences among the population means. We can test the null hypothesis by forming the variance-ratio

$$\mathrm{F} = \frac{4.0}{2.9} = 1.4,$$

(where the larger value of the variance value is always put on top). This is a useful procedure because the sampling-distribution of the F-statistic is known.

If the null hypothesis is true, then these two variances should be equal in the population and the ratio will be 1. But in random samples the estimates will vary and their ratio will follow an F-distribution with, in our case, 2 and 9 degrees of freedom. (The variance estimate based on the means has *two* degrees of freedom since the overall mean 5 has been estimated. The variance estimate based on the individual readings has three degrees of freedom for each sample and hence a total of *nine*.)

From tables of the F-distribution (as given in books of statistical tables, e.g. Fisher and Yates, 1957; Lindley and Miller, 1966) we see that an F-value of 1.4 with 2 and 9 degrees of freedom occurs quite often. So although the three sample means in Table 18.3 differ, we could expect these differences to occur with random sample data, given the quite large variability of the *individual* readings. There is therefore no reason to reject the null hypothesis that the three populations have the same means.

If the null hypothesis were *not* true and the population means *were* different, the variance calculated from the sample means would be greater. Then the F-ratio would be greater than expected from random sampling alone. Thus in our example we would expect an F-ratio above the 5% probability value of 4.5.

The test procedure outlined here is a simple example of the Analysis of Variance. It is called this because it analyses the total variance of the data into separate variance components (here "between samples" and "within samples"). These procedures were first developed by Fisher in the 1920's. Since then they have been greatly elaborated in connection with statistically designed experiments and described in various specialist texts (e.g. Fisher, 1935; Cochran and Cox, 1957; Cox, 1958).

The Problem of Selection. If the F-test is significant, one or more of the population means must be different from the others. Thus if the means in Table 18.3 had been 4, 5, and 10 (with corresponding individual readings), the F-test would have been significant. This is presumably because population C with its sample mean of 10 differs from populations A and B, as common sense suggests.

But such a conclusion is technically difficult to test for its precise degree of significance. Ordinary t-tests or standard error calculations for the difference of two means cannot be applied. These test procedures are not designed for situations where a large difference has been specially selected for testing after inspection of the data, or where *many* differences are tested (e.g. A against B, A against C, etc.).

For example, suppose we have 7 sample means to analyse. We have to compare $(7 \times 6)/2 = 21$ pairs of means. If all the population means were equal, 1 in 20 of the sample results would still be beyond the appropriate 5% limits. If we select the biggest observed difference out of the 21 pairs of means, it may be no more than the *normal* 1 in 20 case. There are additional complications because the 21 comparisons are not all independent. *One* exceptional mean value out of seven would lead to six exceptional *differences*.

Although certain procedures have been put forward for dealing with such problems, none seems to be commonly accepted in practice. The problem is less serious than it might seem because such comparisons are at most needed at early stages of a study. Once the subject matter becomes structured in the light of previous findings, the *uninformed* search for significant differences becomes unimportant.

Correlated Readings

Suppose x and y are paired readings, e.g. heights of brothers and sisters, or readings on the same patients before and after a clinical drug trial, as in Table 18.4. The differences in the sample means m_x and m_y could be due either to the treatment or to the effects of random sampling.

TABLE 18.4 Correlated Readings: Five Patients Before and After Treatment

	Patients					
	A	B	C	D	E	Mean
Before	28	19	19	17	17	20
After	20	15	14	17	14	16
Difference	8	4	5	0	3	4

To test the significance of the differences in sample means for n *pairs* of readings we can use the t-statistic

$$t = \frac{m_x - m_y}{s_{x-y}/\sqrt{n}},$$

with $(n - 1)$ degrees of freedom. Here s_{x-y} stands for the standard deviation of the paired difference $(x - y)$, i.e.

$$s_{x-y} = \sqrt{\left[\frac{\text{Sum}\,\{(x - y) - (m_x - m_y)\}^2}{n - 1}\right]}.$$

In our numerical example, s_{x-y} is $\sqrt{(34/4)} = \sqrt{8.5} = 2.9$, so that

$$t = \frac{4}{2.9/\sqrt{5}}.$$

With 4 degrees of freedom this falls almost exactly at the 5% level of significance (interpolating in Table 18.2). Since the difference in means is larger than would usually occur from sampling error alone, the drug therefore appears to have been effective.

There is positive correlation between the before and after scores (e.g. A is high on both, and E low), and so the assessment of the mean difference, 20 − 16 = 4, is less subject to sampling error variation than if *different* patients had been tested before and after the treatment (where on the same readings the standard error would have been 5.2 instead of 2.9). The gain in sensitivity is typical of the more efficient statistical "design" of the study, using the same patients before and after. The design makes use of the fact that in this case individual differences in response levels before the treatment tend to recur *after* the treatment.

Because the analysis here essentially involved only the readings in the "difference" row, the appropriate degrees of freedom are 4, i.e. five differences minus 1 for the overall mean. (Sometimes the same test procedure is described in terms of the individual before and after readings plus the correlation coefficient between them.)

A More Complex Analysis of Variance

A further question for the data in Table 18.4 is whether the patients really differ *significantly* from each other. Their *apparent* tendency to differ (e.g. A high both before and after, and E low) might only be a fluke due to random sampling variation for the particular 5 patients sampled and not typical of the population sampled.

This can be tested by calculating an F-ratio with a numerator based on the variance of the mean values of figures for each patient (i.e. 24, 17, 16.5, 17 and 15.5) multiplied by 2, since each value is a mean of 2 readings (the σ^2/n effect). The denominator is again s^2_{x-y}. The appropriate degrees of freedom of the F-ratio are 4 and 4. (We could also have tested the before and after treatment effect, $m_y - m_x$, by the F-ratio. In the case of two means, the F-ratio is the square of the t-statistic used in the preceding section.)

These are simple examples of more advanced types of Analysis of Variance. The basic concepts will be outlined further in Chapter 19 in connection with the design of experiments.

Lawlike Relationships

Tests of significance for lawlike relationships are mainly tests of the mean values $\bar{x}$ and $\bar{y}$ of the different sets of data analysed.

An early problem is that we may need to establish whether there is any relationship at all between variables x and y. To do this, one can first test the difference of the means $\bar{x}_1 - \bar{x}_2$ for two (or more) sets of data against the null hypothesis of no difference, along the lines already discussed here. Secondly, one similarly tests the differences of the means $\bar{y}_1 - \bar{y}_2$. If both tests are significant, i.e. significant variation in x and significant variation in y, then the apparent correlation between $\bar{x}$ and $\bar{y}$ in the various samples must be significant.

To establish whether a particular pair of mean values $(\bar{x}, \bar{y})$ based on n pairs of readings differs significantly from a previously established relationship $y = ax + b$, one tests the t-statistic

$$t = \frac{\bar{y} - a\bar{x} - b}{\text{stand. dev.}\,(y - ax - b)/\sqrt{n}}$$

with $(n - 1)$ degrees of freedom. The analysis is similar to that for correlated means because it depends on the differences between x and y. However, here one needs to adjust the scales of measurement of x and y, i.e. working with the deviations $(y - ax - b)$ instead of $(y - x)$.

The same test procedure can be used to establish whether an *individual* pair of readings (x, y) is significantly different from the relationship. Then the values of x and y are inserted in the numerator instead of the means and $n = 1$. The test determines whether the observation differs from the line by more than 95% or 99% of the readings generally do, or whatever cut-off criterion one wishes to use.

Correlation and Regression Coefficients

Tests of significance of correlation and regression coefficients are mostly of the null hypothesis of *zero* correlation or regression.

The standard error of the product-moment correlation coefficient r for a sample of n from a bivariate Normal Distribution with zero correlation in the population can be estimated by the formula $(1 - r^2)/\sqrt{(n - 1)}$, and the sampling distribution of r is then approximately Normal for samples of $n = 100$ or more. For an observed correlation coefficient of $r = .2$ based on a sample of $n = 100$, the estimated standard error is therefore $(1 - .04)/\sqrt{99} = 0.1$; the observed sample value of .2 is therefore twice the standard error from 0 and hence significant at the 5% probability level. For smaller samples, one can use the fact that the quantity $r\sqrt{(n - 2)}/\sqrt{(1 - r^2)}$ is distributed as Student's t-distribution with $(n - 2)$ degrees of freedom, if the population value ρ of the correlation is zero.

To test hypotheses of *non-zero* values of the correlation ρ in a bivariate Normal population, Fisher showed in 1915 that for different sample values r, the quantity $\frac{1}{2}\log_e\{(1 + r)/(1 - r)\}$ follows a Normal Distribution to a close degree of approximation, with a mean of $\frac{1}{2}\log_e\{(1 + \rho)/(1 - \rho)\}$ and a variance $1/(n - 3)$. This sampling distribution can therefore be used for tests of non-zero null hypotheses.

The theory of testing *regression coefficients* has various complexities, especially if one is dealing with time-series as in econometrics or other forms of data where successive readings may be serially correlated. But for Normally distributed independent residuals, the basic standard error formulae for the slope-coefficient $a = \text{cov}(xy)/\text{var}(x)$ in the linear regression $y = ax + b$ is

$$\frac{\text{stand. dev.}\,(y - ax - b)}{\text{stand. dev.}\,(x)\sqrt{(n - 2)}},$$

and the standard error of the intercept-coefficient $b = \bar{y} - a\bar{x}$ is

$$\frac{\text{stand. dev.}\,(y - ax - b)}{\sqrt{n}}.$$

These expressions could therefore be used in tests against a specified null hypothesis for large samples, where the sampling distributions should be approximately Normal.

The Variance

So far we have concentrated on tests related to mean values. These include correlational analyses of how one variable tends *on average* to vary with another. But in studying the scatter of readings about a mean, we may need to test hypotheses about variances. For example, does a particular sample variance s^2, based on a sample of n readings, differ significantly from a hypothesised population value σ^2, or is the observed difference likely to be due only to random sampling errors?

If the readings come from a more or less Normal Distribution with variance σ^2, we test the ratio of the observed to the hypothesised value of the variance:

$$\frac{s^2}{\sigma^2}.$$

If the null hypothesis is true, the sampling distribution of this ratio follows a χ^2-distribution with $(n - 1)$ degrees of freedom. No general test procedure has been developed for variances from markedly non-Normal data.

Two or More Variances

When comparing two or more samples of readings, we want to know not only whether their *means* differ significantly, as drawn out before, but also whether their *variances* are the same. A simple test of whether the variances, s_1^2 and s_2^2, of two samples of n_1 and n_2 readings from two different (Normal) distributions differ significantly uses the variance ratio

$$\mathrm{F} = \frac{s_1^2}{s_2^2}.$$

This ratio follows an F-distribution with $(n_1 - 1)$ and $(n_2 - 1)$ degrees of freedom (using the larger variance as the numerator).

For more than two variances, s_i^2, all having the same number of degrees of freedom, r, one can use Bartlett's index M for "variance homogeneity":

$$\mathrm{M} = r\{\log \bar{s}^2 - \mathrm{Sum}\,(\log s_i^2)\}$$

where $\bar{s}^2$ is the average of the $i = 1$ to q different variances. This index can be tested against a χ^2-distribution with $\{1 + (q + 1)/3qr\}$ degrees of freedom, where q is the number of variances.

If the result is not significantly large, the different samples should come from populations with more or less the same variance (e.g. Bartlett, 1937). However, if the samples come from non-Normal distributions, this test can give apparently significant results even though the different population variances might be equal. Nowadays the test does not seem to be in common use, but there appears to be no adequate substitute.

The general problem of establishing whether different sets of sample data have the same scatter (and whether the observed differences in scatter are significant) can be important. In practice variance heterogeneity is often linked to non-linear relationships, so that the problem is also solved by a non-linear transformation of the variable (e.g. taking logs), which can largely eliminate the heterogeneous scatter.

Goodness of Fit

The scatter of readings about some fitted theoretical model is often arranged in n groupings of readings, as we did in Section 17.4. With sample data we can test the "goodness of fit" of the model by assessing the statistic

$$\mathrm{Sum}\frac{(\text{Observed} - \text{Theoretical Frequency})^2}{\text{Theoretical Frequency}}$$

against a χ^2-distribution with $(n - k - 1)$ degrees of freedom, where k is the number of parameters that had to be fitted in the theoretical model. A non-significant χ^2-value means that the observed deviations from the model are likely to be due to chance alone.

Thus they would not be expected to generalise to other samples or to the population as a whole.

The value of χ^2 must be calculated from the actual frequencies and not from percentages or proportions. The data can be grouped arbitrarily but the grouping must be essentially decided *before* the data are collected (or summarised) to avoid any danger of deliberately biasing the χ^2-value. Another requirement is that the theoretical frequency in any one grouping interval should not be less than 5, otherwise the χ^2-approximation fails to apply. If necessary, adjacent groups can be combined to produce the minimum theoretical frequency and the degrees of freedom reduced accordingly.

The arithmetical calculations for this test are the same as for contingency tables, as is explained below.

Contingency Tables

A special case of goodness of fit problems arise with a "contingency table". This is a two-way cross-classification of *qualitative* data. For example, one classification in a "2 × 2" table might be between male and female, the other between brown-eyed and not-brown-eyed. In general, one may cross-classify k classes one way by m classes the other.

The only null hypothesis that is commonly tested *statistically* is that the classifications are independent of each other, i.e. that in the population, the proportion of men who are brown-eyed equals the proportion of women who are brown-eyed. With sample data the two proportions will usually not be precisely equal, and the question is whether this difference is only due to the sampling errors.

Table 18.5 gives an example of a 2 × 2 table. Two machines produce the same product and each machine tends to yield a proportion of defective items. From a week's production of each machine, 20 and 30 items are sampled at random. When tested, 10 items from the first sample and 5 items from the second are found to be defective. The samples suggest that the first machine, with 10/20 or 50% defectives, is worse than the second, with only 5/30 or about 17% defectives. The question is whether this difference is significant, i.e. was the first machine worse than the other *throughout the week*?

TABLE 18.5 The Incidence of Defectives in Random Samples of Items Produced by Two Machines

	1st machine	2nd machine	Total
Defective	10	5	15
Non-defective	10	25	35
Total samples	20	30	50

On the null hypothesis that there was no difference in the machines' proportions of defective items during the week, the best estimate of this general proportion of defectives is from the total column, that 15/50 or 30% were defective. This leads to the theoretical "expectation" that 30% of the items for each machine should have been defective, 6 out of 20 for the first and 9 out of 30 for the second, as shown in Table 18.5a. (These values tend to be called "expected" rather than "theoretical" because they merely refer to what is expected on the null hypothesis and do not refer to any deep theory or model.)

TABLE 18.5a The Observed and "Expected" Frequencies

	1st machine		2nd machine		Total
	Obs.	**Exp.**	Obs.	**Exp.**	
Defective	10	**6**	5	**9**	15
Non-defective	10	**14**	25	**21**	35
Total	20	**20**	30	**30**	50

To test the hypothesis we calculate a χ^2-statistic as in a goodness-of-fit test. This is the sum of the quantities

$$\frac{(\text{Observed} - \text{Expected Frequency})^2}{\text{Expected Frequency}}$$

for each "cell" of the 2×2 table. Adding these numbers gives

$$\chi^2 = \frac{(10-6)^2}{6} + \frac{(10-14)^2}{14} + \frac{(5-9)^2}{9} + \frac{(25-21)^2}{21} = 6.4.$$

This figure can be tested for significance against a χ^2-distribution with 1 degree of freedom. [Given the marginal totals (15, 35, 20, and 30) in the 2×2 table, only 1 figure in the body of the table can vary freely; the others are determined. For a $k \times m$ table generally, the appropriate degrees of freedom are $(k - 1)(m - 1)$.]

The 1% value of χ^2 is 6.6, so a value as high as 6.4 would occur in only just over 1% of samples. Hence we reject the null hypothesis as being unlikely and accept that there was a difference between the two machines in the week's production sampled.

18.7 Summary

Statistical inference means using the data in a single sample to estimate the values of the population sampled.

In simple cases like the mean, the sample value provides a good estimate of the corresponding population value. But this simple approach may not work as well with other parameters, so more complex estimation procedures have to be devised.

In general, an estimate from a sample will not equal the population value. If the sampling distribution is more or less Normal, the standard error of the estimate provides the basis for its "confidence limits"; this is so for the mean and most other descriptive measures, if the sample size is large enough. For example one would be right 95% of the time in asserting that the population value lies within ± 2 standard errors from the observed sample value. Even if the population value were outside these limits (which occurs for 1 sample in 20), it would not lie *far* outside.

In tests of significance, the question posed is whether the difference between the observed sample value and some hypothesized population value (the "null hypothesis") is due only to random sampling errors, or whether the difference would generalise to other samples and the population as a whole.

The hypotheses tested should generally derive from previous empirical data and should reflect what such prior knowledge has led one to expect. A significant difference therefore means that one's expectation was wrong.

If previous data exist, then extensive *empirical* evidence about the general variability of the data should have been built up. One therefore need not have to rely on *theoretical* sampling theory to provide an inference about the new sample's likely variability.

CHAPTER 18 EXERCISES

Exercise 18A. Significant with a Larger Sample?

An observed sample mean of 5 has a standard error of 3. If the null hypothesis is zero, the sample mean is within the ± 2 standard error limits and hence not significantly different.

With double the sample size, the standard error would be reduced by a factor $\sqrt{2} = 1.4$, i.e. from 3 to $3/1.4 \doteqdot 2$, and a value of 5 would then be more than 2 standard errors away from zero.

Is it right to suppose that the observed sample mean of 5 would have been significant with a larger sample?

Discussion.

No. Another sample would generally have a different mean. And a *larger* sample would (on the null hypothesis) generally have a mean closer to zero. The standard error would be smaller, but so would the observed mean value, so it would generally still not be "significant".

Exercise 18B. The Nature of Statistical Hypotheses

Are statistical hypotheses different from the normal hypotheses of science?

Discussion.

Yes. Examples of scientific hypotheses are Einstein's theoretical deduction that light waves are bent by gravity (a popular example), that a certain drug will reduce people's blood pressure, or that the height and weight of some children will follow the relationship $\log w = .02h + .76$. In contrast, a statistical hypothesis is concerned with whether in random sample data an observed deviation from a scientific hypothesis is real or only due to an error in the sampling. (The question of statistical inference really arises only with small samples. With large samples the standard error of the sample estimate is generally small, so that anything except the smallest overt differences from the hypothesised value will generally be real.)

Exercise 18C. The Full Null Hypothesis

In the example in Section 18.3, the null hypothesis was an average rate of absenteeism of 18 days. Is it possible to deduce a sampling distribution from this information about the population?

Discussion.

In principle, the answer is no; in practice, it is often yes.

To deduce the sampling distribution of the average rate of absenteeism (along the lines discussed in Chapter 17), the null hypothesis must state *fully* what the population is expected to be like. It must give the form of the frequency distribution, e.g. Normal, Poisson, or whatever, and the parameters of the distribution, e.g. its standard deviation as well as its mean μ.

However, certain simplifications arise. Unless the sample size n is very small, the sampling distribution will be approximately Normal whatever the shape of the population distribution. Next, the sampling distribution will have a standard deviation effectively equal to $2/\sqrt{n}$, where s is the observed standard deviation of the *sample*. Hence, for purposes of testing statistical hypotheses about the mean, it is usually unnecessary to specify either the form of the population distribution or its standard deviation (unless the sample is very small). This explains why the null hypothesis is usually described as merely saying something about the expected value of the population mean.

Nonetheless, the analyst will be concerned with the "shape" of the population distribution and the amount of scatter, in order to describe and understand his data. If he has prior expectations or hypotheses about these, he may need to "test" them if the observed sample data look very different.

Exercise 18D. One 95% Confidence Interval or Many?

The 95% confidence limits in the numerical example in Section 18.2 were 14.2 and 15.8. Does this mean that μ lies between 14.2 and 15.8 for 95% of all samples?

Discussion.

No. The calculation of a particular set of confidence limits depends on the particular sample results used. The sample mean of 15 and standard error of .4 gave the 95% limits of 14.2 to 15.8. Another sample from the same population might have a mean of 15.3 and a standard error of .3. This would lead to 95% confidence limits of 14.7 and 15.9.

Probability statements about confidence limits therefore do not refer to a single set of numbers. The situation is more complex. If for any particular sample one can say that the population mean lies between the 95% confidence limits, then for 95% of all samples the corresponding statement for each sample will be correct (i.e. 14.2 to 15.8 for the first sample, 14.7 to 15.9 for the second one, and so on).

This may seem almost intolerably complex. But in practice most confidence limits are numerically similar because most sample means are fairly similar. After all, 95% of them lie within $\pm 2\sigma/\sqrt{n}$ of the population mean μ! It follows that the rough-and-ready interpretation of confidence limits, that the population mean lies in the range 14.2 to 15.8 with a probability of .95, will be close to the truth.

The choice is between making a statement which is true but so complex that it is almost unactionable, and one which is much simpler but not quite correct. Fortunately the *content* of the two kinds of statement is very similar.

Exercise 18E. A 4 × 6 Contingency Table

Table 18.6 shows the incidence of light, medium, or heavy attacks of influenza in random samples from six occupational groups. Does the incidence of influenza in the six populations really differ?

TABLE 18.6 Severity of Influenza in Random Samples of Adults from 6 Occupational Groups

	Occupation						
	White collar	Skilled working	Unskilled working	House-wife	Un-employed	Retired	Total
Severity							
Light	8	15	5	11	6	5	50
Medium	11	30	14	30	5	10	100
Heavy	9	22	16	40	5	8	100
None	32	53	35	99	14	17	250
Total	60	120	70	180	30	40	500

Discussion.

The null hypothesis that the differences in the table are only due to random sampling can be tested by the χ^2-procedure described for a 2 × 2 table at the end of Section 18.6. If the incidence of the various degrees of influenza does not differ among the six populations sampled, the best estimate of their incidence is given by the totals in the right-hand column. Thus 50/500 or 10% of the population would have had a light attack, 20% a medium one, 20% a heavy one, and 50% no attacks at all.

It follows that in the absence of sampling errors the "expected" incidence of degree of illness in each occupational group would be given by applying these percentages to each sample. Thus of the 60 white collar workers, 6 would be expected to have had a light attack, 12 a medium one, and so on.

We now calculate the χ^2-measure of differences between each observed and expected value. Adding for all the items in the 4 × 6 table we have

$$\text{Sum}\,\frac{(\text{Observed} - \text{Expected Frequency})^2}{\text{Expected Frequency}}$$

$$= \text{Sum}\,\frac{(8-6)^2}{6} + \frac{(11-12)^2}{12} + \frac{(9-12)^2}{12} + \text{etc.} + \frac{(17-20)^2}{20}$$

$$= 15.41.$$

Two of the theoretical values are less than 5 (for light attacks among the retired and unemployed). But with a large table as here, this can only marginally affect the χ^2-approximation. [One could combine the retired and unemployed categories, giving a χ^2-value of 13.6 with 3 × 4 = 12 degrees of freedom. The approximation to a χ^2-distribution can also be improved by *Yates Correction*, which consists of reducing the difference between the observed and expected frequencies by half a unit, i.e. $(7.5 - 6)^2/6 + (11.5 - 12)^2/12 + \cdots$). This helps because the observed values are discrete, i.e. whole numbers, and the expected ones are continuous.]

The above quantity will be distributed as χ^2 with $(4 - 1)(6 - 1) = 15$ degrees of freedom for different samples. From tables of the χ^2-distribution it will be seen that a value of 15 is not significant, more than 5 % of possible sample values are greater than 15. The conclusion is that the incidence of influenza does not really vary greatly (if at all) between the different occupational groups in the population sampled.

It should be noted that percentaging each column of figures in Table 18.6 would have shown at a glance that there are no vast differences in the incidence of influenza even in the *sample* data. From this it would seem unlikely that there could be large differences between the populations. Tests of significance are more valuable in establishing whether an apparently dramatic deviation in a sample really represents something real, rather than whether trivial sample differences are real but negligible.

Exercise 18F. χ^2 as a Measure of Correlation

The larger the value of χ^2 (or of any other test statistic), usually the more "significant" the observed result is. Does the value of χ^2 then provide a measure of the "importance" of this result, i.e. of the degree to which one variable varies with the other (e.g. the incidence of influenza with the occupational classification)?

Discussion.

The value of χ^2 is calculated on the basis of the null hypothesis being true (i.e. *no* association). Once this hypothesis has been rejected because of a large χ^2-value, the basis on which this numerical value has been calculated is no longer relevant. In any case, one is usually not concerned with merely measuring the "strength" of the relationship, but with describing its nature.

Exercise 18G. One-tailed Tests of Significance

In a standard test of significance based on the Normal Distribution, a sample mean m is regarded as significantly different from the hypothesised population mean μ if it is more than 2 standard errors *below* or *above* μ. This is a "two-tailed" situation.

Discuss the practical use of "one-tailed" tests of significance, where only sample values lying sufficiently far *above* the mean, say, are regarded as significant.

Discussion.

If the analyst expects that his sample observations will either agree with the null hypothesis or differ in one direction only, a one-tailed test of significance is often advocated. For example, in testing the effect of a drug or a fertiliser one might expect either little or *no* effect, or an *increase* in yield, but not a decrease. Sample values showing a decrease could therefore only occur because of sampling errors and should never lead to the rejection of the null hypothesis.

A one-tailed test is usually more "sensitive" in the sense that a smaller difference in the expected direction will be regarded as "significant". Thus 5% of sample values lie more than 1.6 times the standard error *above* μ, and any such value would then be significant at the 5% level. (With a two-tailed test, $2\frac{1}{2}$% of sample values lie more than 2 times the standard error above μ, and $2\frac{1}{2}$% lie more than 2 standard errors below μ, making 5% in all. The differences from μ generally have to be larger, but in *either* direction, to be significant at the 5% probability level.)

Three warnings about one-tailed tests need to be made. Firstly, the precise level of significance is taken too seriously. Consider a particular sample observation which is 1.6 times the standard error above the mean. This has a 1 in 10 chance in a two-tailed test but only a 1 in 20 chance in a one-tailed test. It is relatively unlikely (but not "impossible") to have occurred by chance whichever way we look at it. (A more extreme observation might have a 1 in a 1,000 chance under a one-tailed test, and a 1 in 500 chance in a two-tailed test; does this difference influence any conclusion?).

Secondly, would the analyst *really* accept all sample results in the unexpected direction (i.e. *below* the mean, say), however large the deviation? Can they *really* only be due to random sampling? What about experimental or computational errors, or some quite irrelevant but real factor (a power-cut, a change in the government)?

Thirdly, if one already knows enough to be sure that a truly negative result cannot happen, is it necessary to test the null hypothesis that *nothing* happened? If a positive result is firmly expected, one should be testing *that* as the null hypothesis.

Exercise 18H. More Complex Estimators

Discuss using estimators from sample data to determine the parameter k of the Negative Binomial Distribution (see Section 12.3).

Discussion.

This illustrates the more complex problems of statistical estimation that occur when one uses measures other than the sample mean.

One approach is to note that the variance σ^2 of a Negative Binomial Distribution with mean μ is given by $\sigma^2 = \mu(1 + \mu/k)$, where k is the second parameter of the distribution. This equation can be rewritten as $k = \mu^2/(\sigma^2 - \mu)$. One can thus estimate k by substituting the observed sample mean m and variance s^2; i.e. by writing $k = m^2/(s^2 - m)$. This is another example of estimating by the "method of moments" (see Chapter 12).

A second approach is to note that the proportion of zeros in an NBD is $(1 + \mu/k)^{-k}$. One can therefore use the sample mean m and the sample proportion p_0 of zeros in the equation $p_0 = (1 + m/k)^{-k}$ and solve for k along the lines referred to in Exercise 12I.

A third approach is to calculate the "maximum likelihood" estimate of k. The maximum likelihood principle is widely regarded as providing the "best" estimates of population parameters from sample data, but the mathematics are very cumbersome for the NBD.

These various estimates, and there are others, will generally give numerically different values for any one sample. For large samples the differences should be small (if the estimators are what is technically called "consistent") as long as the population follows an NBD exactly. But additional problems arise if there are systematic discrepancies, even if small, from the theoretical model.

Blind reliance on any one method should be avoided. Thus the "method of moments" is popular in statistical practice, but it is not very accurate for an NBD with a large proportion of zeros; the distribution is skew, so that the occasional large value markedly affects the variance. In contrast, the estimate using p_0 and m is then a good one because p_0 is observed rather accurately.

Even dogmatic reliance on the maximum likelihood principle is not safe, although is often regarded as the theoretically "best" method. It is widely used in complex situations where the results are not easy to judge, but the principle can give nonsense results under certain circumstances where other procedures work adequately (e.g. Neyman and Scott, 1948; Ehrenberg, 1950, 1951). Statisticians in fact virtually never use the maximum likelihood results in relatively simple situations where "commonsense" judgment *can* be applied. For instance, the maximum likelihood estimate of the variance is $\text{Sum}\,(x - \bar{x})^2/n$ and not the universally used $\text{Sum}\,(x - \bar{x})^2/(n - 1)$. (An apparent exception is the sample mean, which is the maximum likelihood estimate of the population mean. But the sample mean is the "best" estimate of the population mean from every possible point of view.)

PART V: EMPIRICAL GENERALISATION

Successful prediction and scientific knowledge depend on results which are known to generalise. In Chapter 19 we briefly consider the different methods of data collection that can lead to empirical generalisations, and in Chapter 20 the ways in which descriptive generalisations lead to explanation and deeper understanding.

CHAPTER 19

Observation and Experimentation

The different approaches to collecting empirical data can broadly be classified as:

(i) censuses or sample surveys which are statistically representative;

(ii) observational studies, in which the observer selects things to observe or measure; and

(iii) experiments, in which the observer deliberately controls or varies some factors.

The boundaries can be a little fuzzy, such as where selection stops and deliberate variation starts, but an artificial laboratory experiment clearly differs from the mere accumulation of observational records. The crucial element is the amount of control exercised by the observer.

The characteristic feature of scientific results is that they must be either generalisable or repeatable. Any extended study of empirical phenomena must involve more than one set of data and repetition therefore becomes a key element in data collection.

19.1 Repetition

To report that children's heights and weights have followed the equation $\log w = .02h + .76$ in a single study is merely to present an isolated finding. The initial study must be repeated if the result is to gain any scientific meaning.

The first repetition is the most dramatic. If the initial result does not hold again, then we have learned that it cannot generalise, at least not in any simple way. But if the same result *does* hold a second time, we know it could generalise even further.

Establishing the mere possibility that the result is repeatable is clearly only the beginning. But even a few successful repetitions can greatly progress the study if the new conditions of observation are different enough. For example, we needed only three or four studies in Part II to see that the height/weight relationship, $\log w = .02h + .76$, held despite differences in race, sex, age, nation, time, or observers.

No study can be repeated under literally identical conditions. But what matters is that the *result* can be repeated, not whether all the conditions of observation can be duplicated. Indeed, repetition in various *different* situations is required, to determine the range of different conditions or factors under which the result holds. What factors to vary therefore becomes a paramount question.

19.2 Factors to Vary

With a completely new result one usually aims to repeat the initial study with as *few* changes as possible: the same observer, the same apparatus, the same source of material, etc. (although *some* things must change, e.g. time). The purpose is to see easily and quickly whether the same result can be repeated at all.

After this one tries to extend the degree of generalisation. This is best done by changing the conditions of observation as far as possible without making a substantial difference to the results. But one may even go beyond this expected point, since it will either set a limit on the possible range of generalisation or lead to an unexpected, and hence even more dramatic, extension.

For example, one would not follow up a result on the height/weight relationship for white boys with more data on white boys, but with data on white girls or on black boys. Usually more than one factor is varied at a time, so one might go from a study of white boys in Birmingham in 1947 to a study of black girls in Ghana in 1970. If the result holds again, then a single study will have achieved a major breakthrough. One will have found that neither the change of race, sex, nation, time or observer, nor a lot of other non-explicit factors like temperature or seasons, affects the relationship. (There might be compensating factors, e.g. a difference due to sex cancelling a difference due to race; but such details would be covered in subsequent work, where white girls and black boys would also be covered.)

If the ambitious extension of conditions to black girls in Ghana did not work, one would have established one of the limits of the earlier result. To determine why the breakdown occurred, one would have to backtrack on the separate factors. Was the breakdown caused by the difference in time, sex, race, nation, or some other factor like age or nutrition? These further studies would now concentrate on one major factor at a time (but other more exploratory ones could still be varied as well). Varying one factor at a time is only important when the factor actually affects the results.

But even when only one factor is varied, the outcome must still be confirmed by repetition. No single study is all-important. (The "crucial experiment" of science is largely a myth since a study is regarded as crucial only if it has become repeatable.)

19.3 Statistical Surveys

Sample surveys or censuses are, at their simplest, a method of collecting data where no factors are varied at all. Such surveys lead to statistical averages whose validity depends entirely on the data being statistically representative. Thus if 22% of the population lives in Town A, then a nationally representative survey must take 22% of its data from Town A.

Simple statistical surveys are a "last resort" method, used when there is no more structured way of collecting the information. One problem is that, as a technique, the simple representative sample survey cannot cope with the basic requirement of repeating a study under different conditions. For example, taking two samples from the same statistical population is not independent repetition—the results *must* be the same, except for differences due to random sampling errors. On the other hand, taking two samples under different conditions, e.g. sampling from the same set of people on two different days, is no longer mere sampling. One has deliberately changed the empirical conditions of observation, i.e. the populations, and non-statistical control has been introduced.

More generally, control can be introduced into surveys and censuses by stratification. A national opinion survey might show that 40% of the people like the Government. Instead of merely reporting that fact, the results can be broken down to show that it is 40% in the North and in the South, and that it is 60% in large towns and only 20% in small towns and rural areas.

This set of results is more meaningful than the simple national average. By exercising more control over the data, the observer in effect has a number of *different* surveys. Statistical representation is only needed to deal with the uncontrolled variability within each of these smaller surveys. The purely *statistical* element in most structured surveys and censuses is to ensure that little or no systematic error or "bias" has crept into the results.

19.4 Observational Studies

In most observational studies the observer exercises a good deal of control. He selects the variables he will measure and the various conditions under which he will measure them. There is even some interference with the material which may affect the results, since getting people to answer questions or to stand up straight in order to measure their heights is not their normal behaviour. But in an observational study the observer generally does not try to change his material substantially.

Many observational studies are designed so that different factors vary together, instead of just one at a time. For example, a group of patients treated with a certain drug may have recovered better than an untreated group. But aside from the drug treatment the two groups may have differed in the intensity or nature of their illness, their ages, their previous medical histories, the amount of nursing care they received, etc. This makes it difficult to assign causes.

To identify the effects of the medical treatment more closely one has to try to eliminate other possible factors by comparing patients of the same degree of illness, by imposing the same amount of nursing care, etc. The process of eliminating possible factors is piecemeal and certainly not foolproof. It does not positively establish any form of cause-and-effect. This is what makes scientific progress slow and laborious.

But the strength of the observational approach lies in its facility for producing *negative* results: that some factor does *not* matter. A single-observational study showing that the equation $\log w = .02h + .76$ holds for both boys and girls shows that in general sex cannot affect the result.

It may seem naive to base a conclusion on a single observational result, but one is not claiming that all boys are like all girls. One is merely saying that the relationship has been found to be unaffected by sex, the factor did not matter in this study. If in the next study we find that girls *are* different, the cause must be some *additional* factor, like the ages of the girls. Or there may have been something special about the *first* study, if that result never repeats. One does not usually have to rely on a single study for long since the first result will be either confirmed or limited by information from later data.

19.5 Controlled Experimentation

Experimentation can often speed-up the observational approach. Here the investigator deliberately varies or adjusts some of the factors in the situation. If temperature is a relevant factor, he does not wait for the desired temperature level to occur naturally, but does something to make it warmer or colder.

Experimentation has two major roles: deliberate control, to keep specific factors the same in different studies, and deliberate variation, to see what will happen. Thus the typical laboratory experiment is a way of creating *unnatural* observational conditions outside the normal range of variation. Much of the success of science has been due to this power to explore hypotheses by creating artificial situations.

But an experimenter cannot necessarily control all of the factors in a situation. For example, he might select one group of patients and deliberately treat them with a drug and select another "control" group whom he does *not* treat with the drug. He can control the way the drug is administered, try to keep certain background factors like nursing care the same, and generally try to eliminate many of the varying factors that might cause confusion in a purely observational study, where the observer does not even decide which patients are to be treated. But even the deliberate experimenter will usually not wholly succeed in eliminating other factors. His two groups of patients may still not be fully comparable. Consciously or unconsciously, the untreated group may have been selected to have a different chance of recovery, or the drug may have side-effects which required additional treatment and *that* might have caused the difference in recovery rates. When faced with material that is variable and unpredictable, the experimenter's judgment alone is not enough to avoid the possibility of bias.

19.6 The Randomised Experiment

About 50 years ago Sir Ronald Fisher invented a way to control the variability of experimental material through the randomised experiment.

Suppose we have a total of 32 available patients. If we randomly divide them into two equal groups of 16, A and B, we leave the possibility of systematic bias, that the patients in Group A are inherently more likely to improve, literally to chance. (With large enough samples there will only be a very small possibility that the two groups are substantially different. Even with small samples we can estimate the probability of this happening with statistical tests of significance and confidence limits.) If Group A is now treated with the drug but not Group B, we exclude the effects of the variability of the material, subject to the known and usually small degree of statistical risk.

This kind of randomised experiment can greatly cut the time and effort required to eliminate uncontrolled variability in an experiment situation. But the approach still has its limitations.

Firstly, randomised experimentation, and in many cases experimentation of any kind, is often impossible, e.g. in many parts of biology, medicine, sociology, economics, geography, and, until recently, all of astronomy. For example, one cannot randomly allocate children to either "nuclear families" or "communes" and observe the difference. While randomised experiments are helpful, they are in any case not crucial; many scientific results were obtained before Fisher made his discovery in the 1920's.

Secondly, randomisation provides no safeguard that the treatment, the factor deliberately varied, was the only difference between the experimental and control groups. Randomisation only eliminates the inherent variability of the test material itself. The "true" effect of the treatment might still have been due to an impurity and not to the drug itself. Or the shear fact of administering the treatment might have caused some patients to improve: a "placebo" effect.

Thirdly, a randomised experiment will show the effects associated with the treatment (within probability limits), but it does not follow that the result will be repeatable. Randomisation cannot eliminate specific local factors, e.g. that the 32 patients in the clinical trial may have some mineral deficiency due to where they all live, or that the local strain of the virus is perhaps more susceptible to treatment.

As always, the study must be repeated to determine whether the result occurs again under other conditions. But two randomised experiments no longer constitute a controlled experiment. Many of the differences between two studies cannot be controlled, let alone be "randomised away". Thus any sequence of separate experiments degenerates into an observational type of situation. It is therefore the observational approach, with all its complications, that is central to scientific methodology.

19.7 The Design of Experiments

Sometimes a more complex experimental design enables a series of separate experiments to be carried out under more comparable conditions, thereby eliminating much of the ambiguity in interpreting an isolated result. If randomisation can be used in the larger design, the gain in eliminating extraneous forms of variation is even greater.

With the proper design, many different experiments can be conducted simultaneously, whilst still varying only one factor at a time. This is another major advance pioneered by Fisher. Such experiments do not necessarily require increased resources, e.g. increased numbers of readings. Indeed, these "factorial" types of design often reduce the amount of statistical error in the experiment.

Table 19.1 illustrates how a more complex experimental design for a clinical drug test can clarify some of the complicating factors mentioned earlier, like the level of nursing care, the level of dosage, and the possible "placebo" effect of the treatment.

TABLE 19.1 **A More Complex Clinical Trial**

(4 patients in each "cell")

	Dosage				
	High	Low	Placebo	Control	Total
Nursing Care					
High	4	4	4	4	16
Normal	4	4	4	4	16
Total	8	8	8	8	32

We still have a total of 32 patients, but instead of merely dividing them into two sub-groups of 16, one to be treated and one untreated, we have broken them into eight cells of 4 each.

Since level of nursing care was felt to be a possible cause of patients' responses, the patients are evenly divided with two different levels of nursing. Then within each group of 16, 4 patients are given a "high" dosage level of the drug, 4 a "low" level, 4 are a plain untreated "control" group, and 4 are given an inert tablet to simulate the effects of being "treated". The low number in each cell does not necessarily reduce the effective sensitivity of the experiment in establishing the results of the treatment.

Suppose first that (i) the variation in the dosage level makes no difference, (ii) the placebo has no effect (the results being the same as for the control group), and (iii) the variation in the two levels of nursing care makes no difference. We then have several major results from the one experiment: a comparison of 16 "treated" and 16 "untreated" patients just as before, *plus* the knowledge that dosage levels, placebos, and nursing care do not affect this comparison.

Next, suppose that nursing care matters but the medical drug is irrelevant. Then we have a comparison between high and normal levels of nursing, again based on total samples of 16 patients each.

Thirdly, suppose that both the drug and nursing care have an effect but that they do not "interact", i.e. the drug has the same effect at both levels of nursing, and the higher level of nursing produces better results irrespective of the improvement due to the drug. Then we have the result for the drug based still on a comparison of two groups of 16, and the result for nursing based on two groups of 16, two experiments for the price of one. We also have the knowledge that there is no "interaction", i.e. that the drug works in the same way at the different levels of nursing. This is the beginning of empirical generalisation.

Finally, suppose that several of the factors operate and "interact", e.g. that the placebo has a positive effect at the normal level of nursing but none at the high nursing level (where patients are getting enough "attention" anyway). In such a complex case, one needs to compare results based on the smaller sub-samples in the design. These will be statistically less accurate than results based on groups of 16, but usually this loss in statistical sensitivity will be outweighed by the additional information gained about the relative complexity of the situation. Any comparison based on single experimental and control groups of 16 each would have been superficial and potentially misleading and the sooner one knows about this, the better.

Factorial designs are generally easier to interpret if they are randomised, e.g. if patients are allocated to each sub-group at random. Then the initial statistical analysis can often follow the lines of Fisher's Analysis of Variance procedures. (See Section 18.6 and Exercise 19G.) The theory of the design and analysis of such experiments has been vastly elaborated in the last few decades, as described in various specialist texts, e.g. Fisher, 1935; Cochran and Cox, 1957; Cox, 1958.

19.8 Theoretical Norms

Controlled experimentation is largely concerned with comparing a treated group with an untreated control group to determine what would have happened without the treatment. But such a direct empirical check is only required when relatively little is known about the subject-matter. Often enough is known to *predict* the normal response levels.

For example, when we analysed the half-yearly purchases of Corn Flakes and other breakfast cereals earlier, the data had been collected under somewhat unnatural conditions (e.g. only *five* established brands could be bought from the retail outlet in question). One basic question in the initial study was whether these artificial conditions affected the observed purchasing behaviour (see Charlton *et al.*, 1972). But it was not necessary to collect data on normal purchasing behaviour under conditions somehow "matched" with the experimental ones (e.g. same product, same part of the country, same types of household, same season, etc.). It was already known that under everyday conditions the incidence of light and heavy buyers of a brand generally follows the Negative Binomial Distribution (NBD) within close limits of approximation: i.e. for a large variety of different product-fields (food and non-food), for large and small brands, in the U.K. and the U.S., for different lengths of time period, etc. All that was needed was to compare the experimental results with such validated *theory*, as illustrated in Table 19.2.

TABLE 19.2 **Corn Flake Purchases in 24 Weeks**

(From Table 12.9a)

		Number of Purchases											
		0	1	2	3	4	5	6	7	8	9	10	11+
% households buying													
Observed	%	39	14	10	6	4	4	3	2	2	2	2	9
NBD	%	**35**	**16**	**10**	**7**	**6**	**5**	**3**	**2**	**2**	**2**	**2**	**9**

More generally, a doctor dealing with patients diagnosed as having acute appendicitis does not run a controlled experiment ("the operation was successful because half the patients died"). Instead, he already knows from past experience what would happen most of the time if he did not operate.

Again, the physicist measuring temperatures by looking at the length of a column of mercury in a glass tube does not check every reading against "control readings" obtained from beakers of boiling water and crushed ice at the end of his laboratory bench. Instead, his past experience enables him to predict successfully that the mercury would reach the 100 mark for boiling water and 0 for crushed ice. All he need do is check very occasionally that nothing has gone wrong. Similarly, in analysing height/weight data for children earlier in this book we did not compare the numerical data for one group with that for another group. Instead, we only compared the data with the theoretical abstraction $\log w = .02h + .76$ which had held for all the previous data.

One of the most immediate and powerful uses of empirically based theory is to provide such norms based on prior knowledge. It largely replaces the use of statistically designed experiments, because it is easier and more effective than collecting and analysing new empirical data for control or comparison every time. It is the usual procedure in more mature subject areas.

19.9 Summary

The planned experiment, expecially one using randomisation, can greatly reduce the ambiguities in interpreting any finding. More than one factor can be varied in a suitably designed experiment, so that a wide range of generalisations can be established in a single study.

But repeating a controlled experiment under different conditions becomes a form of observational study because the differing conditions cannot be experimentally controlled. Thus the "observational" approach rather than experimentation remains the most basic form of data collection in science.

CHAPTER 19 EXERCISES

Exercise 19A. The Conditions of Observation

An empirical observation is made under numerous conditions. Can these be classified?

Discussion.

Take the measurement of the boiling-point of water as an example. The conditions of observation can be classified as:

(i) conditions used in the analysis, e.g. a correction for the atmospheric pressure;
(ii) conditions recorded but not used in the analysis, e.g. how much heat was applied;
(iii) conditions observed but not recorded, e.g. the time of day;
(iv) conditions that could have been observed but were not, e.g. the humidity;

(v) conditions that could not have been observed, e.g. the phenomenon of super-heating when the relevant concept or measuring procedure was still unknown.

One could record a myriad of factors in any experiment, but one generally ignores those which previous experience has shown to be more or less irrelevant (such as what was going on in the next room, what the observer ate for breakfast, how many measurements he had already made that day, etc.). Only when an unexplained discrepancy occurs may one start digging into such other factors for a possible explanation.

Exercise 19B. Repeating an Experiment

Should we aim to repeat an experiment under identical conditions?

Discussion.

No, for two reasons. Firstly, repeating an observation under identical conditions is impossible (*something* must have changed, e.g. time or place, etc.). Secondly, if identical repetition were possible it would be pointless since we would know beforehand that we must get the same result.

It also follows that a large number of repetitions is pointless if the conditions observed and/or recorded are all very similar. Variety of conditions is what matters.

Exercise 19C. Explaining a Discrepancy

What does a chemistry teacher do when litmus paper turns blue instead of red on being exposed to an acid?

Discussion.

The teacher does not suppose that he has disproved a law of nature, but says "Sorry, something has gone wrong!" and tries again. If the paper still turns blue, he checks whether he has used the right bottle.

All scientific laws only hold under a specified range of conditions and he knows that discrepancies will occur if these conditions are not fulfilled. Occasionally a discrepant observation arises which is outside the "normal" type of exception (Fleming's discovery of penicillin has already been mentioned as a popular example). But most of the time we know (or guess) that discrepant results are due to types of error which are already known about.

In the early stages of studying a topic, exceptions will not yet be well understood, so they require following-up. The various conditions of observation have to be checked and further observations made. Sometimes a discrepancy is non-repeatable. It then remains an isolated exception, with no sort of explanation. But just because it was not repeatable the exception usually becomes increasingly unimportant, a once-only event.

Exercise 19D. The Uncertainty Principle

Is sociology a science?

Discussion.

Qualms are often raised about the scientific study of sociological phenomena, since the act of observation will affect those who are observed. But this also occurs elsewhere. For example, 50 years ago in subatomic physics Heisenberg's *Uncertainty Principle* said that the "quanta" of energy used in trying to measure the position and the speed of an electron are so large relative to the electron itself that they interfere with it too much to measure both position and speed simultaneously.

In fact, everything affects everything else. Lifting one's hand to write affects the gravitational pull of every body in the universe, but mostly only to a trivial extent. The aim of science is to isolate and codify those phenomena which are effectively related to only a few other variables.

There is plenty of experience, both everyday and scientific, to show that there are generalisable regularities in the social sciences, i.e. phenomena which are *not* affected by a myriad of other factors. These phenomena may not always be the important problems which the practical minded person wants to solve immediately, but then the physicist still cannot measure both the position and speed of an electron, nor readily transmute lead into gold.

Exercise 19E. Time as a Condition of Observation

"Forecasting is always difficult, especially when it concerns the future." Discuss.

Discussion.

If the same result has been observed at several different points in time, we know that time as such cannot affect it. We can therefore predict that the result will hold again in the future within the range of conditions already covered, e.g. for different observers, different places, etc.

If such a prediction fails, we know that time itself cannot be the cause. The failure must be due to some more specific factor in the new conditions of observation, e.g. because the temperature or the lighting was different.

Forecasting problems arise because we do not know how factors *other* than time will change, or what their effect on the result will be.

Exercise 19F. The Analysis of a Factorial Experiment

Table 19.3 sets out the average readings of a clinical trial of a drug, designed along the lines of Table 19.1. Discuss the main steps of the analysis.

Discussion.

The marginal averages show three main effects:

(i) treated patients score substantially higher than non-treated patients;

(ii) dosage level makes a marked difference, but there is no placebo effect;

(iii) there is a small difference in favour of the normal level of nursing.

Table 19.3 Mean Values of A Diagnostic Measure

	Treatment Dosage				Average
	High	Low	Placebo	Control	
Nursing Care					
High	82	62	42	50	59
Normal	78	68	62	52	65
Average	80	65	52	51	62

But there is also some indication of "interaction" between the treatment and nursing factors. The results for the two dosage levels differ more at the higher level of nursing than at the normal level, and the placebo group has a substantially lower score at the higher nursing level. Perhaps the normal level of nursing was better in some ways. (The artificially high level of nursing might have been supervised by a tyrannical head-nurse.)

If this was the first experiment of its kind, the interpretation of the observed effects cannot be clear-cut. But the relatively elaborate experimental design allows stronger conclusions about the treatment than would have been possible with just a simple comparison between a treated and a control group of patients.

However, we still do not know to what degree the results have been affected by the allocation of different patients to the various design "cells". This question could have been eliminated by allocating the 32 patients randomly to the 8 "cells".

Exercise 19G. The Analysis of Variance

How would one test the significance of the results in the preceding experiment if the patients had been randomly allocated to the different design cells?

Discussion.

The usual statistical procedure used to establish whether random sampling errors caused the differences in Table 19.3 is the Analysis of Variance (see Section 18.6). This process uses the sampling distribution of the F-ratio and is based on breaking the total variance of all the readings about their overall mean into different components.

The figures below divide the total variance of the data in the clinical drug trial into different components for the four treatment levels, the two nursing levels, the interaction of treatment and nursing, and the pooled residual variance (based on the variances of the four readings in each cell).

	Degrees of Freedom	Variance Estimate	F-ratio
Treatment	3	1,477	10.6
Nursing	1	288	2.1
Interaction	3	208	1.5
Residual	24	140	—
Total	31	281	—

The residual variance is the basic figure used to test the statistical significance of the differences in the observed readings for treatment, nursing and interaction. We are essentially seeking to establish whether the mean values in the different cells differ only because of random sampling errors, or because either the drug or nursing had effects.

We calculate the residual variance by first looking at the variance of the four readings from the mean in each cell. If these eight cell variances are approximately equal, we can pool them into an overall residual variance, giving a value of 140 in our case. (If the cell variances differ markedly, the analysis becomes more complex.) Thus the pooled residual variance is an estimate of σ^2, the variance of the individual patients' readings from the population mean of 62.

We next look at the variances of various sub-group means. Starting with the treatment itself, we have four levels of dosage and therefore four mean values in Table 19.3. The variance of this set of mean values is

$$\{(80 - 62)^2 + (65 - 62)^2 + (52 - 62)^2 + (51 - 62)^2\}/3 = 184.7.$$

On the null hypothesis that the treatment has had no effect, this figure should be an estimate of the sampling variance of the means in each cell. It should therefore equal σ^2 divided by the sample size of each mean, i.e. σ^2/n. Since in our case $n = 2 \times 4 = 8$, the treatment variance multiplied by 8, i.e. $184.7 \times 8 = 1{,}477$, should give an estimate of σ^2 just as the pooled residual variance of 140 did.

The two values in our case, 1,477 and 140, clearly differ markedly. Whether the difference is real (reflecting a treatment effect) or probably only due to random sampling errors in allocating particular patients to the different types of treatment is what one seeks to determine by calculating the ratio of the two variance estimates, $1{,}477/140 = 10.6$.

If the null hypothesis is correct and the treatment has not been effective, this "F-ratio" should follow an F-distribution with 3 and 24 degrees of freedom. (The degrees of freedom reflect the number of independent comparisons in the data. With four treatment levels there are only three such comparisons, one less than the number of readings. There are $8 \times 3 = 24$ degrees of freedom for the residual variance.)

Referring to tables of the appropriate F-distribution we see that an F-ratio of 10.6 is statistically significant. Thus we must reject the null hypothesis and conclude that the treatment has been effective: the *apparent* differences in the last row of Table 19.3 reflect something real.

The same steps can be carried out for the nursing effect and for the interaction between different dosage levels and different nursing levels. (The interaction should normally be tested first.) We will find that neither is statistically significant. This indicates that the large difference between the two nursing levels in the placebo group was due to random allocation of patients to the different design cells. The implication is that a similar difference would occur by chance in most experiments like this with only 32 readings (though not always in the placebo group).

Once we have established that the treatment effect is significant, we still have to estimate and interpret the effects in detail. We tested the treatment on an overall basis. The next step is to determine which specific comparisons matter, i.e. the low level of the drug versus no treatment, the two dosage levels against each other, and so on. As mentioned in Section 18.6, selecting the largest differences by inspection and then testing them for statistical significance leads to technical difficulties. But in our example prior expectations provide specific hypotheses; i.e. whether the treatment is *generally* effective, whether the higher dosage is *more* effective, and whether the placebo has an effect compared with no treatment. Such prior hypothesis can then be tested for statistical significance by standard t-tests or the like.

Exercise 19H. How Many Factors in a Factorial Design?

Could the clinical trial discussed in the two preceding exercises have had more than two factors (treatment and nursing)?

Discussion.

A fully balanced experimental design needs at least one reading for every possible combination of factors. Thus with a total of 32 patients we could have had two additional factors of two levels each, e.g.

four dosage levels = 4

two nursing levels = 2

younger vs. older patients = 2

severe vs. mild cases = 2

This would lead to a $4 \times 2 \times 2 \times 2$ design with 32 test combinations.

By exercising more control on the allocation of different patients to each cell (e.g. two old, two young; two mild cases, two severe cases), the experimenter derives more information from the one study. Normally it is not advisable for *every* factor in an experiment to be expected to produce a striking effect (or at least not a striking interaction with other factors). This would make the results too complex to interpret. Instead it is generally better to include some factors which one merely wants to establish as having *no* effect.

Such fully utilised controlled experiments are obviously beneficial in gaining greater information and saving time, effort and money. In practice they are, however, often difficult to organise.

Exercise 19I. Latin Square Designs

What is lost if there is less than one observation for each combination of factors?

Discussion.

Consider the following "Latin Square" design in an agricultural trial, using nine plots of land.

		Fertiliser level 10 lbs	 5 lbs	 0
Fertility of Plot	High	A	B	C
	Med	B	C	A
	Low	C	A	B

A certain fertiliser is applied at 10 lbs, 5 lbs, and zero rates to plots classified as being of high, medium and low fertility. Three different types of seed, A, B, and C, are also used.

Thus we have three factors, each at 3 levels, making a total of 27 possible combinations, but only 9 plots of land. All possible combinations cannot be measured, e.g. there is no reading for 10 lbs of fertiliser applied to seed B in a high-fertility plot.

However, the average reading for the 10 lbs fertiliser level comes from plots of high, medium, and low fertility using all three seeds, A, B, and C. In the same way, every level of every factor is balanced on all the levels of the other factors. For example, the three *low*-fertility plots are given seeds A, B, and C, and the three fertiliser levels. It is therefore possible to make balanced comparisons of the "main effects", e.g. the effect of fertiliser level as an average across the different plot levels and seeds.

The limitation of the Latin Square design is that it is generally not possible to establish *interactions*. For example, we cannot tell whether 10 and 5 lbs of fertiliser have different effects on high- and medium-fertility plots because different types of seed were used, A and B in one case, and B and C in the other.

This type of experimental design is therefore useful mainly when previous work has shown that such interactions are unlikely, or when one is merely "fishing" to see what *major* effects might occur.

The "Graeco-Latin" Square below is a still more ambitious design.

	10 lbs	5 lbs	0
High	Aα	Bβ	Cγ
Medium	Bγ	Cα	Aβ
Low	Cβ	Aγ	Bα

Here a fourth factor, the use of different fungicidal "dressings" of the seed, say, has been introduced at three levels, α, β, and γ. Each level of each factor is still fully balanced against every level of the other factors. Each column and row of the design has one α, one β, and one γ, and one A, one B, and one C.

The names of these two designs stem from the Roman and Greek letters used in them. They exemplify more advanced cases in the statistical theory of experimental design.

Exercise 19J. The Important Factors

Why do scientists so often concentrate on "academic" questions instead of practical problems?

Discussion.

One reason is that before any problem can be solved it is first necessary to establish the effect of the *major* factors (the ones that most affect the variable in question) even if these factors are uninteresting from the practical point-of-view. For example, we may be interested in the effects of race on children's stature and growth, but first we have to establish the effects of *age*. If we do not know how age affects stature, we will not reach any valid conclusions about the relatively minor differences (if any) related to race.

Exercise 19K. The Choice of the Research Design

There has been a great deal of discussion about the association between smoking and lung cancer. The evidence has largely been based on surveys which examined smoking and lung cancer in human populations.

Discuss the procedures which may be used in such surveys to give rise to the strong presumption of a link (Adapted from a specimen "Question-and-Answer" prepared for the Market Research Society by Mr. Colin Greenhalgh.)

Discussion.

One possible procedure is as follows:

(a) A representative sample of the young adult (or even child) population is recruited. This sample should be large enough for small differences in a rare characteristic, the contraction of lung cancer, to be significantly demonstrated between smokers and non-smokers, probably within sub-groups thought to be relevant (e.g. by sex, occupation, parents' smoking and health history, etc.).

(b) The smoking habits of this sample are recorded over time (e.g. by self-completion diaries or by regular personal interviews).

(c) Their medical history over time is also recorded; specifically, of course, the apparent cause of death for any who are unfortunate enough to die in the course of the survey. Alternatively, the survey could be continued until *all* the informants are dead and age and cause of death are recorded.

(d) A simple breakdown analysis is tabulated of mortality rates, cause of death and/or ailments suffered, between smokers and non-smokers. If the incidence of lung cancer among the smokers is shown to be higher than among the non-smokers, then this is taken as strong evidence to support the suggested association.

Such a simple-minded analysis can be refined not only by *type* of smoking (filter or non-filter cigarettes, cigars, pipe, etc.) but also by *amount* of smoking.

However, this kind of a survey would require a long run of continuous data and take many years before any conclusive association became apparent. Instead, a survey could base its evidence on (probably less accurate) data of past smoking behaviour collected by an interview with each recruited informant.

Yet no matter how well designed and conducted, such surveys would always leave some residual doubt about what was cause and what effect. The cause-and-effect could be the opposite of what is hypothesised; a propensity to contract lung cancer may actually cause a craving for cigarettes. Or the two characteristics may be independently associated with a third, possibly unknown, characteristic. For instance, a certain psychological trait may (i) tend to make its possessors want to smoke *and* (ii) tend to induce cancer. In both these cases one could remove the smoking (by persuasion or even compulsion) and not affect the propensity to contract cancer. The association found in the survey would be true, but *irrelevant*.

Other Methods.

Some of the minor shortcomings in such a survey method can be removed by refinements in the analysis. For instance, the sub-samples of smokers and non-smokers can be checked on other characteristics which *might* be influencing the apparent association: e.g. more city dwellers than rural dwellers might be sufferers and it might be hypothesised that the general atmospheric pollution in the cities was causing the lung cancer rather than the smoking, or there might be a difference in inherited propensity to lung cancer. (Sometimes such analyses are made by *post-weighting* the samples of smokers and non-smokers to "match" on rural/urban and parental background factors, etc. But post-weighting of the aggregate results is pointless unless the analyst has established by analysing the separate sub-groups that the factors matched *do* have any effect, and if so, what it is.)

This procedure can be followed for any characteristics where smokers and non-smokers are shown to differ. (It may be necessary to collect further descriptive data about the informants in order to explore all the suggested hypotheses about possible differences.) These other characteristics can then slowly be eliminated from the association, one by one.

Many doubts about cause-and-effect would be eliminated by organising a controlled *experiment* in which two random sub-samples of the human population (preferably after stratification by other relevant characteristics) were artificially induced to smoke and not to smoke regardless of their "natural" behaviour.

Such a procedure would virtually eliminate all the disadvantages of a typical survey as described since (a) the two sub-samples *must* be matched within the terms in which they were stratified, (b) it is highly probable (and can be tested by replication) that they are matched in other unstratified (and possibly unknown) characteristics, and (c) smoking is the one characteristic for which the two sub-samples were definitely and completely unmatched. Therefore it must be this difference in smoking which *either directly or indirectly* causes any difference in lung cancer incidence between the two sub-samples shown in the experiment (and replicated to establish a degree of generalisability).

One difficulty with such an experimental approach is inducing a human population to adopt a behavioural pattern they do not voluntarily wish to adopt. If the inducement is too artificial, it may affect the conclusions to be drawn from the experiment: "forced" smokers may not be equally prone to contract lung cancer as "natural" smokers. Apart from that, it would in this case be considered improper to force, or even to encourage, a sample of the human population to adopt a form of behaviour that might be injurious to their health.

Therefore such experiments have been conducted on animals, e.g. rats. It is possible, and appears ethically acceptable, to induce animals to "smoke" under reasonably realistic conditions (e.g. by bringing them up in a "smoking-machine"). But such experiments are flawed by the fact that rats are not human beings, physiologically or psychologically, and therefore do not necessarily react in the same way as human beings to any particular treatment. Controlled experiments on rats demonstrating that *their* smoking causes a propensity to contract lung cancer are strong corroborative evidence of the human survey evidence, but still do not prove the cause-and-effect relationship in humans. The main function of such animal experiments is in fact not to "prove" that smoking causes lung cancer, but rather to establish an increasing depth of understanding of the mechanisms involved.

Such understanding is desired partly for "applied" reasons; it could lead to ways of reducing or eliminating any bad effects of smoking without necessarily stopping people from smoking altogether. A more important "basic" reason is that even fully controlled randomized experiments on *humans* would not prove that it is the sheer mechanical act of smoking that causes lung cancer. The cause could be "third factors" which happen to be associated with the act of smoking, e.g. frequent movements with one's hand, the effect of touching cigarette packets, frequent visits to tobacconists, or the psychosomatic fear that smoking will cause lung cancer. Thus depth of understanding, rather than an apparently simple one-to-one relationship, is required.

CHAPTER 20

Description and Explanation

This book has concentrated on the description of data. But the purpose of description is to gain understanding, to find out how and why things happen and hence also to gain some *control* over the phenomena in question.

We now discuss the link between description and explanation. We start by considering the nature of descriptive relationships. For example, what sort of statement is it to say that y varies approximately as $(ax + b)$?

20.1 Descriptive Relationships

We have already noted that scientific laws are not universal. They merely describe how particular observed phenomena behave, with exceptions. For example, Boyle's Law, $PV = C$, describes how pressure and volume vary together under certain conditions. As already discussed, the law does not hold when the temperature changes, when there is condensation or a chemical reaction, etc.

Nor is a scientific law directly *causal*. The relationship $PV = C$ does not claim that pressure causes volume, or even that changes in one variable directly causes changes in the other. The law merely describes the values that Pressure P and Volume V take when some third factor, like the piston in Figure 20.1, is moved from position X to position Y.

If we infer more explanation from Boyle's Law it is because we are aware of a wider range of later knowledge about the movements of molecules inside gases, about Avogadro's Hypothesis, and so on. But a descriptive relationship between two or more variables is never a direct statement of causal effects.

One reason why scientific laws in themselves cannot be directly causal is that they are not exact. They are all deliberate oversimplifications and hence are not strictly true. Science aims to find one generalisable equation that covers a wide range of different phenomena with some known degree of error rather than a great number of different specific equations that give a

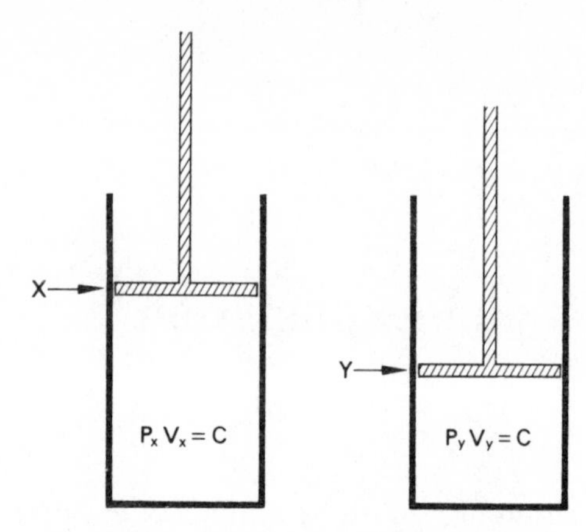

Figure 20.1 $PV = C$ for Positions X and Y of the Apparatus

closer descriptive fit to isolated sets of readings. The result is more useful, but it is also necessarily oversimplified.

We have followed this precept throughout this book. We have forced straight lines onto data which we knew were slightly non-linear, and have fitted Normal Distributions as simplifying approximations to data which could not conceivably be *exactly* Normal, and so on. It is the common process of science.

Generalisations are usually reached by excluding local complicating factors. This is not merely a case of smoothing away irregular errors, but may even be one of by-passing consistent biases, of very *deliberate* over-simplification and reformulation.

For example, the practical usefulness and intellectual glory of Newtonian mechanics is that the same laws cover vastly different phenomena: falling bodies, the swing of a pendulum, the motion of the stars, balls rolling down inclined planes, the movement of the tides, the trajectory of cannon balls, and so on. But it can do this only because the calculations generally ignore complicating factors like friction, air resistance, and the gravitational pull of other bodies. Balls always roll down inclined planes more slowly than is stated in simple mechanics where friction is ignored (and where they would in any case not roll at all, but *slide*).

The size and direction of the errors has still to be established under all the relevant conditions, just like the basic relationships themselves. We have to learn when the theory approximates the data closely enough to ignore the errors, and what correction factors to use when it does not. Once an error is known it is no longer merely an error.

The simplifying approach also shows up in the fact that most well-developed laws in science have no numerical coefficients (other than 1's and certain "absolute constants" like π, e, the absolute zero of temperature $-273°C$, and the velocity of light c in physics). For example, Boyle's Law $PV = C$ also reads $P_X V_X = P_Y V_Y$, without any numerical coefficient. Again, the buyer behaviour relationship in Chapter 10, $w(1 - b) =$ constant, is free of coefficients in the form $w_X(1 - b_X) = w_Y(1 - b_Y)$.

Numerical coefficients are avoided not merely because of the mathematician's delight in the elegance of form ("no coefficients!"), but because a law is only ready for use when it contains no coefficients whose numerical values still have to be established. This applies both to practical and theoretical uses.

20.2 Explanation

If scientific laws merely describe how certain observed phenomena behave, most of us still want to know *why* this behaviour occurs. We are more comfortable with a finding which we can understand than with one based on empirical fact alone: "It is all very well in practice, but how does it work in theory?"

Explanations often take time to develop. Newton was criticised for not having explained why bodies move. He responded to the effect that he had shown *how* they moved, and that should be enough for the moment. In fact he provided much understanding, e.g. by interrelating the movement of the tides and the swing of a pendulum. The same law holds for both. But even today, three hundred years later, there is still no accepted explanation of the *nature* of gravity.

In fact, explanations are often more superficial than is thought. A close look at explanations shows that they are really no more than descriptive linkings of how one thing varies with how another thing varies:

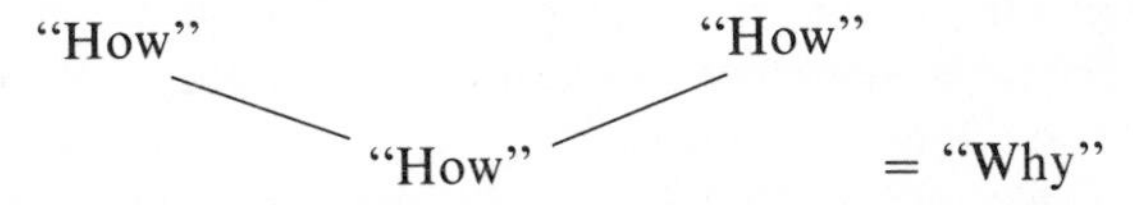

For example, Boyle's Law works because, roughly speaking, the smaller the volume of a given amount of gas, the more often the molecules of the gas hit the side of the containing vessel and create more pressure. The explanation is descriptive.

A typical explanation of a very simple kind is that "Ice floats because it is lighter than water". At first sight this seems little better than the virtual tautology that "Johnny is good at playing the piano because he has a talent for it". But the explanation actually accomplishes two things.

Firstly, it relates a very specific phenomenon—ice floating—to a larger generalisation which we already know about, the notion of specific gravity, that things generally float if they are lighter than water. Hence the phenomenon has been "explained", because it has been linked to something which we know and accept. (Similarly, in Chapter 8 the cube-root formulation of the relationship between children's heights and weights was more comfortable than the logarithmic one. It linked up with, or was "explained" by, our more general knowledge that weight varies as volume and that volume is three-dimensional.)

Secondly, the explanation rules out all other mechanisms which might have accounted for ice floating. For example, it excludes the Aristotelian doctrine that objects float or sink essentially according to their *shape*.

Many explanations are deeper than this because they touch on the mechanics involved. An example is that "hydrogen burns in air because it combines with oxygen". But saying that hydrogen combines with oxygen is not a blow-by-blow causal explanation. None of us knows what actually happens when two atoms of hydrogen meet one atom of oxygen to make water. Similarly, we accept the molecular explanation of Boyle's Law even though we do not fully understand it.

We seek explanation and understanding. But this is never simple because there are always "third factors", and it is never complete because there are always "black boxes" that elude our grasp. And when we use laws and relationships for theoretical work and practical applications we do so only because they actually describe what happens. We know that if we press a light switch, the light will generally come on. We do not need to know more until the light does not work and we have to search for a cause: a blown fuse, a burnt-out bulb, a power-cut, or an unpaid bill? Even then we usually do not know all that much about the precise mechanism, there are still "third factors" and "black boxes". We may not know how fuses work or what actually happens when we pay our overdue bill. These things can function without our understanding them.

Explanations often do not even take the form one might have expected. There would have been nothing obvious to Sir Robert Boyle about the molecular explanation of his law. The mere idea that gases consist of molecules moving around at high speed was effectively unknown in his time. More parochially, consider the equation $w(1 - b) =$ constant, that large brands are bought more frequently by their buyers than small brands are bought by theirs. The explanation might have been sought in terms of people's incomes, the sizes of their households, and their exposure to advertising, but the result actually follows from (or is "explained by") quite different considerations, as was outlined in Chapter 10.

20.3 From Facts to Theory and Back Again

The development of scientific knowledge depends on the interaction between facts and theory. Basically it has to start with facts. Fact-collecting must of course be preceded by some hypothesis, as Popper (1959) has stressed. But initially this only needs to be some very loose, imprecise notion of which facts it might be interesting or relevant to collect and which to ignore. At the beginning these ideas are largely based on ignorance. When they turn out to be wrong it matters little except to the investigator.

Having found some low-level patterns or generalisations among the early facts, one tries to generalise them further and also to account for the exceptions that will arise. Interrelating the different findings leads to the beginning of *theory based on facts.* This in turn often suggests hypotheses, i.e. ideas about new ways of looking at the available facts or about new kinds of facts to collect. Hence we have the continuing interplay between fact and theory and back again.

Theory here is empirically based. But nowadays it has become rather fashionable to formulate behavioural or mathematical models *before* looking at the facts. These explanatory models are based on abstract ideas of how nature might work, using various intuitively reasonable assumptions. ("Reasonable" is a telling word here. It is generally used only when no reasons can in fact be given.)

There is nothing like a few facts to eliminate any number of speculative assumptions. For example, the old view was that the planets move in circles. This was logical enough since the circle is the most perfect geometrical shape. God is good, and therefore planets have to move in circles. Similarly, it is not difficult to think of reasons why black children have a different shape from white children, or why the leading brand of breakfast cereals appeals particularly to heavy users of the product. But just a few facts show these statements are simply not true.

Given the rehabilitation of speculative theorising in recent decades, it is not surprising that theory has earned itself a bad name, as something which does not work. Despite the extraordinary success of science in the last few hundred years, words like "theoretical", "academic", and "intellectual" are nowadays tinged with pity. But there is good theory and bad theory, good scientists and lesser ones.

Theory and applied mathematics are not supposed to be substitutes or precursors of empirical knowledge. Their function is to help model and explain the known facts, but the facts have to come first. Einstein is often quoted as the supposed counter-example to this. (People usually pick on dramatic extremes which they do not understand.) But Einstein's work concerned minor discrepancies in results that were based on centuries of detailed empirical observation and analysis (e.g. that light rays generally travel in straight lines, except, Einstein predicted, very fractionally near the sun). As he himself put it (Einstein, 1949): "... before a theory explaining a process can be tested that process must be known".

There is a straightforward test for the modern mind-over-matter tendency to build mathematical models and theories before any facts are known:

> "Take away the mathematical language and what generalized factual knowledge of the process in question remains? If the answer is none, the mathematical symbol for *that* is very simple."

The pay-off of Sonking, the Scientification of Non-Knowledge, is only in the model-builder's neo-cartesian self-regard: "I sonk, therefore I am". The question is not what the theoretician thinks, but what he *knows*.

Science means knowledge. But particular subject matters, like physics or psychology or economics, may be in very different states of development. Three main stages of knowledge can usefully be identified, depending on the existence of a generally accepted conceptual frame-work or "paradigm" (Kuhn, 1970).

First there are the initial fishing-expeditions: trying to find some facts, some patterns, and some generalisable notions. Subject areas at this early pre-paradigmic stage of development usually lack stable foundations of knowledge. They are often distinguished by intense concern with techniques and apparently endless arguments over methodology. But this is unnecessary: this is where the low-level procedures discussed in this book mostly apply, the notion of empirical generalisation, of seeing whether the results in one study agree with all the previous ones, an approach which is low-key but nonetheless exacting.

The second stage of development is "normal science", firmly based on acknowledged past achievements where its practitioners share some general theoretical viewpoint. The greater part of scientific effort takes place at this level: filling in gaps, linking up pieces, puzzle-solving. This work can be very exciting, like atomic physics in the 1920's and 30's or the discovery of the structure of DNA.

The third stage is the occasional revolution, where some accepted scientific view-of-the-world or paradigm is turned upside down. For aeons people thought the sun moved round the earth. That was the way it looked, a view which could not be changed by any directly discernible evidence (Wittgenstein's "What would if have looked like if it had looked as if the earth rotated?"). That revolution had to wait for Copernicus' creative insight and nerve.

These latter qualities are also needed at the level of this book. It takes more insight and courage to report an average or an abstract relationship than merely to present all the facts.

20.4 Summary

A scientific law is a theoretical abstraction, but inherently it still remains only a descriptive generalisation of the facts. Understanding and explanation are continuous processes that grow as different laws are interrelated.

CHAPTER 20 EXERCISES

Exercise 20A. The Main Conclusion

What is the main point this book makes about the analysis of data?

Discussion.

That data need to be *summarised.*

The criteria of a *good* summary are that it be

(i) succinct,

(ii) complete (i.e. that the original data could be reconstructed, within the stated or implied limits of approximation),

(iii) usable (i.e. that the results can readily be used when analysing further data).

Exercise 20B. Statisticians as Their Own Customers

To what extent do statisticians in fact use previous results when analysing new data?

Discussion.

The statistical literature contains little explicit discussion of the use of previous results in analysing new data. (An apparent exception is the Bayesian approach to statistics mentioned in Chapter 13, but this is not widely practised.)

A small-scale check of statisticians' use of previous results was therefore carried out a few years ago (Ehrenberg, 1969). A number of eminent past presidents of the Royal Statistical Society and editors of its Journal were asked whether they or some nominee could prepare a methodological paper or case-history about using the results of previous statistical analyses for a statistical conference on "Consumer Satisfaction". The question was whether statisticans were satisfied customers of their own results.

The approach was unsuccessful, as is illustrated by the following replies (quoted with permission):

(a) "Customer Satisfaction in any sense of the words is not really me, I think. Although the idea you suggest is an interesting one, I am not really competent to speak on it."

(b) "The proposal to have a session of the kind you describe seems to me an excellent one. However, I feel it essential that speakers should at some time have been in a position where quantitative prior data were available for use in the analysis of new material. As I have never really had such experience, though often felt the lack of it, I am afraid that I could not make any useful contribution to the discussion."

(c) "No, not I; nor yet any other. The job is not one for a statistician *per se*, but for a science-historian."

(d) "I have put out a few feelers in the department [in agricultural statistics] but nobody seems to have a case-history which they think would make a sufficiently interesting paper. I wonder whether the accumulation of prior information would be more regular in an *industrial* environment ['The grass on the other side of the fence...']. I agree that it would be a fascinating topic with the right case-history to initiate the discussion."

(e) "Incorporating previous data? This just hasn't come my way."

(f) "I think the proposal you make is a very interesting one, but I shall not be able to take it on as this is clearly one which would require a good deal of thought and preparation."

(g) "Participating in the conference at Sheffield will not, I am afraid, be possible. I have nothing useful to conribute to this subject."

(h) "We use other people's *conclusions*, but these are qualitative results. We don't use other people's *figures* very much, if at all, in the sense of other people's estimates of parameters. We don't combine our statistical test results with theirs either, in a numerical way. The reason is I think that we are never sure that parameters have not varied or, more generally, that our statistical set-up is strictly comparable with that which went before. There is also the question of *authority*. I don't think I would trust more than 20% of statisticians that I know not to miss the obvious. Too many are so blinded by their theory that they don't look at the data. And as for the statisticians I don't know...."

And so on and so forth. Apparently one must never use previous results because things *might* have been different (so one will never find out whether they were or not). Statisticians with 20 or 50 years' experience claim never to have used previous results! This is wrong. Why *produce* statistical results, if even oneself never expects to use them?

Exercise 20C. Description Without Explanation

Can relationships which are merely descriptive and have not been explained be valid and useful?

Discussion.

The history of science shows that explanations often come long after a result has been descriptively well-established, e.g. the gap of more than one hundred years between Boyle's Law and Avogadro's molecular explanation.

Farmers have used fertilisers since time immemorial without knowing how they work. Are they absorbed into the plant (and then what happens?); or do they stimulate soil bacteria that break down inorganic nitrogen compounds; or do they work in some other way completely? Agricultural scientists and manufacturers now know more about fertilisers, but they still do not understand many of the detailed mechanisms.

Exercise 20D. Assumptions Without Justification

Is it proper to make an assumption without direct justification or explanation?

Discussion.

Except in speculative theorising, assumptions must be founded in fact. However, the empirical justification does not have to be direct, and it may come much later. Nor does the assumption have to be fully understood.

As long as the assumption links up *other* empirical results, it performs a useful descriptive function. (A typical example of this process was provided by the justification of the Poisson–Gamma theory which underlies the NBD model of repeat-buying, as outlined in Exercise 13P.)

Exercise 20E. Testing an Explanatory Theory

To what extent do theories have to be tested against the facts?

Discussion.

Many theories develop organically by piecing together empirically well-established but low-level relationships. The basic results therefore do not require testing, as they are already empirically established. But the process of linking different results together and the development of explanations involves working assumptions and hypotheses that do require testing.

In contrast with such theories which start with a basis of factual generalisation, there are the speculative theories that are developed as possible explanations or models of certain phenomena before it is empirically known just how these phenomena actually behave and what there is to explain. Most writers suggest that such models need to be tested against the facts, but they frequently miss the right emphasis, as in the *Principle of Empirical Viability* (Montgomery and Urban, 1969, p. 89).

> "It seems reasonable to require that a model be demonstrated to be empirically viable for at least one set of data. That is, the model should be consistent with (fit) at least one set of data."

Saying something twice does not make it any more true, but requiring a model to hold for *two* sets of different empirical data would at least serve to establish whether it could in fact generalise. (Many theories can hold for one selected set of data, but are they known to apply to a sufficiently wide range of circumstances to be of any practical or academic interest?) The real problem is that the intending model builder here did not know anything by way of empirical generalisations before he started building the model. All he intends to do is to "test it" against some new and undigested data.

Exercise 20F. The Timing of an Explanation

Could explanations be arrived at earlier than they are?

Discussion.

In principle the explanation of some new result could often be reached earlier, but in practice it takes time for the new result to become familiar enough for its explanation to be "seen".

Often an explanation requires additional kinds of findings, and then a delay is inevitable. (This happened with both the illustrations referred to in Exercises 20C and D.) Sometimes one can speculate theoretically on the new kinds of findings that might be needed to explain the given result, and this may speed the process. However, creatively guessing at new explanatory factors is often difficult. Part of the problem is that empirical models are not *exact*. Possible connections cannot therefore be deduced by logical or mathematical arguments alone. Seeing the nature of the necessary approximation in the argument usually requires a major imaginative jump.

Consider a simple result, e.g. that a particular plant in the garden has died. One looks for an explanation such as lack of water, too much heat, certain pests or disease, mere age, or whether someone trod on it. But if none of these appear to apply, some new kind of explanatory factor has to be found. To do so requires creative insight as well as technical knowledge.

Exercise 20G. Knowing the Causal Direction

The precise form of certain descriptive analyses is sometimes said to depend on the causal direction. What is the justification for this?

Discussion.

An example occurs in regression analysis (Chapter 14). To apply this technique the analyst has to choose between the two possible lines, y on x or x on y. To do this he often makes a prior assumption about the direction of the "causation", e.g. that x influences y and not the other way round.

But he is then claiming that he knows the causal direction even though he does not know the coefficients in the equation, whether the equation is really linear, or even whether it exists. (He usually puts much effort into testing the null hypothesis of zero correlation!) This seems untenable, an analyst's half-baked ideas about causal connections determining the results of the analysis. (Not surprisingly, therefore, this approach does not seem to have led to any lasting results.)

Exercise 20H. Correlation Is Not Causation

It is commonly said that just because two variables are correlated, this does not necessarily mean that there is causation.

Can you think of an example to the contrary, where a correlation between two variables *does* mean that one variable directly causes the other?

Discussion.

No. None of the laws of science seem in themselves to tell us directly about what causes what. At best there always seem to be "black boxes" or "third factors" involved.

The medical drug Thalidomide is known to be related to the incidence of malformed babies. An obvious explanation is that the drug causes the deformities.

But it is also known (i) that a proportion of human foetuses are naturally malformed; (ii) that many of these are rejected by the mother's body during the early stages of pregnancy; and (iii) that Thalidomide helps to suppress the body's rejection of strange or foreign tissues. We could therefore form the alternative hypothesis that the correlation is due to a *negative* effect, that Thalidomide prevents the natural abortion of naturally deformed babies.

These different hypotheses about the correlation cannot be elucidated in any simple manner, but only by interrelating the descriptive results of many different types of studies (e.g. the *kinds* of deformities observed under different conditions, the length of time for which the drug is taken, etc.).

Exercise 20I. The Causal Direction

Is it possible to establish the causal direction between two variables?

Discussion.

It is generally very difficult to make a precise, watertight statement about the causal connection between any variables occurring in a quantitative scientific law. An electric current running along a wire does not cause resistance, nor vice versa. The current does not even "cause" a difference in voltage; if anything, we might think of a difference in voltage causing the current to flow, but it is actually due to a third factor or black box mechanism (e.g. a battery or dynamo), and this does not appear in the equation at all.

Among logicians and philosophers, the notion of causation is at best controversial. Among scientists it is hardly used, except as a loose and superficial shorthand, such as that a charged battery, or a dynamo when driven and suitably connected, will cause a voltage gradient to exist and a current to flow. But this is not a causal interpretation of the explicit relationship between the observed variables R, V, and I in Ohm's Law $V = RI$.

For another example, it is often thought that people's expenditure depends on their income. You have to *have* money in order to spend it. But in fact people's earnings are often influenced by what they need to spend. And some people spend more than they earn, thus ruling out any *simple* cause-and-effect relationship.

As a further example, it is widely thought that price determines the volume of sales. But often it is also the other way round, prices for high-volume goods can sometimes be reduced because of economies of scale.

Again, it might be thought that a product's sales level is influenced by the amount spent on advertising. But in practice advertising budgets are frequently set as a proportion of sales income, and advertising is generally the first item of expenditure cut if sales drop. The causal direction is not simple and unambiguous.

In driving a motor-car, it might seem that turning the steering-wheel causes the road-wheels to turn. But if the car is running in a rut, the opposite occurs. If there is a bend in the road, then the road-wheels need to follow it and this usually causes the steering-wheel to be turned. (In any case, few drivers understand the steering mechanism in their cars, rack and pinion, etc., or what happens to the causal chain when the steering fails.)

Nevertheless, it can be helpful in everyday terms to say that a current flows if there is a difference in voltage or that, by and large, a car is directed by turning the steering-wheel. At least this describes what generally happens!

Exercise 20J. Spurious Relationships

How can one distinguish spurious relationships from the proper laws of science?

Discussion.

It is common to deride so-called "spurious relationships", like the correlation between the number of pigs and the production of pig-iron over the years.

But the association between pigs and pig-iron is no less real than that between the pressure and volume of a gas. It has happened over and over again, and typifies a growth-pattern that is desired by almost every developing nation ("more of everything"). The relationship seems humorous partly because we know of circumstances where it does not work: e.g. that nothing much happens to the number of pigs if the government interferes with the production of pig-iron, or if there is a strike of foundry workers. But the same is true of Boyle's Law; there are circumstances where pressure and volume are not related.

Nor is the association between pigs and pig-iron merely a blind correlation. We actually know a good deal about the underlying mechanism, that the growth in the number of pigs and in pig-iron production is related to the growth in the population and the growth in productivity. The association is therefore due to "third factors". So is that in Boyle's Law, the piston in Figure 20.1 and what the experimenter does to it.

It is easy to think of examples of relationships which are spurious, and difficult or impossible to think of one which is not. In this sense then, all lawlike statements are "spurious". They are in themselves merely descriptive generalisations.

Exercise 20L. Empirical Generalisations

Few statistical texts refer to the idea of empirical generalisation, let alone emphasise it. Why all the fuss about it in this book?

Discussion.

A result which only holds for one set of data remains an isolated historical event.

List of Exercises

References

Aitcheson, J. and Brown, J. A. C. (1957). *The Lognormal Distribution*. Cambridge: Cambridge University Press.

Anscombe, F. J. (1967). "Topics in the Investigation of Linear Relations Fitted by the Method of Least Squares." *J. Royal Statist. Soc. B*, **29**, 1–52.

Bartlett, M. S. (1937). "Some Examples of Statistical Methods of Research in Agriculture and Applied Biology." *J. Royal Statist. Soc. Suppl.*, **4**, 137–146.

Bartlett, M. S. (1949). "Fitting a Straight Line when Both Variables are Subject to Errors." *Biometrics*, **5**, 207–212.

Bird, M. and Ehrenberg, A. S. C. (1966). "Intentions-to-Buy and Claimed Brand-Usage." *Operational Research Quarterly*, **17**, 27–46, and **18**, 65–66.

Bird, M. and Ehrenberg, A. S. C. (1970). "Consumer Attitudes and Brand Usage." *J. Market Res. Soc.*, **12**, 233–247; **3**, 100–101, and 242–243; **14**, 57–58.

Bortkewitsch, L. Von (1898). *Das Gesetz der Kleinen Zahlen*. Leipzig: Teubner.

Brookes, B. C. and Dick, W. F. L. (1969). *Introduction to Statistical Method*. London: Heinemann.

Bureau of the Census (1972). *Pocket Data Book: USA 1971*. Washington DC: U.S. Government Printing Office.

Business Week (1971). "Figures of the Week". *Business Week* (*April 24*), New York: McGraw-Hill.

Central Statistical Office (1972). *Facts in Focus*, London: Penguin Books.

Chakrapani, T. K. and Ehrenberg, A. S. C. (1974). "The Pattern of Consumer Attitudes". Aapor-Wapor Conference, Lake George, New York.

Charlton, P., Ehrenberg, A. S. C. and Pymont, B. (1972). "Buyer Behaviour under Mini-Test Conditions." *J. Market Res. Soc.*, **14**, 171–183.

Chatfield, C. (1969). "On Estimating the Parameters of the Logarithmic and Negative Binomial Distribution." *Biometrika*, **56**, 411–414.

Chatfield, C. and Goodhardt, G. J. (1970). "The Beta-Binomial Model for Consumer Purchasing Behaviour." *Applied Statistics*, **19**, 240–250.

Chatfield, C. and Goodhardt, G. J. (1973). "A Consumer Purchasing Model with Erlang Inter-Purchase Times." *J. Amer. Statist. Assoc.*, **68**, 344–351.

Clements, E. M. B. (1953). "The Statute of British Children." *British Medical Journal*, (October 24), **4842**, 897–902.

Clements, E. M. B. (1954). Private Communication.

Cochran, W. G. (1963). *Sampling Techniques* (2nd ed.) New York: Wiley.

Cochran, W. G. and Cox, G. M. (1957). *Experimental Designs*. New York: Wiley.

Collins, M. A. (1973). "The Analysis and Interpretation of Attitude Data." Market Research Society Course on Consumer Attitudes, Cambridge, March 1973.

Corlett, T. (1963). "Ballade of Multiple Regression." *Applied Statistics*, **12**, 145.

Cox, D. R. (1958). *Planning Experiments*. London and New York: Wiley.

Deming, W. E. (1960). *Sample Design in Business Research*. New York: Wiley.
Ehrenberg, A. S. C. (1950). "Estimation of Heterogeneous Error Variances." *Nature*, **166**, 608.
Ehrenberg, A. S. C. (1951). "The Unbiased Estimation of Heterogeneous Error Variances." *Biometrika*, **37**, 348–357.
Ehrenberg, A. S. C. (1959). "The Pattern of Consumer Purchases." *Applied Statistics*, **8**, 26–41.
Ehrenberg, A. S. C. (1963). "Bivariate Regression is Useless." *Applied Statistics*, **12**, 161–174.
Ehrenberg, A. S. C. (1967). "Where Were You in the Revolution?" *Admap*, **3**, 247–250.
Ehrenberg, A. S. C. (1968). "The Elements of Lawlike Relationships." *J. Royal Statist. Soc.* A, **131**, 280–302 and 315–329.
Ehrenberg, A. S. C. (1969). "Statisticians as Their Own Customers." Royal Statistical Society Conference on 'Consumer Satisfaction', Sheffield, April 1969.
Ehrenberg, A. S. C. (1972). *Repeat-Buying: Theory and Applications*. Amsterdam and London: North-Holland; New York: American Elsevier.
Ehrenberg, A. S. C. (1974a). "Repetitive Advertising and the Consumer." *J. Advertising Research*, **14**, 62–71.
Ehrenberg, A. S. C. (1974b). *Aviation Fuel Contracts*. Cambridge: Marketing Science Institute.
Ehrenberg, A. S. C. and Goodhardt, G. J. (1968). "The Nature of Brand-Switching." *Nature*, **220**, 5764, 304.
Ehrenberg, A. S. C. and Goodhardt, G. J. (1969). "A Model of Multi-Brand Buying." *J. Marketing Research*, **7**, 77–84.
Ehrenberg, A. S. C. and Pyatt, F. G. (1971). *Consumer Behaviour*, London and Baltimore: Penguin Books.
Ehrenberg, A. S. C. and Twyman, W. A. (1966). "On Measuring Television Audiences." *J. Royal Statist. Soc. A*, **130**, 1–54.
Einstein, A. (1949). *Out of My Later Years*. New York: Philosophical Library.
Elderton, W. B. (1902). "Tables for Testing the Goodness of Fit of Theory to Observation." *Biometrika*, **1**, 155–163.
Fisher, R. A. (1935). *The Design of Experiments*. Edinburgh: Oliver and Boyd.
Fisher, R. A. (1950). *Statistical Methods for Research Workers*. Edinburgh: Oliver and Boyd (1st edition 1925).
Fisher, R. A., Corbett, R. S. and Williams, C. B. (1943). "The Relation Between the Number of Species and the Number of Individuals in a Random Sample of an Animal Population." *J. Animal Biology*, **12**, 42–48.
Fisher, R. A. and Yates, F. (1957). *Statistical Tables for Agricultural, Biological and Other Research Workers*. Edinburgh: Oliver and Boyd.
Foy, F. C. (1973). "Annual Reports Don't Have to be Dull." *Harvard Business Review*, January–February, 51–58.
Gatty, R. (1966). "Multivariate Analysis for Marketing Research: An Evaluation." *Applied Statistics*, **15**, 157–172.
Geissler, A. (1889). "Beiträge zur Frage des Geschlechts-verhältnisses der Geborenen." *Zeitschrift des K. Sächsischen Statistischen Bureaus*, **35**, 1–22.
Golde, R. A. (1966a). *Thinking with Figures in Business*, Reading: Addison-Wesley.
Golde, R. A. (1966b). "Sharpen Your Number Sense." *Harvard Business Review*, July–August, 73–83.
Goodhardt, G. J. (1966). "The Constant in Duplicated Television Viewing." *Nature*, London, **212**, 5070, 1616.

Goodhardt, G. J. and Chatfield, C. (1973). "The Gamma Distribution in Consumer Purchasing." *Nature*, London, **244**, 5414, 316.
Goodhardt, G. J., Ehrenberg, A. S. C., Collins, M. A. and Doyle, P. (1975). *The Television Audience*. (In preparation.)
Greenwood, M. and Yule, G. U. (1920). "An Enquiry into the Nature of Frequency Distributions Representative of Multiple Happenings, with Particular Reference to the Occurrence of Multiple Attacks of Disease of Repeated Accidents." *J. Royal Statist. Soc. A*, **83**, 255–279.
Haight, F. A. (1967). *Handbook of the Poisson Distribution*. New York: Wiley.
Hansen, M. H., Hurwitz, W. N. and Madow, W. F. (1953). *Sample Survey Methods and Theory* (*Vol. I* and *II*). New York: Wiley; London: Chapman and Hall.
Healy, M. J. R. (1952). "Some Statistical Aspects of Anthropometry (with Discussion)." *Royal Statist. Soc. B*, **14**, 164–184.
Hinshelwood, C. N. (1967). "President's Anniversary Address." *Proc. Roy. Soc. B.*, **148**, 5–16.
Kendall, M. G. (1951 and 1952). "Regression, Structure and Functional Relationship, Parts I and II." *Biometrika*, **38**, 11–15; **39**, 96–108.
Kendall, M. G. (1961). "Natural Law in the Social Sciences." *J. Royal Statist. Soc. A*, **124**, 1–19.
Kish, L. (1965). *Survey Sampling*. New York: Wiley.
Kpedekpo, G. M. K. (1970). "The Height and Weight of Ghanaian Children." *J. Royal Statist. Soc. A*, **133**, 86–93.
Kpedekpo, G. M. K. (1971). "Pre-School Children in Ghana—The Use of Prior Information." *J. Royal Statist. Soc. A*, **134**, 372–373.
Kuhn, T. W. (1970). *The Structure of Scientific Revolutions*. Chicago: University of Chicago Press.
Lindley, D. V. (1947). "Regression Lines and the Linear Functional Relationship." *J. Royal. Statist. Soc. B*, **9**, 214–244.
Lindley, D. V. and Miller, J. C. P. (1966). *Cambridge Elementary Statistical Tables*. Cambridge: Cambridge University Press.
Lovell, H. G. (1972). "Heights and Weights of West Indian Children." *J. Royal Statist. Soc. A*, **135**, 569–578.
Medewar, P. (1952). *The Art of the Soluble*. London: Penguin Books.
Montgomery, D. B. and Urban, G. L. (1969). *Management Science in Marketing*. Englewood Cliffs: Prentice-Hall.
Moore, P. G. (1969). *Principles of Statistical Techniques*. Cambridge: Cambridge University Press.
Moser, C. A. and Kalton, G. (1971). *Survey Methods in Social Investigation*. London: Heinemann.
Neyman, J. and Scott, E. L. (1948). "Consistent Estimates Based on Partially Consistent Observations." *Econometrics*, **16**, 1–12.
Pearce, S. C. (1969). Data From a Classical Apple Experiment. (Unpublished.)
Popper, K. R. (1959). *The Logic of Scientific Discovery*. London: Hutchinson.
Scott, J. R. (1973). "A Method of Predicting Trial Performance of Single Screw Merchant Ships." *Transactions of the Royal Institute of Naval Architects*, **115**, 149–171.
Shewan, J. M. and Ehrenberg, A. S. C. (1955). "Volatile Bases and Sensory Quality-Factors in Iced White Fish." *J. Science Food and Agric.*, **6**, 207–217.
Shewan, J. M. and Ehrenberg, A. S. C. (1957). "Volatile Bases as Quality-Indices of Iced North Sea Cod." *J. Science of Food and Agric.*, **8**, 227–231.
Sillitoe, A. F. (1971). *Britain in Figures*. London: Penguin Books.

Sprent, P. (1970). "Some Problems of Statistical Consultancy." *J. Royal Statist. Soc. A*, **133**, 139–165.

Theil, H. (1971). *Principles of Econometrics.* Amsterdam and London: North-Holland; New York: John Wiley.

Wald, A. (1940). "Fitting of Straight Lines if Both Variables are Subject to Errors." *Ann. Math. Stat.*, **11**, 284–300.

Yates, F. (1960). *Sampling Methods for Censuses and Surveys* (3rd ed.). London: Griffin.

Name Index

Subject Index